# AMERICAN ZOO

# AMERICAN ZOO

## A Sociological Safari

## David Grazian

PRINCETON UNIVERSITY PRESS

Princeton and Oxford

press.princeton.edu
Jacket art © Kairos69/Shutterstock.
"Holocene,"was written by Justin DeYarmond Edison Vernon and published
by Chris In The Morning Music LLC. Administered
by Kobalt Music Publishing America, Inc.

ISBN 978-0-691-16435-9

Library of Congress Control Number: 2015934773

British Library Cataloging-in-Publication Data is available

This book has been composed in Electra Lt std

Printed on acid-free paper. ∞

Printed in the United States of America

1 3 5 7 9 10 8 6 4 2

# Contents

*For my family*

# AMERICAN ZOO

# The World in a Zoo

**WALKING THROUGH A GRAND METROPOLITAN ZOO** before opening hours tickles the mind with sights both beautiful and strange. African elephants swing their massive trunks. Bengal tigers bathe themselves while perched on elevated rocks. Arctic polar bears frolic in the sun. The cacophonous soundscape of lion roars, rhinoceros snorts, and spectacled-owl hoots forms a magical orchestra, a feast for the ears. Of course, one's senses become most immediately attuned to the earthy smells of the zoo, pungent and dark, the musky scent of nature itself.

Yet a step closer into the bowels of the zoo reveals even more astonishing discoveries. In a small kitchen, zookeepers slice open cantaloupes and pull heads of romaine for a peccary's lunch, and boil eggs for a spotted skunk. Volunteers prepare a salad of chopped apples, shredded carrots and yams along with a spoonful of mealworms for an armadillo's breakfast. Others prepare a cougar's meal with raw beef soaked in buckets of blood and topped off with laxatives, or carefully weigh out horse meat and count out frozen chicks or rats for a hungry peregrine falcon. Nearby, education volunteers spray designer perfume on colored toys designed to stimulate the senses of a blue-and-gold macaw or hedgehog. In the yards, keepers shovel bison dung and cart away wheelbarrows full of dirty hay while everyone gossips about how the latest volunteer recruit mistakenly wore shiny white pants on her first day of zoo work.

At the entrance gates the first visitors eventually arrive in groups, mostly children accompanied by their mothers, and elementary school classes on field trips with students all adorned in identical T-shirts as a safety precaution. Toddlers chase the free-range peacocks left to roam the zoo grounds and nibble on dropped pretzel crumbs and hotdog buns. Older kids approach wildcat enclosures and bang on their glass walls, hoping to awaken their families of slumbering felines. Others imitate the gorillas and chimpanzees, sensing in them sparks of intelligence and a common ancestry. Occasionally a romantic couple strolls in, hand and hand, oblivious to the nearby primate screams of howler monkeys and crying human infants. Adults glance among the zoo's many placards explaining the life cycles of monarch butterflies, winged migrations of plovers and warblers, and endangerment of poison dart frogs. Multimedia spectacles at the zoo invite audiences indoors to watch animated penguins dance on a giant screen in climate-controlled theaters instead of enjoying the very real ones waddling outside.

Some zoo-goers have already set up their tripods and cameras decked out with light sensors and high-definition telephoto lenses to capture photographs of captured animals. Their close-ups may reveal the man-made technologies that zoo exhibits hide in plain sight: electric cables designed to look like sagebrush; hollow logs containing caches of kibble; palm trees made of fiberglass and steel. Cold-blooded lizards rest on artificially heated rocks molded out of concrete. Gray wolves are expertly trained to receive vaccinations on command. Prerecorded soundtracks of the rainforest and savanna accompany the chirps and howls provided by the zoo's actual wildlife.

These sounds reverberate against the chatter of seasoned workers and opinionated visitors swarming among the zoo's garden paths. Keepers tell dirty jokes while feeding pronghorn antelope, and surprise their giraffes with special treats on their birthdays. Zoo educators warn of global warming and the vulnerability of Alaska's polar bear population, while climate change skeptics discourage their children from believing such nasty realities. Outside the zoo gates, animal rights activists protest the captivity of African elephants and great apes.

These social elements of the American metropolitan zoo—its man-made environments, landscapes of meaning making, caregiving communities, and an uncanny ability to generate moral dispute among participants—all emphasize how zoos are not merely cordoned-off refuges of animal life in the city but repositories of culture, living museums that chronicle how we humans make sense of the natural world. Zoos are therefore perhaps best understood by focusing attention on their most advanced primates—the zookeepers and animal trainers, educators and volunteers, exhibit designers and landscape architects, audiences and animal advocates that inhabit their gardens and surrounding social ecologies. Indeed, it is the human life of zoos that best animates what we might think of as *the culture of nature*, the ways we attach shared meaning and sentiment to our environment and all its creatures, great and small.

## The Culture of Nature

In our postmodern world we have begun to tear down all sorts of traditional boundaries in recognition of their feeble foundations as manmade artifacts carved from social convention. Yet the one boundary that remains fixed in our collective imagination as taken for granted and unchanging may be the binary distinction between culture and nature, as well as its logical parallels—the city and the country, civilization and wilderness, humans and animals. Urban dwellers flee asphalt cities in droves during the dog days of summer to seek solace in countrified or naturalized landscapes from mountainside resorts to ocean beachfronts to national parks. We revere remote islands of wilderness as if the last refuges of unspoiled frontier, the literally undiscovered country—Australia's Great Barrier Reef, the Amazon River Basin, the icy deserts of Antarctica. We organize social life itself according to further refinements of this primary culture/nature distinction: modern and primitive, mind and body, love and lust, order and chaos, artifice and authenticity.[1]

Notably, the distinction between culture and nature has historically served as a dominant organizing principle for the development of American city zoos and continues to guide their operations today. Most of the nation's first and most prominent zoological societies and accompanying gardens and parks were established in the late nineteenth and early twentieth centuries during an age when industrialization and urban development made Americans feel ever more distant from the natural living world.[2] In fact, the first U.S. zoos were specifically designed to serve as idyllic oases of nature protected from surrounding downtown business districts and immigrant neighborhoods in metropolitan cities expanding in population, urban density, and economic growth. These zoos were sited in the great urban parklands of that era: New York's Central Park Zoo, Chicago's Lincoln Park Zoo, the Philadelphia Zoo in Fairmount Park, the Buffalo Zoo in City Park, and the Baltimore Zoo in Druid Hill Park. (The development of these early parkland zoos was followed by the founding of the Atlanta Zoo in Grant Park, the original San Francisco Zoo in Golden Gate Park, the Denver Zoo in City Park, the Pittsburgh Zoo in Highland Park, and the San Diego Zoo in Balboa Park.) Zoos emerged as part of the urban landscape just as everyday life was becoming increasingly alienated from the rural countryside and routine exposure to farm animals and wildlife.[3]

Today, zoos continue to order themselves on successive iterations of the culture/nature divide.[4] For instance, not only are zoos thought of as places of nature in the city, but within zoos themselves there are workers and departments considered more directly aligned with *animal* care (zookeepers, veterinarians, diet volunteers) and others that handle more social and cultural matters concerning the zoo's *human* participants: educators, guest services personnel, media relations liaisons, caterers, and cashiers. Likewise, American zoos and their visitors consider exotic African, Asian, and Australian animals to be more *wild* than the *domesticated* U.S. farm animals such as the roosters, ponies, sheep, goats, and cows typically segregated in so-called children's zoo areas. (Of course, *all* zoo animals are under some domesticated regime, given that they are fed scheduled meals and regularly receive veterinary care.) In aquatic zoos and marine parks like SeaWorld San Diego, dolphins, orcas, sea lions, and other highly intelligent mammals are given pet names—Seamore, Clyde, Shamu—and trained to mimic *human* gestures like "waving" before large crowds with their gigantic fins. Less anthropomorphic fish and other sea life that populate aquarium tanks remain nameless as anonymous representatives of their *animal* phylum and subspecies.

Yet however sensible such schemas may seem, binary distinctions between culture and nature are as fabricated as any other set of collective beliefs. We may contrast human civilization and its cities to greener climes, but the predominance of biodiverse urban habitats such as the Los Angeles parks where mountain lions and coyotes roam in search of prey, the Manhattan alleyways where brown rats copulate, and the Chicago four-star hotels infested with bedbugs tell a different tale. Likewise, the apelike *Homo sapiens* is an organic specimen, an amalgam of bones, cartilage, DNA strands, blood vessels, neurotransmitters, finger-nails, and skin. Rife with human bodies both living and dead (to say nothing of New York's feral pigeons and red-tailed hawks, Philadelphia's raccoons, Houston's armadillos and opossums, and Tucson's javelinas), the city is a jungle.[5]

Nature itself is a cultural construction organized by human imagi-nation and experience. Just as the Earth's topography does not feature longitudinal lines painted across an actual equator, our world's perma-nent state is one of elemental disorder, a swirling soup of biological and inorganic matter from which we as language-generating humans selec-tively name and define as "pollen," "rivers," "hurricanes," "meadows," "antelope," and all else that we collectively categorize as nature. Of course, this does not mean that those empirical forces that we have earmarked as distinct and identifiable—gravity, photosynthesis, electro-magnetism, plate tectonics, lunar tides—do not shape the Earth's ambi-ent environment and its inhabitants. (Ask anyone who has ever been injured in an earthquake or tsunami, caught malaria, or fought off a leopard.) Rather, it means that distinctions between culture and nature have always gained their symbolic power through habits of mind rather than ecological realities. Chimpanzees share 98.4 percent of their genes with humans, yet as nonhuman animals they are considered categori-cally closer to wombats or gerbils than people.[6] Even more arbitrarily, during the Holocaust the Nazis quite famously honored nonhuman animals as moral beings deserving of compassion, protection, and even cult worship while Jews were regarded as filthy "lower" beasts likened to parasites and vermin, as expendable as laboratory rats.[7] Today we associ-ate nature with quiet peacefulness and solitude, majestic yet fragile—something to preserve and protect. Yet until only a few hundred years ago, Western civilizations thought of the natural "wilderness" as dark, desolate, and dangerous, something to be feared if not avoided alto-gether—just ask Little Red Riding Hood or the wandering Israelites.[8] Colonial settlers in North America also fantasized that forested sites inhabited for thousands of years by indigenous people represented an untapped New World of "virgin" wilderness.[9] Even though what we call nature is quite literally all around us, we can render it meaningful only through the prism of human understanding, the culture of nature.[10]

Nature is a social reality constructed by humans, and in fact land-scape features that we characterize as natural are often *literally* con-structed by humans as well. Although the Global North has tradition-ally thought of the Amazon as a pristine Garden of Eden unadulterated by human hands before European conquest, in fact evidence now suggests that Mesoamericans actually planted and cultivated the Amazon rainforest for millennia prior to Columbus's accidental voyage to the New World. As Charles C. Mann observes in his book *1491*, the Amazon is "a cultural artifact—that is, an artificial object." Elsewhere in his book Mann quotes Penn archeologist Clark Erickson, whose research demonstrates that "the lowland tropical forests of South America are among the finest works of art on the planet." In this sense, most Neotropical landscapes are quite literally man-made environments.[11]

Indeed, even a sight as glorious to behold as Niagara Falls and its surrounding scenery was shaped by landscape architect Frederick Law Olmstead in the late nineteenth century, and then redesigned in the 1920s to divert additional water over the American Falls and augment the water level in its rapids below, all for "the betterment of the spectacle by using water to greater scenic advantage."[12] Most areas of North American "wilderness" like California's Yosemite National Park and Alaska's Denali National Park and Preserve are protected spaces, which means they are the product of government legislation, surveying, mapping, management, and other human activities. As UCLA professor N. Katherine Hayles wisely asks, "What counts as natural? Can we consider Yosemite National Park an embodiment of nature? If so, then nature is synonymous with human intervention, for only human intervention has kept Yosemite as a nature preserve. Asked for a definition of wilderness . . . Richard White offered the following ironic observation: wilderness is managed land, protected by three-hundred page manuals specifying what can and cannot be done on it."[13]

Jon Krakauer poignantly illustrates the literal indivisibility of culture and nature in his nonfiction book *Into the Wild*. Interspersing his reporting with autobiographical tales of his own intrepid outdoor adventures, Krakauer investigates the strange life and tragic death of Chris McCandless, a young Emory graduate who attempted to escape civilization to "live amongst the wild" in Alaska's Outer Range in 1992 and eventually succumbed to starvation and died, all alone. Yet despite McCandless's urge to lose himself in the snowy wilderness, his body was discovered within a six-mile radius of no fewer than four cabins, nearby a gauging station built by the U.S. Geological Survey, and less than sixteen miles away from a well-traveled road in Denali patrolled by the U.S. National Park Service. No fewer than three separate parties found his corpse rotting in an abandoned Fairbanks City Transit System bus.

As Krakauer observes, "In coming to Alaska, McCandless yearned to wander unchartered country, to find a blank spot on the map. In 1992, however, there were no more blank spots on the map—not in Alaska, not anywhere."[14]

The whittling down of places on Earth untouched by mankind has been rendered absolute in the current era of human-induced global warming and climate change. Since the Industrial Revolution, anthropogenic activities—notably massive population growth, urbanization, rainforest clearing and woodland destruction, intensive farming and livestock production, toxic pollution, and unabated fossil fuel consumption—have rendered every inch of the planet's soil, ocean water, and atmosphere transformed, perhaps for millennia. In its fifth assessment report released in 2013, the Nobel Peace Prize–winning Intergovernmental Panel on Climate Change (IPCC) details the outsized mark that industrialized civilizations have made to the Earth's biosphere, notably growing atmospheric concentrations of carbon dioxide, methane, and nitrous oxide. IPCC experts have conclusively identified the rapid intensification of these greenhouse gas emissions as the primary contributor to the recent warming of the Earth's surface, atmosphere, ocean, and climate system; increased ocean acidification; rapid polar ice loss and glacier melt; and global rising sea levels.[15] As just one of the many devastating symptoms of our new environmental reality, the journal *Science* reports that recent nonlinear increases of species extinction exceed prehuman baseline levels by a factor of up to *one thousand*—a crisis that scientists refer to as the sixth major extinction event in the history of life on our fragile planet.[16]

We have also begun to see how anthropogenic climate change will affect human societies the world over: extreme storms, punishing droughts, and other weather-related catastrophes; ruined agricultural harvests for a growing population; decreased access to water in unstable regions of the Global South; and the flooding of coastal cities from New York City to New Orleans. All this confirms that there is no meaningful distinction between *natural* history and *human* history.[17] Even hardnosed geologists have considered officially renaming the late Holocene period the "Anthropocene," arguing that humans have impacted the planet "on a scale comparable with some of the major events of the ancient past," changes that may be seen as "permanent, even on a geological time-scale."[18] In a 2011 article published by the Royal Swedish Academy of Sciences, a team of scientists argues,

The Anthropocene implies that the human footprint on the global environment is now so large that the Earth has entered a new geological epoch; it is leaving the Holocene, the environment within which human societies themselves have developed. Humanity itself has become a global geophysical force, equal to some of the "great forces of Nature" in terms of Earth System functioning. . . . The Anthropocene provides an independent measure of the scale and tempo of human-caused change—biodiversity loss, changes to the chemistry of atmosphere and ocean, urbanization, globalization—and places them in the deep time context of Earth history. The emerging Anthropocene world is warmer with a diminished ice cover, more sea and less land, changed precipitation patterns, a strongly modified and impoverished biosphere and human-dominated landscapes.[19]

Admittedly, the Anthropocene is as creatively rendered as the Earth's longitudinal lines. Distinctions of time are marked only by our collective imagination, not by some celestial stopwatch that beeps three times when a new geological epoch begins.[20] Still, the cultural rise of the Anthropocene provides a moment for us as a society to acknowledge that humans are as much a part of what we call nature as ocean currents, jet streams, erosion, and other planetary forces. Of course, the Earth itself does not care whether rising extinction rates are due to the spread of deadly viruses, cyclical changes in the planet's temperature, or decades of carbon emissions from fuel-inefficient trucks. Again, the boundary between culture and nature, human civilization and the wild, is an imaginary one, important only to us.

In this book I argue that zoos best reflect how humans construct the natural world, both literally and figuratively. With its carefully curated animal collections, audiovisual entertainment media, educational and conservation programming, and staged encounters with wildlife, the zoo provides a fitting model for how humans distill the chaos of the outdoors into legible representations of collective meaning and sentiment that project our prejudices and desires, past and present. At the same time, zoos are monuments to human domination and folly, designed and constantly transformed by the ingenuity of architects, horticulturalists, and zookeepers, and the occasional whims of zoo trustees and elite donors. Their lifelike habitats, curated gardens, and captive creatures therefore help to illustrate humankind's recurring impact and reconstitution of the Earth's geology and biosphere in the age of the Anthropocene. Perhaps more than any other cultural attraction in the contemporary American city, zoos cannot help but represent nature as a human creation, a product of both imagination and hubris. My hope is that by understanding zoos we might rethink our preconceptions and priorities in a world already beset by mass species extinction, suffocating heat waves, unquenchable forest fires, record-breaking droughts, and the poisoning of the world's oceans.

# Put Me in the Zoo

As a tenured professor at the University of Pennsylvania, I usually spend my time giving lectures to classrooms of college students, advising doctoral candidates on their dissertations and comprehensive exams, and slogging through committee meetings with university faculty and administrators. So how did I wind up at the zoo in the first place? As a cultural sociologist and urban ethnographer by training, I had earlier in my career written two books about urban nightlife, the first on Chicago's blues scene, the second on the growing nightlife economy of restaurants, nightclubs, and cocktail lounges in and around downtown Philadelphia. These projects required me to spend long hours at late-night music venues, corner taverns, nouveau-fusion eateries, martini bars, high-end speakeasies, dance palaces, and corner taverns until the wee morning hours. (Poor me.)

After my wife gave birth to our son (whom I will henceforth refer to as Scott), this nocturnal lifestyle no longer seemed tenable nor all that appealing. But at a very young age Scott developed an acute fondness for our neighborhood's menagerie of leashed puppies, wandering housecats, and backyard chickens. This enthusiasm eventually brought us to the local Philadelphia Zoo nearly every weekend to take in its far more exotic elephants, red pandas, marmosets, pumas, gorillas, and giraffes. From the helium-filled Channel 6 Zoo Balloon that rises hundreds of feet up into the sky to its intricate naked mole-rat exhibit down below, we took it all in: the unforgettable sights, sounds, and smells of the nation's oldest zoo. (The smells were the most difficult to forget.) With each visit, our curiosity grew about the zoo's strange creatures, and what makes them tick—and chirp, moo, growl, honk, quack, roar, and squeal.

Yet as Scott and I continued our visits to the zoo, the sociologist in me couldn't help but wonder about its strange allure. Zoos share a number of apparent similarities with nightclubs and cocktail lounges, and not only because they all feature the public display of uninhibited mating rituals. Both city zoos and nightlife scenes are cultural worlds of urban entertainment, tourism, and popular culture. They also serve as theatrical stages for the performance of authenticity and fantasy, and enveloping zones of rich experience and meaning making for a range of consuming audiences. Both zoos and nightlife scenes promise to serve their surrounding communities by bringing people together in a spirit of sociability and celebration of public life, yet often slip into the trap of commercialization and superficiality. Perhaps most of all, both zoos and nightlife environments force their participants to reflect on their relationship to a larger social world, and their place within it.

I therefore traded in my tweed jacket, bookshelf-lined office, and other professorial trappings for a zoo uniform of khakis, work boots, green polo shirt, and accompanying nametag, and got my hands dirty (quite literally) by volunteering for a total of four years at two urban zoos.[21] At one institution that I call City Zoo I worked primarily in an outdoor children's zoo. There, I was responsible for cleaning enclosures and exhibits, preparing and distributing zoo-prescribed diets to birds of prey and small mammals, managing children in a petting yard filled with goats and sheep, and providing behavioral enrichment to a variety of animals. Along the way, I shoveled cow manure and chicken dung, goat pellets and duck droppings. I scrubbed owl and macaw cages and lined them with old issues of USA Today and the Wall Street Journal, clipped a ferret's toenails, and once got locked inside a bird's double-caged enclosure. I picked horse and donkey hooves, stuffed frozen feeder mice with vitamin E capsules, bathed tortoises, and exercised overweight rabbits.

At a second zoo that I call Metro Zoo I worked as a docent, or volunteer educator. I handled and presented a variety of small live animals to a range of zoo audiences, including families, school groups on field trips, children's birthday parties, and busloads of local nursing home residents. I learned to handle Arizona desert king snakes, ball pythons, chinchillas, fat-tailed geckos, screech owls, black vultures, millipedes, giant Flemish rabbits, and an American alligator. I also regularly helped prepare diets for most of the animals in the zoo's collection, which included giraffes, jaguars, howler and saki monkeys, cougars, river otters, peccaries, golden lion tamarins, bison, wolves, and Jamaican fruit bats. Much of the food was expired (but still safe) meat, fish, and produce donated by local supermarkets and grocery stories, including whole strip loins, boxes of oranges and kale, and odds and ends of raw salmon and squid for the otters. Notably, keepers and support staff went all-out when preparing animal diets. While at Metro Zoo I watched one retirement-age volunteer teach another how to chiffonade romaine lettuce for an iguana, while another regularly grated carrots to make a slaw for the zoo's spotted skunk and armadillos.[22]

While on and off the job at both zoos I experienced some of the daily labor expected of zoo workers, observed social life as it unfolded, and freely conversed with keepers, veterinary staff, educators, volunteers, animal curators, guest services personnel, administrative staff, and hundreds of zoo visitors. (Fortunately, no animals were ever seriously harmed on my watch, although I myself endured bites, scratches, and other humiliations from several domestic rabbits, a bearded dragon, an African gray parrot, and at least one goat.) Of course, I cannot claim to have worked nearly as rigorously nor under the same precarious conditions of low-wage employment, danger, and stress as full-time zookeepers regularly do every day without complaint. Still, I managed to log more working hours of animal husbandry and fecal cleanup than your average Ivy League sociology professor, and learned a great deal while doing it. I introduced live boa constrictors, blue-tongued skinks, and tarantulas to hundreds of children and their parents; gave tours of a greenhouse packed with flittering butterflies; and illustrated to camp counselors the visible differences between male and female Madagascar hissing cockroaches. Although I had never in my life displayed any prior kinship or rapport with animals, even the neighbors' pets, I eventually grew attached to the zoo animals under my care—and not only the cute rabbits (which can actually be quite vicious and nasty), but less popular critters as well, including black rat snakes, chuckwallas, and, yes, even the cockroaches.

Both City Zoo and Metro Zoo are highly respected institutions accredited by the nonprofit Association of Zoos and Aquariums, or AZA. To give a sense of the exclusivity of the AZA, according to the most recent publicly available figures the Animal and Plant Health Inspection Service of the U.S. Department of Agriculture licenses 2,764 animal exhibitors, ranging from roadside attractions to high-tech breeding centers. In comparison, as of September 2014 the AZA had accredited only 228 current zoos and related facilities, or fewer than one-tenth of all U.S. animal exhibitors. From the Alaska SeaLife Center to Zoo Miami, AZA-accredited institutions include 138 conventional zoos, 43 aquariums, 10 hybrid zoo/aquariums, 18 safari and theme parks, 15 science and nature centers, 2 aviaries, and 2 butterfly houses. More than half of these accredited zoos, 54 percent, are private nonprofit institutions (such as the San Diego Zoo), while 35 percent are public (Smithsonian National Zoo) and 11 percent are for-profit entities (Disney's Animal Kingdom, SeaWorld Orlando).[23] The AZA claims its zoos attract more than 181 million annual visitors, teach 12 million students a year who visit through school field trips, support 142,000 jobs, and contribute $16 billion annually to the U.S. economy. These accredited zoos contain a total of 6,000 species and 751,931 individual animals, and are part of an elite network of institutions that can loan and breed their animals with one another. They may also participate in the AZA's Species Survival Plan (SSP) programs that manage the propagation of threatened species in attempts to ensure the continued survival of endangered creatures like the Panamanian golden frog and the Siberian musk deer.[24]

In addition to my hands-on research at City Zoo and Metro Zoo, I traveled to twenty-six AZA zoos (or 12 percent of all AZA-accredited U.S. zoos) across the country, including some of the nation's most prominent metropolitan zoos, aquariums, and marine mammal parks, from New York's Bronx Zoo to SeaWorld San Diego to the Monterey Bay Aquarium in Northern California. My visits included backstage tours at a number of facilities as well as extended periods of public observation at selected exhibits, animal shows, and other attractions. (I also visited a number of nonaccredited roadside zoos and aquariums, just for good measure.)

It is perhaps unsurprising that men in their early forties who wander the zoo alone while staring at strangers in public, jotting down notes, and taking photographs of crowds that may include small children are often seen as strange or even dangerous among zoo-going families.[25] Therefore, in order to more comfortably blend into such family-oriented surroundings (and also because I desperately wanted the company), for much of this secondary research I was accompanied by my aforementioned son Scott, whose age spanned from three to seven years during the course of the research. To be sure, my attempts at blending seamlessly into these zoo environments would not have been possible without his agreeable participation as an adventurous fieldwork companion.[26] I also relied on the watchful eyes of a team of trained research assistants from Penn who conducted over one hundred total hours of public observation among visitors at a number of zoos around the region.[27] I followed up this research by meeting with a wide variety of key informants and stakeholders in the zoo world, including keepers, veterinary technicians, educators, and volunteers; media relations executives and other administrative personnel; zoo architects and exhibit designers; and anti-zoo animal advocacy activists.[28]

# The World in a Zoo

On a beautiful fall morning, Scott and I arrived at the Philadelphia Zoo especially early in order to get in line for the zoo's hot-air balloon ride. After waiting our turn, we climbed aboard its basket and began our speedy ascent: a four-minute ride through the blue sky to a spectacular height of four hundred feet. After our flight upward the balloon sat aloft, tethered to a cable that could support 98,000 pounds—ten times stronger than necessary. The cable was lined with small red flags to make it visible to low-flying aircraft.

With my acute fear of heights, I found none of this reassuring. I was unsure whether to be more terrified about the possibilities of the balloon plummeting back down to the concrete parking lot below, or simply breaking off from our tether and flying away. Still, I knew I had to put on a brave face for Scott (he was only four at this point), but he seemed too much in awe of the view to notice my furrowed brow. Looking down through a small window in the basket, he breathlessly exclaimed, "Daddy, you can see the whole world from up here." He said this because he spotted what he thought was "the edge of the Earth," which I explained was really the horizon. But this was a fitting metaphor, given that Scott and I could look straight down to the zoo's exhibits and attractions, a scaled-down model of the world with its artificially replicated ecosystems in miniature—the African savanna, the Australian outback, the Arctic icecap.

This book similarly presents a pocket-sized depiction of a much larger place, the social and cultural world of the American metropolitan zoo. Think of each of its chapters as the different staged areas of a zoo safari, perhaps one where waterbuck and hippos stomp in tall grasses behind hidden electrified wires. First we visit the zoo's roaring animal exhibits, driving past simulated jungles where captive gorillas thump their chests and forage for breakfast cereal scattered about their enclosures. As built environments, zoo exhibits and their facades rely on the stuff of culture, both synthetic materials (fiberglass, acrylic, epoxy) and symbolic fictions (canonical myths, idealized representations, moral narratives). Just as the Earth itself, the zoo's environment is a result of human engineering and its unintended consequences, with its mountains and ice floes made from inert cast-in-place concrete, and bamboo stalks made of inflexible steel.

From here we turn to the families and other audiences who wander the grounds together in small groups, attaching sentiment and symbolic value to the zoo's resident animals. Like nature, the culturally filtered meanings we attribute to animals may be the product of imagination and myth, yet such fictions are no less powerful to those who believe them.

Our sociological safari then takes us behind the scenes, where we will be introduced to the community of zookeepers whose dedicated efforts keep zoos afloat by keeping their resident creatures alive and healthy. Birds of a feather, keepers forge loving relationships with zoo animals and each other, collectively animating their work with shared meaning and moral purpose. More like hospitals than prisons, zoos employ responsive caregivers who assign helpless animals pet names when they are born, tend to them throughout their lives, and solemnly mourn them when they pass away, just as if such animals were needy human patients.[29]

Zoos are like hospitals in another way as well. In a reversal of historical trends, today women constitute the majority of American zookeepers.[30] Female zoo recruits also tend to be younger and more educated than the retired men they typically replace. Yet although 82 percent of today's zookeepers have four-year college degrees in biology, zoological science, or other related fields, most keepers are vastly underpaid, just as are female caregivers in other occupational contexts.[31] Nevertheless, the camaraderie and selfless commitment displayed by zookeepers speak volumes, and provide a model for the contemporary demands of environmental stewardship over the Earth's growing numbers of endangered inhabitants. We will meet these brave zookeepers as well as their important counterparts: the educators who provide cultural interpretations of the zoo's landscape and its resident creatures to visitors and the public, and in doing so attach human meaning to animal life. The attentiveness and drive of these teachers illustrate the vital need for disseminating knowledge about basic science and an appreciation for the Earth's biodiversity in the wake of the current environmental crisis. They will share with us both the enormous pleasures and difficult challenges involved in educating the general public about these issues.

Our safari will then head uphill to a more expansive view of competing zoo priorities, as places of entertainment and amusement on the one hand, and centers of environmental protection and species conservation on the other. We will visit zoos and aquariums erected in some of the world's most visited tourist destinations, including Orlando's Disney World, SeaWorld San Diego, and the Las Vegas Strip, stopping by their shark tanks, killer whale spectacles, amusement park adventures, and gift shops. We will also see how the nation's most esteemed zoological institutions similarly traffic in themed environments and branded diversions, as well as fanciful myths about the Global South.

We will then take a critical look at how zoos attempt to fulfill their missions as environmental stewards and conservation educators, with an eye toward recognizing the challenges they face in both confronting the threat of the environmental crisis and effectively communicating the urgency of the problem to the public. Our safari eventually comes to a head in the zoo's beating heart, a hub of contestation and debate surrounding the morality of zoos and the exhibition of captive creatures. In the social world of the zoo, keepers and educators, media relations staff, zoo visitors, and animal rights activists publicly wrestle with complex issues surrounding the domination and display of animals—what I call the *captivity question*. We finally conclude our scenic jaunt by reflecting on the zoo as a repository of culture, and consider some suggestions derived from my research for reimagining zoos as sanctuaries for protecting wildlife, schools for educating the public about the environmental crisis, and showcases for modeling future possibilities for dealing with urban life in the anthropogenic age.

*"Daddy, you can see the whole world from up here."* Well, maybe not the whole world, but certainly *our* world, a socially constituted reality as reflected by the zoo and its culture of nature—its geographies of meaning and moral sentiment, regimes of conservation and captivity, landscapes of learning and entertainment, and communities of caregiving. We begin our sociological safari of the American metropolitan zoo by exploring the artificial environments of authentic animals.

Chapter 1

# Where the Wild Things Aren't

Exhibiting Nature in American Zoos

**THE AIR WAS HEAVY** with humidity and dewy mist as I traipsed through the rainforest. Through the thick vegetation I could make out shocks of bright orange and red as a distant cackle grew louder. And WHOOSH—a fairy bluebird swooped in and I ducked. Glancing downward, I caught a glimpse of a Victoria crowned pigeon strutting to my right, and up in the tall trees I spotted the rest of them, almost all at once. First a rhinoceros hornbill, followed by a green toucan, and then an honest-to-god Micronesian kingfisher! Funny, I thought they went extinct in the wild. But of course, this was not the wild—it was the artificial tropical rainforest of the Philadelphia Zoo's McNeil Avian Center, where visitors can enjoy watching birds from Africa, South America, and the Pacific Islands that would never actually mingle in nature, at least not on the same continent.

As the Earth's most biodiverse ecological habitats disappear along with untold species of fauna and flora in the age of the Anthropocene, we rely ever more heavily on distinctly man-made environments to experience what we have traditionally considered the natural living world. But whereas modernist zoo exhibits of the past emphasized functionality through cheerless and antiseptic living spaces adorned with barred steel cages and easily washable concrete floors, recent decades have seen the emergent popularity of what has been called the "new naturalism" in zoos, whereby exhibit designers envelop animals and audiences together in landscaped gardens that simulate the wild bush, all while providing realism based on scientific field research on wildlife environments and their topography.[1]

To this end, the best American zoos today use scenic backdrops, lifelike props, soundscapes, and other elements of stagecraft to furnish seemingly authentic and realistic depictions of the Gobi Desert, the Everglades, Malaysia's mangrove forests, and other habitats of life on Earth. Just as contemporary audiences experience the spirituality and enchantment attributed to consecrated wilderness spaces by consuming nature documentaries and other popular cultural entertainments, naturalistic zoo exhibits similarly capture visitors' imaginations by transporting them to exotic landscapes of sublimity around the Earth's biosphere, if only in their mind's eye.[2] As feats of both engineering and imagination, these exhibits best represent how zoos attempt to both literally and figuratively construct the natural world according to human desire and will, much like the Earth itself. In doing so, zoos fuse together modern science with the craft of storytelling and the aesthetics of visual art, just as paleontologists create colorful and seemingly lifelike models of dinosaurs no human has ever actually seen. As the biologist Edward O. Wilson observes, "The role of science, like that of art, is to blend exact imagery with more distant meaning, the parts we already understand with those given as new into larger patterns that are coherent enough to be acceptable as truth."[3]

Yet the fabrication of the idealized natural world is always a complicated proposition—especially for zoos, given the challenges inherent in staging live animal exhibits that offer authentic-looking depictions of wild habitats, endangered or otherwise. Zoo visitors gravitate toward African lions to observe them at play and hear their ferocious roars, yet in the wild lions rest approximately twenty hours a day.[4] Audiences delight in watching furry zoo creatures frolicking about in sociable groups, but many captive animals are solitary creatures by nature, while others resort to dangerous games of hierarchical dominance when placed in close proximity to one another. American zoo-goers revere the stately bald eagle, our national totemic symbol, and yet at feeding time some visitors might find its diet of thawed rats or mice unsettling, or simply gross. On this first leg of our safari we will explore how the built environment of the zoo and other idealized representations of *nature* ultimately rely on the synthetic and symbolic materials of *culture*, the man-made tools of imagination and art. Given how often Americans today experience the outdoors in highly controlled environments that manage audience perception and meaning—national and state parklands, wildlife and game preserves, themed amusement parks, natural history museums, even shopping malls—the zoo is an ideal social world for examining the cultural construction of nature in the age of the Anthropocene.[5]

## Zoo Design as Nature Making

First things first—nothing could be more self-evident than the staged or manufactured authenticity exhibited in zoo displays, no matter how naturalistic. No visitor approaching an Amur tiger's enclosure at the Philadelphia Zoo would somehow mistake its glass-walled habitat for the Siberian tundra. Yet as audiences we nevertheless expect zoo exhibits to prominently display landscapes that we at least associate with the natural environment, however imaginary and romanticized such renderings might be. Like moviegoers and theater buffs who give themselves over to the virtual realities provided by the special effects of cinema and the stage despite their obviously engineered quality, zoo visitors enjoy lush, immersive habitats that evoke a sense of authenticity, however illusory.[6]

For this reason, the biggest impediment to exhibiting nature in metropolitan zoos is not technical but organizational. Staging naturalistic exhibits requires that zoo designers and staff negotiate among a variety of competing institutional priorities. Zoos simultaneously serve as providers of animal care, promoters and funders of zoological research and endangered species conservation, and purveyors of both recreational amusement and scientific education. Therefore, zoo architects must build aesthetically alluring and sufficiently entertaining animal displays that offer enough scientific realism that they can serve an educational purpose, while also providing safe and comfortable living quarters for the zoo's resident species.

Negotiating these competing priorities—the cultural expectations of audiences, the educational mission of zoos, and the practicalities of managing live animals—gives rise to a particular bundle of stagecraft techniques, a kind of collaborative team performance that I call *nature making*.[7] Nature making in zoos requires adherence to a set of aesthetic conventions regarding what audiences collectively imagine the natural world to look and sound like, as illustrated by the popularity of attractively landscaped immersive exhibits densely overrun with lush vegetation, and frosty polar bear and penguin habitats filled with snow and ice. In such exhibits, designers attempt to hide all visible signs of artificiality, man-made technology, and human domination over displayed animals, just as modern societies erect mental barriers between human settlements and the surrounding environment, however imaginary and elusive such borders may be.[8]

American zoos engage in nature making not only to create delightful and aesthetically pleasing cultural attractions but to provide educational opportunities for the public as well, which the zoo industry celebrates as one of its paramount goals.[9] According to the Association of Zoos and Aquariums (AZA), two-thirds of all grown-up zoo visitors accompany children to the zoo, and these adults are primarily women, specifically mothers aged twenty-five to thirty-five.[10] Within both the zoo industry as well as among middle-class families, specifically those attentive to parenting strategies that emphasize what my Penn colleague Annette Lareau refers to as the "concerted cultivation" of their children through participation in enriching leisure activities, scientific education provides a loftier warrant for the existence of zoos than mere entertainment.[11] In keeping with their educational priorities, zoos must therefore present aesthetically *attractive* animal displays to the public while adhering to scientifically *accurate* and thus edifying renderings of the environment and its ecological realities. This can be a tricky feat, because although zoo visitors might think they crave realism in naturalistic zoo displays, it is often an idealized realism void of unpleasantries such as animal feces and regurgitated prey. As James Parker observes in the *Atlantic* in an essay on nature-based reality television shows, "Nature unobserved, unsentimentalized, unpolluted with our delusions, is just a bunch of stuff eating itself."[12]

Nature making in zoos also requires balancing the aesthetic tastes shared among visitors with the health and safety of the animals in their collections. This includes designing and maintaining secure yet comfortable enclosures for a variety of living creatures, feeding them individually calibrated diets, and providing veterinary care, exercise, behavioral stimulation, and opportunities for reproduction. While in the past zoo collections consisted of healthy specimens taken from the wild, today zoos usually rely on animal populations either born and bred in captivity, abandoned by negligent owners, or else injured in the wild and subsequently rescued and rehabilitated.[13] These managed populations therefore completely depend on their caretakers for their fitness, sustenance, and protection, and indeed many zoo animals could not survive in the wild at all without such assistance.

Sensitive to the treatment of captive animals, audiences themselves often try to evaluate the naturalism of zoo exhibits on behalf of the creatures they display. Of course, what nonexperts deem accommodating to zoo animals does not always correspond to their genuine needs. As animal scientist and advocate Temple Grandin argues,

> Zoos have gotten sidetracked by bad ideas about animal welfare. One of the most common is the "back to nature" approach where the goal is to make the enclosures as close to the animal's natural habitat as possible. That sounds logical until you stop to think that "nature" in a zoo is nothing like nature in the real world. Real nature means predators or prey, disease, hunger, and danger. "Zoo nature" doesn't have any of those things except disease, and a sick zoo animal gets immediate attention from a veterinarian. The result is that some zoos have spent a lot of money building fancy enclosures that appear natural to people, but are just as boring and painful for the animals as a barren concrete cage. I remember one tiger exhibit that looked really pretty with lots of rocks molded from concrete. There was absolutely nothing for the tiger to do. The enclosure was visually stimulating for people, but it was a barren environment for the tiger.[14]

Staging naturalistic exhibits in zoos therefore requires careful sensitivity to the concerns of resident animals, the goals of zoo educators and other stakeholders, and the cultural expectations of audiences. Given the organizational complexities and tensions involved in portraying nature to the public under such conditions, the diverse and multidisciplinary network of zoo participants that I call *nature makers*—the exhibit designers, landscape architects, horticulturalists, animal curators, veterinarians, animal care keepers, and zoo educators responsible for bringing naturalistic zoo exhibits to life—must work collaboratively to solve routine dilemmas encountered in the everyday functioning of zoos.[15]

## Dilemmas in Exhibiting Nature

In many ways, the social and collective nature of zoo work does not really differ from the production of art and popular culture more generally.[16] Perhaps this is why renowned zoo designer Jon Coe draws on the metaphor of the stage to advise contemporary nature makers in zoos to conjure up a "scenario that fully describes the exhibit context, just as a cinematic or theatrical scenario sets the scene for a performance."[17] Yet even under the most favorable of circumstances, nature makers run into difficulties when creating naturalistic zoo exhibits. Consider the design of animal enclosures. Most obviously, managing captive zoo animals requires the use of elaborate technologies designed to prevent escape, for the safety of both the animals as well as the public at large, and for good reason. On Christmas Day in 2007 Tatiana, a Siberian tiger, escaped from its exhibit at the San Francisco Zoo and killed Carlos Sousa, Jr., a teenage boy, while in March 2011 a venomous Egyptian cobra escaped from its Reptile House enclosure at the Bronx Zoo and evaded capture for nearly a week.[18]

It also bears remembering that barriers not only keep zoo animals *inside* their enclosures, but also keep dangerous animals *out*, whether they be stray dogs, foxes, raptors, or even other escaped zoo animals. And lest we forget, zoo animals can find themselves visited upon by the least predictable trespassers of them all, *Homo sapiens*. In July 1999, twenty-seven-year-old Daniel Dukes snuck into an orca exhibit at Sea-World Orlando and was drowned by a five-ton killer whale. In November 2011, a group of territorial spider monkeys from the Sorocaba Zoo in São Paulo, Brazil, viciously attacked an inebriated zoo visitor after he foolishly climbed into their enclosure. Human intruders occasionally sneak into lion and other wildcat enclosures both during and after zoo hours, only to be devoured or at least badly injured. While such instances are statistically rare, human recklessness often prevails nonetheless, as when the *Guardian* reported in July 2014 that among drunken visitors partying at Friday night events at the London Zoo, one man poured a beer over a tiger while another stripped off his own clothing and attempted to enter the penguin pool.[19]

But while enclosure technologies may remain crucial to the display of live zoo animals, most of us regard traditional zoo cages lined with iron or steel bars as antiquated and inhumane, just as audience research shows that zoo visitors dwell longer at exhibits unobstructed by visual obstacles such as fences or screens.[20] (There are therefore stark differences between the mammal exhibits that zoos present to the public and the offstage holding cells hidden from audiences, behind the scenes. For instance, although Zoo Atlanta displays its massive collection of over twenty gorillas—the largest in the nation—in a spectacularly verdant rainforest habitat of an acre and a half, at night these apes sleep in small indoor cages reminiscent of a medieval dungeon.)[21] It would therefore seem that nature makers ought simply to replace remaining old-timey zoo cages with either less conspicuous technologies of captivity, such as razor-thin electric wiring or large panes of transparent glass, or else naturalistic moats that allow for unobstructed and thus more seemingly realistic depictions of zoo wildlife.

Yet each of these technologies creates a new set of problems that runs counter to the aforementioned priorities established by zoos. While thin electric cables, or "hot wires," can seem practically invisible even at close range, they are no panacea, and only most obviously because electric shocks can be hazardous and painful to zoo creatures. In fact, they often do not work at all, as thick-skinned animals such as rhinoceroses can withstand their high voltages. In one instance, chimpanzees in a New Mexico facility seemed oblivious to the hot current produced by an electrified fence, while an orangutan at Seattle's Woodland Park Zoo appeared entertained as an electrified windowpane sent shocks coursing through its arms. Elsewhere, animals have learned to short-circuit the cables with sticks, while elephants have done so with their tusks.[22]

Meanwhile, other seemingly invisible enclosure technologies such as glass can be just as conspicuous as iron cages or fences. For safety reasons, protective glass-paneled enclosures must be constructed thicker than some walls, much less fencing. According to Jeffrey Smith, an architect at a leading American zoo design firm,

> We always look for the magic barrier where there's "nothing" there. . . . We joke about a "force field" where the animals can't get out—you're always trying to get rid of the barriers, or not *see* the barriers, to get this interaction with the animals. Glass, obviously, is one of the best ways to do that. So a lot of thinking goes into it, as you can imagine. We hire glazing engineers, glass engineers that we say, *"Okay, we have a four-hundred-pound gorilla that could be running at this thing at twenty miles an hour, can you help us out? And how thick of a glass does that need to be?"* Usually this exhibit's glazing is several pieces of glass laminated together, so it may be two or

three (usually we don't get above three layers), but those layers of glass can be anywhere from a quarter-inch thick to a half-inch more thick with a membrane in between them to help adhere it, and actually the membrane gives it a little bit of strength as well. Glazing engineers help us come up with that sandwich, the appropriate thickness of that, as well as how much the edge frames need to grab the glass to give it strength on both sides.

Because of the thickness of these laminated, multilayered glass sandwiches some zoo architects argue that glass-paneled enclosures actually *amplify* the psychological barrier between the animal and the human viewer, as they not only present a physical (if transparent) barrier but also mute animal sounds and smells, all while reflecting sunlight glare during daytime hours.[23] Moreover, glass enclosures invite visitors to *bang* on their translucent walls when attempting to get the attention of sleeping lions or otherwise occupied gorillas or tigers, a point of stress for animals and keepers alike.

Given these limitations, moats would seem to provide a more perfect solution, alleviating the need for steel bars, panes of thick laminated glass, or any other physical impediments to visibility, scent, and sound on the part of visitors. The wide berth they provide can potentially help protect zoo animals from pathogens spread by humans (and vice versa), as well as inappropriate or even toxic foods that visitors might surreptitiously try to feed them, and projectiles that ill-mannered visitors might throw at them (a complaint made by staff at both City Zoo and Metro Zoo).[24] Of course, since moats generally increase the spatial distance between exhibited animals and the public, they make it more difficult for visitors to actually *see* such creatures, much less experience feelings of empathy and intimacy for them or learn about their anatomy and behavior. At least one published study of audience behavior at zoos reveals that the greater the distance between visitor and animal, the less likely visitors will stop by the exhibit.[25]

More important, moats can take up much-needed space in a zoo exhibit that can otherwise be appropriated for larger living areas for the animals on display. As Smith explained to me,

Moats and pools are very aesthetically pleasing, and you can have dry moats that form the same kind of earth barrier. But those often take up usable habitat for the animals . . . you're taking away space where these animals can go. So you have to weigh that back and forth. . . . If their space is more limited, they may decide not to go with the moat and do the barriers. . . . Ideally, if it was just aesthetics, you would not see any barriers and have lots of space and do everything you want, but oftentimes the zoo may only have two or three acres to do an elephant habitat. Do you really want to take up a third of that with moats that are unusable by animals?[26]

In addition to their drain on zoos' spatial resources, moats and similarly designed pools also pose certain risks to animal safety. Chimpanzees and other great apes have been known to drown in deep moats filled with water.[27] Likewise, although moats can prevent certain kinds of animal escapes, they sometimes cannot stop feral creatures from encroaching *into* zoo animals' living spaces. At Metro Zoo, bald eagles enjoy an open-air exhibit, since as injured birds rescued from the wild, they pose few flight risks. On the other hand, regional flocks of wild black vultures regularly swoop into their cageless enclosure to pilfer the eagles' carefully calibrated rations of meat prepared by keepers. In such cases, the distinction the zoo makes between captive and free birds of prey obviously means very little to the birds themselves.

Nature makers face a second set of dilemmas when creating zoo exhibits. In the past several decades zoos have increasingly relied on what exhibit designers call *landscape immersion* to ensconce both their captive animals and audiences in settings that dramatically simulate natural environments, whether Indonesian rainforests, Kenyan savannas, or the Mojave Desert. Nature making therefore relies as much on the botanical and horticultural sciences as it does on zoology and animal behavior. In keeping with the educational aims of AZA-accredited zoos, some of the most sophisticated immersive exhibits envelop zoo animals and visitors alike in a maze of trees, grasses, and other flora indigenous to the animals' native ecosystems, and zoo educators use them as backdrops designed to teach guests about biogeography as well as habitat preservation and other environmental issues.

For example, Zoo America in Hershey, Pennsylvania, exclusively exhibits North American regional habitats that include indigenous plant life along with native animal species. In the zoo's Eastern Woodlands habitat, local trees such as white oak, Eastern hemlock, and American beech shade the black bears and bobcats on exhibit. Gray wolves and peregrine falcons reside in its Northlands exhibit alongside Alaskan hardwoods such as black spruce, lowbush blueberry, and paper birch. Handy nameplates help visitors identify the different botanical species growing in each enclosure by the shape of their leaves, needles, and flowers. Meanwhile, the Arizona-Sonora Desert Museum in Tucson takes this a step further by exhibiting only local plants, animals, minerals, and fossils indigenous to the Sonoran Desert region of the Southwest.

Yet attempts on the part of zoos to present ecologically accurate portrayals of wild habitats for both aesthetic and educational purposes can be all too easily undermined by the realities of animal management, and occasionally Mother Nature herself. Most obviously, while zoos showcase exotic animals from all over the world, the plants indigenous to their native habitats simply cannot survive outdoors in zoos located in radically different geological or climatic zones. The Detroit Zoo is home to black-and-white ruffed lemurs and tree boas indigenous to the forests of Madagascar off the African coast, yet native plants such as the tropical species of palm endemic to that island country cannot easily be transplanted to the zoo's outdoor exhibits, especially with its frigid Michigan winters.

Nature-loving visitors may scoff at one possible solution to this problem: swapping out living trees for synthetic plants. Yet ironically, live, organic trees can be hazardous to zoo animals since they attract disease-spreading microbes and insects, and unlike in the wild, captive creatures cannot easily migrate to less infested habitats. Similarly, although wood may seem like more of an aesthetically "natural" material to construct exhibit walls than fiberglass or epoxy, wood walls might also provide shelter to mice, cockroaches, and other pests. (The irony here is that while cockroaches and mice may be considered pests in this context, some zoos simultaneously *exhibit* such insects and rodents as part of their curated collections.)[28] Moreover, zoo animals such as apes can destroy the elaborate plantings of even the hardiest of landscaped environments. According to the famous field primatologist Dian Fossey, wild gorillas living in the forests of Rwanda tear up their surrounding vegetation to build new nests in different locations on a *nightly* basis, and in her landmark book *Gorillas in the Mist* Fossey suggests that such behavior may be innate, even among zoo apes born in captivity.[29]

Exemplified by this last example, a third set of dilemmas faced by nature makers concerns the biological appetites and inclinations of zoo animals. The educational mission of AZA-accredited zoos encourages adherence to ecologically accurate renderings of the natural world, and zookeepers, educators, and volunteer docents at zoos across the country regularly offer live interpretive presentations to visitors concerning the native habitat, current stage of development, diet, and endangered status of the animals in their collections. Nevertheless, zoos sometimes consider the realities of animal life inappropriate to reveal to young children, and aesthetically or morally unpleasant to their parents as well. For example, zoos often struggle with the question of how much information about animal mating habits and their sexual practices ought to be explained to small children and preteens. Of course, when zoo animals copulate or simulate intercourse in public view, there is often little that can be done to stop them *in flagrante delicto*, regardless of who may be watching. Indeed, sometimes nature simply calls, just like when goats or sheep urinate on unlucky children or their parents during visits to petting barnyards in family-oriented zoo areas.

The diets fed to zoo carnivores may similarly repulse visitors of all ages, especially the sorts of foods that approximate what these animals would naturally hunt, ravage, and eventually consume in the wild: (live) mealworms and crickets; and (dead) rats, mice, quail, rabbits, chicks, and guinea pigs. As for feeding meat-eaters live game more substantial than mealworms, the expectation of audience outrage generally prevents zoos from serving living mammals and birds to carnivores, to say nothing of the suffering such prey would experience. (Live prey can also present dangers to zoo animals themselves, particularly when they attack their would-be devourers.)

A somewhat related dilemma in nature making concerns how the relative sociability of certain animals should be represented in zoo exhibits. On the one hand, evidence suggests that visitors gravitate toward zoo mammals exhibited in social groups, especially when they include infants.[30] Then again, some creatures, such as the ocelot (or dwarf leopard) and the coati (a member of the raccoon family), live alone in the wild, solitary by nature. The sociability of these mammals varies dramatically according to life cycle phase, and some adult carnivores cavort only when mating.[31] At the same time, many social animals, such as wolves, live in intensely hierarchical groupings in the wild and therefore risk being attacked by members of their own species if housed in the same zoo exhibit. This is a particular concern for children's zoos that exhibit bunny rabbits to toddlers, as their world-famous cuteness masks their rather vicious disposition. (I have the scars to prove it.) Zoo personnel must measure the popularity of group displays of mammals against the need to create both safe and zoologically accurate depictions of the wild.

A final set of dilemmas encountered by nature makers concerns the use of behavioral stiumulators. Life in a zoo has its advantages: a reliable, safe supply of food for life; consistent veterinary care; protection from the elements; and freedom from fear of being eaten alive by predators. The downsides are equally obvious, including spatially constrained living quarters; a general lack of privacy; limits on mobility and autonomy; and bouts of irrepressible boredom. To compensate for the dullness of captive living, AZA-accredited zoos provide their animals with behavioral (also called environmental) *enrichment,* or additions to their habitats that will hopefully bring out their animals' "naturally" occurring behaviors while providing them with sensory and cognitive stimulation. Zookeepers might hide a bobcat's allocated portion of ground beef and bones in the crevices of a hollowed-out tree stump, or feed a polar bear frozen fish or blood popsicles. Since noncaptive gorillas spend up to 70 percent of each day foraging for food in the wild, keepers will scatter raisins or breakfast cereal about the grounds of their exhibit.[32] The introduction of behavioral or environmental enrichment in zoo exhibits can not only encourage activity in otherwise inactive creatures, but also help reduce abnormal stereotypic behaviors symptomatic of boredom, stress, anxiety, frustration, or fear among animals, such as pacing, excessive nodding, fence licking, or other repetitive movements.[33]

Enrichment often consists of playing and climbing apparatuses, swimming pools, scratch posts, tree branches, hollow logs, loose browse, nest boxes, and toys of all kinds—bowling balls, puzzle feeders, colored wooden blocks, and so on. Zoo personnel also provide enrichment by altering the temperature, ambient sound, or even odor of an animal's habitat. To accomplish the latter, keepers may introduce designer fragrances to their environments. The African wild dogs at the Philadelphia Zoo enjoy Chanel No. 5, while the San Diego Zoo's giant pandas prefer Ralph Lauren Polo for Men. The San Francisco Zoo's snow leopards go wild for Old Spice, and keepers at the San Diego Zoo Safari Park supply their tigers with deer antlers sprayed with Calvin Klein's Obsession. With spring in the air, Metro Zoo keepers once sprayed a donated sixty-dollar bottle of Trésor in Love eau de parfum by Lancôme on the insides of a coati's enclosure.

Enrichment also involves employing strategic animal management practices that provide mental stimulation while discouraging stereotypic behaviors, as in the aforementioned example of providing novel foraging opportunities for gorillas. One autumn morning before visitors arrived, Metro Zoo animal keepers served the zoo's two gray wolves a freshly killed deer carcass shot by a local bowman, thus providing the carnivorous canines an opportunity to experience predatory behaviors while in captivity. Contemporary enrichment strategies also include innovations in animal rotation in which different species alternate among a variety of exhibit enclosures, offering zoo animals new opportunities for exploration outside the confines of their everyday habitats. In 2011 the Philadelphia Zoo unveiled its Treetop Trails, a $1.5 million experimental zoo enrichment system in which small mammals traversed the zoo on their own through an elaborate system of overhead bridges, walkways and lookouts made of flexible, stainless-steel mesh. Golden lion tamarins, white-faced sakis, red-capped mangabeys, and other exotic primates climbed through 700 feet of elevated paths among the trees, free to take in a bird's-eye view of local wildlife as well as neighboring zoo attractions. In later years the skyway system, designed in collaboration with Jon Coe, expanded to over 1,735 feet, and could accommodate larger apes like orangutans. Eventually the Philadelphia Zoo added Big Cat Crossing, a 330-foot extension that allowed its massive tigers, lions, pumas, jaguars, and snow leopards to wander the grounds from fourteen feet above the heads of awestruck visitors.[34]

Research shows that among zoo gorillas and orangutans the presence of behavioral enrichment objects, especially movable objects like coconuts or trashcans, may do more to encourage activity than even an enlarged enclosure. Moreover, enrichment augments not only the physical and psychological welfare of zoo animals, but also the popularity of zoos themselves. Physically active animals impress zoo visitors far more than their sleepy counterparts, and in fact studies show that audiences spend twice as much time viewing energetic animals than those same creatures lying dormant. Meanwhile, captive animals displaying repetitive stereotypic behaviors (such as pacing) naturally disturb zoo audiences, particularly those concerned with animal welfare more generally. Behavioral enrichment can therefore increase visitor satisfaction as well as provide stimulation and therapeutic relief for captive animals.[35]

However, while enriched zoo animals may exhibit behaviors common to their brethren in the wild, many successful enrichment objects—tractor tires, oil drums, ping-pong balls—clearly do *not* appear in idealized natural landscapes, and may therefore interfere with the illusion of authenticity otherwise suggested by more naturalistic environments.[36] At City Zoo, orangutans swing from Kevlar-lined fire hoses, as their keepers find them more difficult to shred than vines or ropes. Turkeys and guinea fowl peck at broom heads. Rabbits play with wicker balls stuffed with carrot bits, along with blue icepacks on hot days. At Metro Zoo, cougars occasionally stalk an empty beer keg and enjoy venison served out of big plastic blocks. Jaguars play with old sneakers while woodchucks play inside old cartons of Woodchuck cider. Cheerios boxes stuffed with newspaper adorn an indoor enclosure featuring a green iguana, cotton-top tamarin, and red-footed tortoise. Meanwhile, at the San Diego Zoo Safari Park keepers give a male lion giant phone books it can rip to tatters, and at the Milwaukee County Zoo orangutans communicate via Skype and play Fish Farm, Magic Piano, and Flick Kick Football on donated Apple iPads.[37]

## Nature Making as Stagecraft

Given the dilemmas and tensions inherent in creating naturalistic zoo environments, nature makers rely on specific strategies of stagecraft designed to overcome these challenges. One set of exhibit design strategies involves controlling the sight lines and visual experience of both animal and visitor alike through the manipulation of physical space. An obvious impediment to building even simulated wilderness in zoos is that animals live in massively spacious habitats in the wild, especially nomadic creatures such as polar bears and wolves. Nature makers therefore manipulate sight lines to create an illusion of limitless space and total animal control over the environment. The Philadelphia Zoo's Amur tiger exhibit features a variety of angled windows that hinder visitors' ability to see the entire habitat at once, thus making the enclosure seem deceptively larger than it actually is. Occasionally, the illusion confuses visitors who believe that each new vantage point reveals yet another tiger when all the while they are simply seeing the *same* tiger from alternate points of view.

This illusion of vastness is heightened in zoos that cleverly employ sight lines to appropriate the trees, hills, mountain ranges, and other surrounding elements of nearby landscapes as theatrical backdrops for animal displays.[38] While the Brandywine Zoo in Wilmington, Delaware, sits on only 12 acres, its scenic surroundings include a 178-acre state park thick with foliage that envelops the zoo grounds, providing an impressive mise-en-scène for its outdoor exhibits. The Arizona-Sonora Desert Museum (ASDM) in Tucson accomplishes this illusion better than perhaps any other American zoo. Like the city of Tucson itself, the zoo is situated in a Sonoran Desert valley surrounded by five minor mountain ranges, a dry sandscape where saguaro cactus, agave, and aloe grow under a big blue canopy of Arizona sky. With no visible skyscrapers or railway lines to obstruct one's view, all one sees beyond the zoo's desert-themed exhibits is this spectacular ecological landscape. The viewer then sees the zoo's animal enclosures and expertly planted cactus gardens as part of a continuum with the surrounding desert environment itself. Together these combined elements depict a panorama that allows ASDM audiences to imagine the zoo's captive animals living freely amid the windswept layers of the enveloping natural wilderness, rather than trapped behind simulated rock formations made of poured concrete.[39]

The ASDM accomplishes this feat of staged authenticity by attending to the production of nature at every level of crafted exhibition. First, the zoo displays only a limited number of animals (about 230), all native to the southwestern corner of the United States, if not the Sonoran Desert proper. This means not only that the exhibits resemble similar habitats, but also that there is little to no dissonance *among* exhibits—as there is at the Philadelphia Zoo, which displays Arctic polar bears within view of African giraffes and white rhinos. The constantly maintained and reconstructed gardens feature 1,200 indigenous plant species (56,000 individual specimens) arranged in appropriate proportions, with a purposeful randomness and irregularity that allows them to seamlessly blend into the surrounding terrain.[40] Finally, the desert surroundings themselves are emphasized by the open-air nature of the zoo. This stands in contrast to many metropolitan zoos where tall trees and other plantings intentionally *obscure* sight lines to *hide* elements of the surrounding urban terrain, whether train lines or the tops of towering buildings.

At the same time, the nonnative tourist (like myself) experiences the Sonoran Desert ensconced in an immersive dry heat that gave me a decidedly authentic case of dehydration and sunburn during my field visit. Leaving little to chance, the zoo promotes this kind of visceral authenticity as well, as a placard instructs passersby:

> Get Ready for a REAL Desert Experience. This trail is hot, dry, dusty, and bumpy—a half-mile long and uphill on the way back. . . . The Desert Loop Trail is our attempt to give you a real outdoor desert experience. You will see fascinating desert plants and enjoy beautiful vistas. The exhibits are quite large and blend into the natural desert, so you might not see all of the animals on a single visit. Be on the lookout for both javelina and coyote. You may also see many kinds of lizards and birds. In fact, you never know what you might see. Remember, just like hiking in the wilds, a little effort and patience will always be rewarded.

In ASDM's messaging, visitors should experience the zoo as if they were trekking through the open desert itself: inhaling its arid and dusty air, spotting feral as well as exhibited animals. Unlike many zoos that emphasize their *distinctiveness* from their exterior environments, the ASDM consciously draws on its surrounding landscape to *blur* the boundary between inside and outside, domesticated and the wild. In a similar fashion, aquariums are often built on waterfronts in order to best emphasize their visually scenic qualities, from maritime cultural artifacts to marine life and even the water itself. In Chicago, the John G. Shedd Aquarium's Abbott Oceanarium overlooks the rough waves of Lake Michigan, while Baltimore's National Aquarium juts out alongside the docks of the city's Inner Harbor. Adventure Aquarium in Camden, New Jersey, sits along the Delaware River and its waterfront attractions, including a decommissioned World War II battleship and a fifty-slip marina. At the Monterey Bay Aquarium in Northern California, visitors enjoy brilliant outdoor deck views of the bay and its U.S. National Marine Sanctuary, where humpback whales, dolphins, harbor seals, sea otters, and pelicans dance along its kelp-filled waters. Such attractions that deliberatively blur man-made boundaries between the zoo or aquarium and its surrounding ecology can be far more effective as educational tools than the typical blizzard of signage that so often overwhelms visitors with small children in tow.

Other nature making practices similarly employ strategic sight lines to bolster the visitor's naturalistic experience of the zoo. Architects and designers strategically elevate exhibits above public viewing areas on the ground, thereby placing visitors in a spatially subordinate position relative to the animals on display. In the grassland habitats of Metro Zoo, spectators feel the majesty of massive elk that peer down at them from the top of their steeply sloped enclosure. The San Diego Zoo's Lost Forest immersive exhibit requires visitors to gaze up at Malayan tigers perched at terrific heights. According to Jon Coe, "the simple procedure of locating the animal in a position or location superior to the viewer may relatively predispose the viewer to want to learn from the animal, be more attentive to it, and perhaps be even more respectful of it."[41] As Catherine Brinkley, an international zoo consultant with doctoral degrees in veterinary medicine and city planning, explained to me during a walk through the Philadelphia Zoo, zoos ought to keep "the people on a lower level than the animals. I always think that's really important, especially with predators. You want people to have a healthy respect and fear for the animals, and if you are constantly towering over them, as you would in most moat exhibits or bear pits, you don't get that sense, and there's more of a tendency to throw stuff, to throw food or pennies or whatever, into the exhibits."[42]

Nature making also involves the strategic placement of behavioral enrichment designed to coax otherwise shy animals into public view. One common trick involves placing heat-emitting sources, such as hot rocks, in the front of reptile exhibits in order to maximize visitor views. Like all cold-blooded creatures, reptiles such as iguanas, geckos, turtles, and crocodiles rely on environmental sources of heat because they cannot regulate their body temperatures internally, unlike humans and other warm-blooded mammals. For this reason zoo personnel deliberately position natural-looking thermal rocks in reptile enclosures so as to lure them toward the parts of their exhibit most visible to audiences.

Finally, sight lines at the zoo are as much about what viewers do *not* see, especially with regard to non-naturalistic enrichment items used behind the scenes. Keepers at City Zoo treated program animals who resided in backstage holding pens to handmade enrichment toys made from PVC drainpipes stuffed with newspaper, cardboard toilet paper rolls, and empty boxes of Coke and Diet Pepsi, Hostess Mini-Muffins, Cheerios, Trix, and Eggo Waffles, all occasionally spiked with Italian seasoning, dill weed, cinnamon, or other spices. Meanwhile, a spectacled owl that lived off-exhibit regularly watched children's animated movies on a TV/VCR, including *Robin Hood*, *Mulan*, and *The Brave Little Toaster*. Apparently, this is not uncommon. At Metro Zoo keepers frequently showed movies to animals as part of their enrichment regime. The zoo's collection of videocassettes included not only animal-centric children's classics such as *Lassie*, *Lady and the Tramp*, and *The Lion King*, but also more risqué fare including *Austin Powers*, *The Dukes of Hazzard*, and *Fight Club*.

## Where the Wild Things Aren't

A second set of strategies employed in naturalistic zoo design makes use of theatricality and stagecraft to simulate the wild. Natural-looking immersive landscapes rely on man-made materials, including artificial rocks and termite mounds made of concrete, epoxy, and fiberglass, to say nothing of the perfectly sculpted aquarium glass windows, hi-fidelity sound systems, and advanced parasite control technologies used in zoos. In part, this is because of the difficulties involved in using organic materials in zoo environments. As noted above, zoo animals such as nest-building apes can easily destroy the planted vegetation in their landscaped exhibits. In such instances, designers use synthetics to simulate the natural world. At the San Diego Zoo, orangutans clamber up artificial trees made of steel. The Francois' langur exhibit at the Los Angeles Zoo includes five outdoor pine trees, also made from steel. The company (revealingly named NatureMaker) that designed and fabricated those fire-retardant pines promises that each of its trees will be "87 percent botanically accurate," and advertises its products and services in its promotional materials by tapping into the adoration for naturalism felt among zoo audiences and staff alike:

> Idyllic. Majestic. And perfectly imperfect. Exactly as Mother Nature intended. . . . Bug holes. Fungus, moss and decay. Twisted, knobby, contorted and distorted trunks and limbs. Faithful reproductions for hands-on discovery, but are they realistic enough to convince the animals at the zoo? Absolutely, but don't tell any of them. They've been living comfortably among NatureMaker Steel Art trees for years.
>
> Look closely and you'll see painstakingly-sculpted impressions of real bark with all its natural anomalies. Step back and sense the grandeur of the tree's presence. What could be more important to educators, museum administrators, librarians—anyone whose vocation is to educate? The curious always inhabit the confluence where art and nature intersect. NatureMaker truly sees the forest for the trees.

Steel trees help zoos simulate environments that neither elephants, rhinoceroses, orangutans, nor gorillas can easily destroy, all while saving on expenses such as pest control, irrigation, and drainage. Yet as noted above, nature makers must protect plantings from not only powerful animals, but basic climate and weather patterns as well. Again, zoos frequently display animals from all over the world, yet the plants indigenous to their native habitats simply cannot survive outdoors in zoos located in inappropriate biogeographical environments. Given the enthusiasm among zoos and their audiences for authentic-looking immersive landscaping, nature making in North American zoos requires the use of "plant simulators," domestic plants that perform the role of other plants from South America, Asia, and Africa. For example, according to Zoo Atlanta horticulturalist Donald Jackson, plants found in West African tropical rainforests have "large leaves with a relatively smooth, waxy upper surface and long 'drip tips,' two features that help shed excessive rainfall in the humid topics." Therefore, credible plant simulators for a western lowland gorilla exhibit in a American zoo, for instance, could include smooth and staghorn sumac, tulip poplar, royal paulownia, northern catalpa, and a variety of magnolia, including sweetbay, cucumber, big leaf, and umbrella, among others.[43] As Jeffrey Smith, the aforementioned zoo architect, explained to me, "In Africa, tropical plants have one big single leaf, so we look for plant species like hostas, for example—a very common perennial that has a large leaf that looks very jungly. In a setting where you can't grow big tropical leaf plants, you might be able to grow a hosta, because they are very cold-hardy in northern climates, for instance, so that's one African rainforest simulator." Zoo exhibits depicting forest habitats also appear more authentic when their trees are planted at varying angles, since trees do not typically grow perfectly straight in natural ecosystems.[44]

Similarly, an American zoo exhibit featuring zebras, lions, giraffes, African elephants, or Thomson's gazelles, all indigenous to the arid savannas of Tanzania and Kenya in East Africa, would require trees and shrubs with "thorns or spines and small leaves to simulate the native flora's need to conserve water," like the acacias of the Serengeti.[45] Yet again, as Smith explains, "You can't really grow those in many parts of North America. So we use the honey locust; it's a good simulator, with its similar leaf look and form as an acacia." Plants native to the United States that would also serve as successful East African simulators might include mimosa, yaupon holly, cockspur hawthorn, pampas grass, and small soap weed.[46] In addition, nature makers need to take differences in climatic zones *within the United States* into account when appropriating plant simulators for immersive zoo exhibits. Again according to Smith, "If you're doing an African or jungle-type thing, palm trees grow great in Florida, but in Denver they're not going to do so well."

Of course, zoos simulate not merely plantings as they might appear in the wild, but the animals themselves. Although humans care for them daily, pampered zoo creatures frequently play the role of wild beasts, sometimes with an assist from keepers and other staff. As Metro Zoo's docent handbook advises all volunteers performing live animal presentations, "A clear and consistent message of conservation and education must be a pertinent part of every program presented by Metro Zoo. It must also be made very clear that our program animals are NOT pets and are to be presented as captive representatives of the wild population." Docents were therefore instructed during their training that because zoo animals ought to be regarded *as if* they were wild species, they were not to caress the animals as if they were pets, nor address them using baby talk when in the presence of visitors, even though such behavior was fairly common among zoo staff and volunteers behind the scenes.

The display of zoo animals as wild creatures also extends to their physical appearance. For instance, in the wild flamingos feed on brine shrimp, which contain beta-keratin; it is for this reason that their plumage is famously bright pink. (Beta-keratin is the same organic compound that makes carrots orange and tomatoes red.) However, at zoos flamingos enjoy a more managed diet of grain that often does not include shrimp, which would therefore make their feathers exhibit a pale pink or even white coloration—if not for the fact that many zoos now feed their flamingos pellets with extra beta-keratin, simply so that they will retain their pinkish hue. In this sense, the pigmentation of captive flamingos in zoos functions as a kind of aviary costume for their public performances as "wild" pink birds.[47]

## Birds Do It, Bees Do It, Even Educated Fleas Do It

A final set of stagecraft strategies employed in naturalistic zoo exhibition involves the censoring of certain animal behaviors and husbandry practices from public view in the interests of masking some of the more aesthetically unpleasant or uncomfortable aspects of zoo work and biological reality. In their attempts to create an entertaining landscape for audiences—particularly families with children—zoos will sometimes purposely avoid trespassing sensitive social boundaries that may alienate some visitors, despite the potential contribution that revealing such realities to the public might make toward their educational mission.

First, during live presentations zoo educators often mute any allusions to animal mating or sexuality, especially when in the presence of small children. Tyler, twenty-three, one of two Metro Zoo education volunteer coordinators, warned a classroom of docents during a training session, "Don't talk about sex," even *nonhuman* animal sexuality, when in the presence of small children.[48] Staff usually communicated these warnings in a joking if nevertheless sincere manner, as when Tyler humorously advised during a training session, "For example, it would be inappropriate to say, 'A *blue whale has a ten-foot penis*,' in front of a bunch of six-year-olds." As he explained,

I've told a lot of people that the [mating] butterflies are "dancing" with each other. . . . It's a nice and inoffensive way to skirt the issue, but it's not true. You're purposely feeding the guest incorrect information. . . . We'll do shows where you talk a lot about a lot of the specifics of breeding without actually talking about sex. I did a Mother's Day show about animal moms. When you talk about marsupials, it gets harder and harder to explain what a marsupial is without explaining its birth. . . . And then the Father's Day show is even harder to do. . . . A lot of interesting facts about animal dads usually involve their actual sexual performance, and you can't talk about that, as fascinating as it would be.[49]

At City Zoo, the turtles and tortoises that resided in the reptile exercise yard were probably the most notorious public exhibitionists on display—so much so that the zoo's educational staff specifically instructed its volunteers to respond to curious children using strategies of misdirection and evasion, along with a litany of fun animal facts. For example, if a young visitor should ask, "What are those tortoises doing?" zoo staff advised volunteers to respond with a question: "What do *you* think they are doing?" or "What does it *look like* they're doing?" If the child answered, "It looks like they are playing leapfrog," volunteers were told to give a tactfully evasive response, followed by a quick change of conversation topic: "Yes, it does *look* as though they are playing leapfrog. . . . Say, *do you know how much food a rhino can eat in one day?*"[50]

Second, while exceptions exist, it is not uncommon for American zoos to hide the feeding of prey (live or dead) from public view by spatially or temporally sequestering such practices from audiences. Again, this is especially the case in areas designated for young visitors. At City Zoo, keepers and volunteers working in the children's zoo prepared and distributed small birds and rodents to owls, hawks, and other large raptors on a daily basis. These carnivorous diets included the bleeding carcasses of thawed frozen mice (referred to in the industry as "pinkies," "fuzzies," or "hoppers," depending on their size), white rats, quail, and yellow chicks, some stuffed with additional helpings of raw ground meat. Zoo personnel regularly doled out these morsels to zoo animals about fifteen to thirty minutes before closing time. By then most young visitors had already exited the park grounds, saved from exposure to such a ghastly menu.[51] (Meanwhile their parents were invited to purchase sanitized pellets of grain for their children to feed to the zoo's goats and ducks.) Although carnivores quite obviously eat other animals in the wild, at zoos this fact is often kept from small children whom some parents fear might be confused by such a discovery, or at least disturbed by the sight of dead prey. In these instances, zoo visitors care less about experiencing the authenticity associated with what the nineteenth-century British poet Alfred, Lord Tennyson once described as "Nature, red in tooth and claw," as much as an airbrushed depiction of nature as a tidy cultural artifact, free of blood and bones. Of course, it is not only visitors who experience this mediated image of nature, but the animals themselves—after all, predators in the wild hardly feed on measured-out portions of ground beef cleaned of fur, much less disease-free breeder mice. According to the field biologist George Schaller, African lions feast on zebras, warthogs, impala, buffalo, and waterbuck in the wild—all animals more likely to be exhibited by zoos than served to their captive wildcats for brunch.[52]

Many American zoos sanitize their animals' diets for the sake of visitors. At one accredited zoo in the southern United States, a keeper in an African birds exhibit explained to me,

> We have feeding platforms out here [in public] . . . and then we will also put some food upstairs [out of public view]. One of the things we put upstairs is for the meat-eating birds, because most people find it a little distasteful when they see one picking up a pinkie [a small mouse] or something like that. . . . They are all dead: they are frozen, the ones we feed them. We also feed them commercial-type meat that we mix the insectivore pellets in. So people are okay with that—but they generally don't want to see them eating mice. We feed that upstairs [out of view], and the birds will take those to their nesting area.
>
> If you have someone around that can explain that owls eat mice, I think that's acceptable. But it is not something you want to throw in everybody's face. You know, sometimes the birds will drop the pinkies on the ground, and it is kind of weird.
>
> It just seems to be human nature that you can feed fish; people don't seem to have a problem with that, as long as they're dead—you don't want to see anything flopping around. Anything with fur, they seem to have issues with. . . . I have done some raptor programs where we fed large mice to raptors, and the people who were interested watched, and they understood that that is what these birds eat in the morning. But we can't always explain it that way.

In order for zoo visitors to experience nature in a cultivated or sanitized manner, the presentation of animal diets is often edited for public viewing. At City Zoo, carnivore food bowls and trays always remained under aluminum foil, never to be shown to visitors—even when asked. At Metro Zoo, prey animals would sometimes be partially dismembered before being fed to exhibited animals on display. During the aforementioned episode in which keepers at Metro Zoo served a deer carcass to its gray wolves before visiting hours, a staff person explained that the deer's head and limbs had been removed prior to feeding, lest zoo visitors "think of Bambi" should they arrive before the wolves finish devouring their meaty breakfast.

In addition to animal sexuality and diet, a third biological reality commonly censored in zoos is the ubiquity of death among resident animals. While zoos devote an abundance of resources to veterinary care and animal safety, demographic research shows that captive zoo animals actually exhibit the same mortality rates as animals in the wild.[53] In fact, while researching this book many resident species died at both City Zoo and Metro Zoo—alligators, mountain lions, jaguars, peccaries, greater rheas, pygmy hedgehogs, hognose snakes—the list goes on.

Nevertheless, keepers and other zoo staff painstakingly hide animal deaths from visitors when possible, especially young children. Shortly before I began my fieldwork at City Zoo, two feral foxes entered the chicken coop in the barnyard exhibit one night and ravaged about twenty chickens. The staff managed to clean up the mess before any children arrived. The timing of the cleanup was very deliberate—I asked Phil, one of the zookeepers, what would have happened if any children had witnessed the aftermath of the incident. He admitted that it probably would have scared them, or made them cry. Tony, another zookeeper, assured me that "it would never happen," that they would never expose children at the zoo to such gore. If any kids did happen to find feathers strewn about the yard, he said the keepers would sooner *lie* to young visitors, and make up a story about how the chickens lost some of their feathers by fighting one another, rather than reveal the savagery of the animal kingdom.

At Metro Zoo, keepers hid news of unexpected animal deaths from guests by issuing a "10:19" when communicating by radio. According to Heather, a zookeeper and veterinary technician, "10:19 is radio code for dead. So, instead of saying *'dead'* out loud on the radio, you say, *'I have a 10:19 hourglass tree frog.'* It's just so you don't say it over the radio for a guest to overhear it." (I first discovered this when Stanley, one of the zoo's two resident peacocks, was found dead under an old repurposed caboose near the bison yards and someone wrote in blue marker on a whiteboard in a restricted area, *"Stanley is 10:19."*) Also at Metro Zoo, staff personnel sometimes needed to surreptitiously kill predator insects such as dragonflies in the butterfly house, in order to protect the monarchs and moths on exhibit. As Tyler advised the zoo's volunteers, "If you need to, squash it with your foot. . . . And if nobody's there to see it, then *it didn't happen.*" Staff also instructed docents to euthanize injured butterflies with broken legs or wings, but only to do so "when no kids are around."[54]

## The Enchantment of Culture

Representations of nature always rely on the materials of culture, and in zoo exhibits they include man-made synthetics, animal management practices, and live animal encounters and performances by zoo staff and volunteers. Nevertheless, a variety of competing institutional priorities at zoos—the expectations of audiences, the educational mission of zoos, and the practicalities of ministering to live species—all shape the context in which nature makers construct naturalistic zoo exhibits.

However, there is an ongoing debate within the zoo industry regarding whether zoo exhibits or enrichment need look naturalistic in the first place, as some argue that such design elements should more transparently reflect the conditions of zoo animals in captivity. According to this postmodern logic, even exhibits that merely evoke nature by using simulated materials like artificial logs can seem dishonest to visitors. This would suggest a move toward presenting guests with *non*-naturalistic stage sets that draw attention to *human*-dominated environments—call it *cultural* immersion.

This can be seen in a variety of contemporary zoo exhibits that represent man-made settings where animals may dwell in real life (or not). The Philadelphia Zoo displays its Sumatran orangutans and western lowland gorillas in an emphatically non-naturalistic environment—a cleverly designed stage set intentionally created to look like a repurposed industrial site where apes climb on crane rails and multistory metal scaffolding. (During one visit to the exhibit, a baby orangutan climbed the scaffolding to the second floor where I was standing, and then swung over to smear a handful of poop on the glass, inches in front of my face.) At KidZooU, the Philadelphia Zoo's children's pavilion, a domestic rat exhibit simulates a scientific laboratory where zoo rodents perform the role of lab rats running along an obstacle course of ramps, rope ladders, running wheels, and plastic tanks full of pet litter. While signs explain the rats' natural origins (Asia), they also note laboratory animals' relationship to humans. "We train rats and other animals by giving them rewards—their favorite foods—when they perform a behavior correctly." "My French cousin Hector [a rat] became famous when he was launched into space in 1961." "Rats can be affectionate with people who treat them nicely."[55] At both the Philadelphia Zoo and the San Diego Zoo, avian exhibits featuring domesticated pigeons depict them living in simulated rooftop coops where bird handlers might fly them over urban neighborhoods in New York or other metropolitan outposts.[56]

Meanwhile, on the West Coast the Monterey Bay Aquarium displays some of its marine wildlife—walleye surfperch, strawberry anemone, vermillion rockfish—in its Wharf habitat, that blurry border space between terrestrial human settlements and the salty ocean where sea creatures swim on tides full of trash that they sometimes transform into makeshift homes. In an underwater display depicting the ecosystem below a harbor's docks, a fish takes shelter among rope-wrapped wooden pilings in the ankle of an old white rubber boot. Nearby placards read,

> *The sea meets bustling business at the wharf.* Though the wharf's not a natural habitat, many animals find it the perfect place to settle down. The water's calm, and food comes in on the tide—or from above, when fishermen dump fish scraps after their catch of the day.
>
> *One man's junk is anemone shelter.* Natural cracks and crannies aren't the only places animals live under the wharf. Castoff junk provides hiding places for so many creatures.
>
> An old shoe, rusty can or bottle sometimes becomes a fringehead's home. [The onespot fringehead is native to California's waterways and the Pacific Ocean.] Tucked snugly into the junk with just his head poking out, a male guards his castle, ready to charge at trespassers.

The Monterey Bay Aquarium also showcases an Art for the Environment exhibit, in which artists use plastics and other toxic synthetic materials that wind up polluting our seas to create works that attempt to help visitors "re-envision our relationship to the world we live in," particularly in the Anthropocene.[57] Scientists estimate that 5.25 *trillion* particles of plastic weighing nearly 269,000 tons float atop the world's oceans.[58] Much of that drifting debris concentrates around five major garbage patches, accumulations of bath toys, tires, polystyrene foam containers, bags, bottles, and other waste that oceanic winds, currents, and tides draw together in swirling masses of trash. Taking inspiration from that thickening ring of sludge and debris known as the Great Pacific Garbage Patch, Chris Jordan's 2009 photo collage *Gyre* incorporates toothbrushes, buttons, combs and other plastic items, in total "2.4 million pieces of plastic—the number of pounds of plastic estimated to enter the world's oceans every hour." Such artworks poignantly illustrate how human activities reconstitute every inch of the Earth's environmental reality. While zoos and aquariums are themselves landscapes of human domination, they can also serve as sites for contemplating the scale of our impact on the planet and (hopefully) potential solutions to our current environmental predicament.[59]

Similarly, the Monterey Bay Aquarium features models of a leatherback sea turtle and a humpback whale, both constructed out of blue recycled plastic by artist Sayaka Ganz. The signage surrounding *Humpback Whale* echoes the urgency of Jordan's *Gyre*:

> *Lost fishing gear traps humpbacks.* Each year, tons of commercial fishing gear is lost at sea. As these plastic ropes, lines and nets drift with currents and drag along the seafloor, they continue "ghost fishing" for animals for years. Tangled masses of lines and nets trap and injure migrating humpbacks.
>
> *Every effort makes a difference.* Fishermen, universities, government agencies and other organizations are teeming up to collect lost fishing gear in the oceans. What can we do? Continue to zero in on plastic trash—and clean it up before it makes its way out to sea.

Along with these strangely compelling plastic models of biological creatures, Ganz's most disturbing piece of artwork, *Laysan Albatross*, uses excavated artifacts from the Anthropocene to bolster her plea to protect the oceans from human-produced polymers. Drawing on the cultural connotations and ecological impact of man-made toxins and consumer waste, Ganz shows us a bowl brimming with bottle caps and other plastic debris, all discovered inside the stomach of a Laysan albatross chick found on Midway Island in the central Pacific Ocean. Just like the staged zoo exhibits discussed in this chapter, *Laysan Albatross* melds together organics and synthetics, rupturing long-standing myths about the distinction between culture and nature while sounding the alarms of the environmental crisis for us all.[60]

<center>~~~~~~</center>

Nature is a cultural invention, and perhaps more than any other contemporary urban attraction, zoos best reflect how humans distill the elemental muddle of the ambient world into a curated universe of distinctive landscapes and habitats of varying social sentiment and status. At the same time, the literal construction of the zoo's built environments—its ersatz tropical rainforests, polar icecaps, grassland savannas—illustrate how humans have always worked upon the Earth, and not only to raise agricultural crops and farm animals, but to cultivate fertile rainforests, renovate scenic waterfall vistas, and preserve dedicated lands in national parks and refuges. The zoo's fiberglass palms, artificially pink flamingoes, and man-made panoramas swiped clean of animal corpses and the other nasty bits of the natural world also remind us that we typically experience nature when it is shrouded in romanticism, as purple mountain majesties. Yet as the Monterey Bay Aquarium's luminous artwork illustrates, zoos (aquatic or otherwise) can also serve as centers of environmental education with the potential to mobilize audiences around issues of great import, from ocean pollution to climate change. In the Anthropocene, zoos are not only collections of man-made environments—steel trees and all—but models for understanding life on a planet whose history has become forever entwined with our own.

In the next chapter, our safari draws us in for a closer view of the American alligators, Siberian tigers, squirrel monkeys, and giant pandas that live in our zoos, and the symbolic power we collectively attach to various members of the animal kingdom. Although wildlife exists all around us, we can make sense of such creatures only through cultural lenses that shape our shared understandings of such beasts according to the fashion of our deepest prejudices, wildest dreams, and pop-cultural fantasies. To best illustrate how we socially construct nature by investing its inhabitants with meaning and moral value, we must return to the public grounds of the zoo.

# Chapter 2

---

# Animal Farm

## Making Meaning at the Zoo

**THE DESIRE TO ANTHROPOMORPHIZE** zoo creatures is so great that visitors often incorrectly assume that animals displayed in groups must be organized into idealized nuclear families. At Metro Zoo I assisted visitors who had paid five dollars apiece to hand-feed lettuce leaves to a lovely pair of reticulated giraffes named Mwangi and Ayo, ages four and one, respectively. Guests always assumed they were mother and child, especially given their differences in height and age, when in fact both were males and unrelated to one another—they were even on loan from different zoos, and thus perfect strangers.

Sociologists and anthropologists have long been concerned with how humans invest animals with emotional value and symbolic significance. In 1912 the French sociologist Émile Durkheim proposed in *The Elementary Forms of Religious Life* that ancient tribes of aboriginal Australians and Native Americans adopted the names and images of particular animal species—the wedge-tailed eagle, the black cockatoo, the buffalo, the wolf—to serve as collective symbols of kinship, badges of group belonging. During rituals of worship, celebrants decorated costumes, paintings, and other ceremonial objects with these sacred emblems of social solidarity. Given their exalted significance, these totemic animal species themselves were often endowed with special magical powers and charisma.[1]

Ancient civilizations tended to derive their totems from animals found close at hand, given the proximity of their societies to the surrounding *terra firma* and its wild inhabitants. But given emerging environmental threats of mass species extinction, habitat destruction, and biodiversity loss, it may be instructive to ask how we modern humans assign meaning to animals in a more urbanized context, especially given the remoteness with which most of us experience what we have traditionally considered the natural environment. (This may even be the case for those ecological habitats immediately within our reach, like city parks or our own backyards.) Zoos are repositories of culture and imagination, and by observing how groups of visitors—primarily families with children—react toward those creatures residing behind their gates, we can better understand how humans socially construct the natural world by investing the animal kingdom with shared meaning and moral sentiment, no matter whether we express such collective sentiments as enchantment, empathy, dismissiveness, or disgust.

## Animals as Totems

As a respite from the virtual realities of the digital age, Americans seek out the authenticity commonly associated with consecrated spaces of wildlife—the Sequoia National Forest, the Great Barrier Reef, the Serengeti—as well as their most famous cultural reproductions, as illustrated by the continued popularity of Ansel Adams's landscape photography, IMAX-formatted nature documentaries, and, of course, the naturalistic zoo exhibits described in the last chapter. The sociologist James William Gibson celebrates this heightened emphasis on the authenticity and awe that we humans attach to ecological habitats and their wildlife as a renewed "culture of enchantment" surrounding the natural environment. At the same time, our media-saturated landscape of eco-awareness incites everyday pop-culture consumers to experience their own postmodern reveries of nature as sublime or even spiritual—even as such experiences of nature occur in the mediated environments of metropolitan zoos.[2]

According to the biologist Edward O. Wilson, scientists have identified and categorized approximately 1.9 million species of organisms.[3] Naturally, zoological parks exhibit only less than 1 percent of those creatures. But even among the roughly 6,000 species displayed by AZA-accredited institutions, zoos and aquariums ultimately rely on a very small handful of charming creatures to attract guests, members, and donors. While taking in the exhibits at the world-renowned San Diego Zoo, I asked a tour guide to recommend the zoo's five most popular animals. Her list included gorillas, polar bears, elephants, koalas, and, San Diego's star attraction, its giant pandas. At City Zoo, the development office maintains what they call their Big Twelve animals that attract the most curiosity from donors: again polar bears and koalas, in addition to chimpanzees, giraffes, rhinoceroses, lions, and hippopotamuses.

What explains the magnetism or charisma of one zoo animal over another? Zoo research suggests that among visitors, size matters. Large mammals, specifically pachyderms, great apes, bears, and big cats, hold the attention of audiences longer than their less hefty counterparts.[4] The zoo industry characterizes these massive exotic animals as *charismatic megafauna*, just as they do hippos, giraffes, zebras, rhinos, and a host of large marine mammals, including whales, dolphins, seals, walruses, and seal lions.

Size matters not just at the zoo. When it comes to protecting threatened or endangered species in the wild, economists have shown that these decisions tend to be more correlated with the physical size of the animal than its *actual endangerment status in the wild*. The protection of the North American brown bear receives among the highest levels of government spending relative to other animals, yet the World Wildlife Fund ranks its endangered status at the *lowest* possible level. As George Orwell reminds us in *Animal Farm*, "All animals are equal, but some animals are more equal than others."[5]

Because of their seductive allure among the public, charismatic megafauna also frequently serve as symbolic totems for zoos themselves, adorning membership cards, public signage, and promotional materials featuring the institution's logo, just as name brands rely on animal mascots—Lacoste's alligator, the MGM lion, the NBC peacock—to enliven their corporate images.[6] At the entrance to the Philadelphia Zoo stands a ten-foot granite and stainless-steel reproduction of the zoo's logo, which features a silhouette of a wildcat patrolling its grasslands. At Adventure Aquarium, kids take turns climbing into the smiling jaws of Gill, the aquarium's official shark mascot, to have their photo taken while posing in between his sharp-pointed lower teeth.

Further down the Delaware River, staff members at the Brandywine Zoo in Wilmington, Delaware, wear polo shirts with a tiger's face displayed under the zoo's name. In keeping with its feline iconography, for twenty years Brandywine exhibited a celebrated Siberian tiger named Ashley. When she had to be euthanized in 2010, the zoo was left without a living institutional ambassador, its literal corporate face. (The crisis was alleviated the following year when the zoo acquired another Siberian tiger, three-year-old Zhanna, from the Saint Louis Zoo as a replacement.) When a city's zoo loses its most cherished animals to other institutions, its citizens experience the loss as a symbolic death, a hit to local pride and metropolitan status. When the Philadelphia Zoo transferred their elephants to larger facilities in central Tennessee in 2007, the public outcry was as deafening as the near-constant local news coverage. City locals signed farewell cards, while others lamented, "I just don't understand how you can have a zoo without elephants."[7]

While some zoos and their cities revere their most prized large mammals as trophies of status and distinction—just as ancient Chinese emperors, Indian princes, Persian kings, and Egyptian pharaohs expressed their wealth, power, and prestige through their palace menageries, full of exotic creatures from conquered lands—others do so with a lighter touch.[8] For example, animal naming practices at zoos run the gamut from cute to clever to cliché: many a zoo names its ball python Monty, and its barred owls Shakespeare or Ophelia. (Why? Say "barred owl" aloud.) The San Diego Zoo's orangutan and gorilla exhibits display celebrity headshots of their great apes labeled with assigned names normally attributed to people, thus blurring the distinction between human and nonhuman primates. Many of the orangutans' designations—Satu, Indah, Unkie—are Indonesian or Malaysian names designed to encourage visitors to associate them with their species' native countries of origin, even though Satu and Unkie were both born in American cities (San Diego and Miami, respectively).

Meanwhile, since wild gorillas hail from only a small set of habitats in sub-Saharan Africa, most of the San Diego Zoo's captive gorillas have been allocated Swahili names: Mandazzi, Membe, Ekuba, Imani, and Maka.[9] (Only Membe was born in the wild.) Mandazzi is also known by his assigned Americanized nickname, the Comedian. The gorilla's pinup poster further anthropomorphizes him as having "boundless energy, a show-off, a real cutup." It reads, "He used to be known as the 'little terror' by his keepers due to his rambunctious nature as a youngster. Now, he's a great older brother. . . . He is playful, intelligent, and has developed quite a personality." Strangely, one zoo that does *not* share the pet names of its animals with the public is Disney's Animal Kingdom—which is ironic, since the very foundation of Disney's success has been in transforming animals into animated characters with human names and personalities: Mickey and Minnie Mouse, Donald Duck, Jiminy Cricket, and whatever animal Goofy is supposed to be.

Like Disney's menagerie, sometimes zoo celebrities become cultural sensations, thanks to both the synergy of the corporate media industry and the fierce brand loyalties of American youth. Shortly after she was born, Zoo Atlanta named its newborn baby panda Po, after the lead character in the 2008 animated film *Kung Fu Panda*. Hardly a coincidence, the stunt was an outcome of a cross-promotional alliance with DreamWorks, the studio that produced the film. Jack Black, the actor who lends his voice to Po in the movie, formally announced the panda's name at the zoo in February 2011. The event coincided with the release of the film's much-hyped sequel later that year.

The past several decades have been marked by the exponential rise in the production of mass media and advertising pitched to American kids through children's books, television shows, animated films, video games, and even breakfast cereals. The typical first grader can identify nearly two hundred brands, and a 2000 study revealed that two-thirds of mothers confessed that their own children were brand conscious at three years of age, while one-third discovered this awareness in them by age *two*.[10] Flamboyant cartoon characters, whiz-bang toys, junk food, and kid-friendly brands regularly promoted to children by corporate behemoths such as Disney, Viacom, Hasbro, and McDonald's all populate the vast symbolic repertoire available to the preadolescent American child.[11]

Children reference these cultural touchstones when socializing with one another in peer group settings, and collectively learn to invest branded consumer goods and pop icons with meaning and value in school, on play dates, and also at the zoo.[12] For this reason, Adventure Aquarium dedicates one of its tanks to displaying only two live underwater species: *Amphiprion ocellaris*, the clown anemonefish, and *Paracanthurus hepatus*, the Pacific regal blue tang. Children of all ages flock to this centrally placed tank but never refer to these fish by their scientific names—only as Nemo and Dory, the lead characters of the 2003 blockbuster Pixar film *Finding Nemo*. I first discovered this during a visit to the aquarium in 2009, when my son Scott immediately identified the clownfish in question as Nemo. He was three years old.

Public observations also reveal the pervasive power of popular culture as a means of organizing collective meaning *within* families, especially since shared pop-cultural references help to lubricate interactions between children and their knowing parents. At the Great Cats exhibit at the Smithsonian National Zoo in Washington, D.C., a mother directed her daughter's attention to a lioness and exclaimed, "Look, there's Nala!" referring to one of the lead feline characters in the animated Disney film *The Lion King*. In City Zoo's reptile room, a father pointed out one of the snakes and joked with his daughter, "This one looks like the snake from *Harry Potter*, right?" perhaps remembering an early scene from *Harry Potter and the Sorcerer's Stone* in which a boa constrictor speaks to the young wizard Harry during a zoo visit and gives him a wink before escaping from its tank and slithering off in search of its native Brazil. The father then asked his daughter, "Can you speak Parseltongue?" referring to the language of snakes associated with the Dark Arts in the *Harry Potter* world.

His daughter laughed. "No, Daddy, I'm not in Slytherin; I'm in Gryffindor."[13]

As middle-class urban and suburban dwellers, American zoo visitors typically have limited experience with the feral animals that live in their own backyards, much less the exotic African, Asian, South American, and polar creatures that reside in zoos across the country. It is therefore no wonder that they resort to familiar pop cultural touchstones in order to simply have something to say in the presence of such strange critters. "Hey, hey, we're the Monkees!" a mother sang nearby a troop of Central American squirrel monkeys at City Zoo. A woman in her mid-sixties noticed that at least one of the zoo's endangered mammals was "from Madagascar, like that movie." On more than one occasion, guests asked me if City Zoo's Caribbean giant cockroaches were the same ones fed to contestants on the NBC reality show *Fear Factor*. Multiple visitors to the Cape May County Zoo on the New Jersey Shore referred to one of its colorful Neotropical birds as Toucan Sam, the Kellogg's breakfast cereal mascot. At the Smithsonian National Zoo at least one visitor compared the massive Asian elephant enclosures to the maximum-security dinosaur cages in Steven Spielberg's *Jurassic Park*.[14]

Sometimes guests' visions of the animal kingdom are so warped by mass media they have trouble distinguishing actual zoo animals from their movie-screen counterparts. At City Zoo, a woman in her midtwenties approached the gorilla enclosure while one of the apes appeared fixated on a baby in its stroller on the public side of the glass.

Referring to the gorilla, the woman asked, seemingly in disbelief, "Is that real?" The other visitors assured her that the gorilla was, in fact, real. She responded, "That's so creepy. Have you ever seen *Planet of the Apes?*"[15]

In fact, when zoo visitors *do* reference their familiarity with actual live animals, they often cite their own pets rather than more exotic beasts from the wild kingdom. One summer at Metro Zoo I set up an educational station in front of the cougar exhibit and invited visitors to touch a soft piece of wildcat fur from an ocelot. As passersby caressed the ocelot's pelt, mothers asked their children if it resembled the fur of their housecats.

"Does it feel like Gypsy?"

"Does it feel like Rex?"

"This feels just like Sammy."

One mother told her son, "It's like Moose," who isn't a cat but their pet yorkiepoo, a cross between a Yorkshire terrier and a toy poodle. I also presented a crushed and tooth-pocked stainless-steel food bowl that the zoo's jaguars mauled one day, and a mother asked her child, "Could you imagine if Cinder did that?" (Cinder is a black Labrador.) Even the cougars themselves attracted such comparisons, as when a mother asked her daughter, "It's like a really big kitty, huh?"

Wildcats like cougars and jaguars are clearly totemic among zoos, just as visitors associate such creatures with the sublime. On an autumn day at City Zoo, an elderly man walked through its exhibit of lions, tigers, and pumas, and turned to his grandson, "I think these are the most beautiful cats God ever created." Another variety of charismatic megafauna that enchants visitors, especially children, includes the most mega of them all—dinosaurs. Obviously, zoos do not display live prehistoric specimens (although a mother visiting Metro Zoo once asked me if we did, without irony), but many do rely on the mystery and allure of dinosaurs to enthrall audiences. In addition to tigers and lions, the Great Cats exhibit at the Smithsonian National Zoo features a life-size *T. Rex* skull and footprint. A placard adds emphasis: "The biggest, scariest and coolest of all! Lions and tigers may have fierce reputations, but Tyrannosaurus Rex ranks as one of the most ferocious creatures of all time." (The zoo also boasts a life-size fiberglass triceratops that appeared as Uncle Beazley in the 1968 television adaption of the Oliver Butterworth children's novel *The Enormous Egg*.) Disney's Animal Kingdom in Orlando, Florida, an AZA-accredited zoo in its own right, features the DinoLand USA micropark that displays live reptiles consid-

ered to have evolved from dinosaurs. The placard at the American crocodile exhibit reads, "All modern alligators and crocodiles are descendants of the crocodilians—an ancient family of reptiles that flourished
in the Late Cretaceous. Found throughout Central America and the
northern costal areas of South America, the American Crocodile preserves many primitive characteristics from that era." DinoLand USA
also exhibits a live alligator gar, a predatory fish whose progenitors also
date back to the Late Cretaceous period, along with Tyrannosaurus and
Triceratops.[16]

Among species of charismatic megafauna, what other specimens do
zoo visitors find most enchanting? Research shows that zoo audiences
lavish much longer viewing times on exhibits featuring animal infants
compared to those that do not.[17] (No wonder American zoos obsessively
promote animal births by featuring images of newborn cubs and baby
calves on their brochures and other publicity materials.) One September morning at City Zoo, a group of mothers watched a family of Sumatran orangutans (one of the zoo's Big Twelve animals) from behind their
glass enclosure. "Look, it's a baby!" one exclaimed as the baby orangutan climbed into its mother's arms. The mothers all swooned—
"Awwwww!" as their six toddlers pressed their faces against the glass.

Orangutans also share certain evolved physical similarities, or "phylogenetic relatedness," with humans, which might explain why gorillas
and other great apes also engage audiences for longer periods of time
than other animals.[18] This is especially the case for animals possessing
what physical anthropologists refer to as *neotenic* traits that resemble
not just humans but specifically human *babies*, which explains much of
the visceral magnetism and iconic status of the giant panda, with its
rounded body, circular face, forward-facing eyes (along with black
patches that make its eyes look enlarged), floppy arms, clumsy gait, and
ability to sit up.[19] Neoteny also explains the popularity of *stuffed* panda
bears among children, which in turn further explains why they find *live*
pandas so enchanting—they appear to resemble the make-believe plush
toys made in their image.

Of course, pandas are also critically endangered in their natural
habitat in the mountain forests of southwestern China and thus excep-
tionally rare here in American zoos, which further adds to their appeal.
On loan from the People's Republic of China, these furry, two-toned
ambassadors reside in only four U.S. zoos: the Smithsonian National
Zoo, San Diego Zoo, Memphis Zoo, and Zoo Atlanta. In 2011 I visited
the San Diego Zoo's popular Panda Canyon exhibit to see them for

myself, although I found it difficult to get a close look without having my view obstructed by the endless waves of tourists simultaneously video-recording and snapping photographs of pandas Gao Gao and Yun Zi on their digital cameras and smartphones. (They behave just as tourists at the Louvre in Paris do upon encountering iconic works of art. Gao Gao and Yun Zi are the *Mona Lisa* of the San Diego Zoo.)[20]

Later that year I visited Lun Lun and Yang Yang, two of the giant pandas residing at Zoo Atlanta (which at one time paid the Chinese government $1.1 million *annually* for the rental of its pandas), and I discovered yet another explanation for the magnetism and charm associated with these magnificently plump creatures. The giant panda essentially spends the majority of its waking hours doing one thing: eating bamboo. Although pandas are technically carnivores, bamboo makes up approximately 95 percent of their diet. Since bamboo stalks and leaves do not have much nutritional value, they therefore have to eat a lot of it, up to 40 percent of their body weight *daily*. (According to keepers, Zoo Atlanta's giant pandas must be fed five or six times a day.) Since research suggests that zoo audiences spend more time viewing the exhibits of animals actively engaged in animated activity, this voracious munching occupies visitors as much as the pandas themselves.[21] By observing restless animals as bodies in motion, audiences are reminded that zoo animals are living, sentient beings—in other words, they are *real*—rather than inanimate plush toys, or cartoons, or stuffed pieces of taxidermy. When one considers the flood of simulations that saturate the zoo landscape, it is no wonder that visitors are drawn to some exhibited animals for simply being alive.

Given the charisma that audiences associate with active animals in constant motion, the fluttering of exotic birds attracts nearly as much attention as large mammals do. (In fact, just like charismatic megafauna, birds also enjoy high levels of conservation spending by federal and state governments: avian causes célèbres include the northern spotted owl, Florida scrub jay, red-cockaded woodpecker, and bald eagle.)[22] Zoo audiences are especially drawn to birds with vibrant plumage—especially peacocks, since zoos often allow them to strut around unencumbered by cages or netting. These free-range zoo animals would occasionally confuse very young children. At City Zoo, a three-year-old boy walked by a peahen and asked his mother, "Are we in the wild?" Two seemingly identical peacocks regularly strolled the grounds at Metro Zoo, and upon hearing a keeper greet one of them—"Hi Stanley!"—I wondered how staffers told the two birds apart. (I eventually figured out that the zookeepers had named *both* peacocks Stanley.) Children frequently chase after the peacocks, collecting the blue tail feathers they shed along the zoo paths. Whether found on zoo grounds or purchased at a gift shop, bird feathers serve as lucky talismans for young visitors, and in fact the keepers at City Zoo make a point of saving colorful turkey, parrot, and macaw feathers to donate to the Pueblo Indians of the American southwest to be repurposed for their traditional Native American religious ceremonies and rituals.

Older female visitors admire zoo birds while occasionally making passing references to *wearing* bird feathers as fashion accessories, as was stylish among American women during the late nineteenth century. Two seniors visiting City Zoo, both women, pointed out a flock of passing peacocks. One said, "Look at the beautiful peacocks. How pretty they are."

"Yes," the other woman replied. "They are gorgeous. I have a brown hat with peacock feathers the same shade of blue. It's one of my favorites. I wear it with a brown suit and blue sweater the same color as the feathers. I always get complements when I wear it."[23]

Guests made similar comments when I exhibited reptiles at Metro Zoo, perhaps owing to their lack of familiarity with these animals in any other context besides their own wardrobes. One spring day a woman approached me as I publicly presented Hero, a spiny-tailed lizard, along with a brown monitor lizard skin for visitors to touch. She pointed to the skin and whispered, "You should cover [Hero's] ears, but *this would make a nice pair of boots.*"

Later that summer I displayed Puff, a bearded dragon, alongside the skin of a tegu, another kind of lizard. A mother with blonde hair remarked, "I have a purse that looks just like that," while another agreed that the tegu skin would "make a nice handbag."

Meanwhile, a grandfather seemed slightly more circumspect. "Is that tanned? Yeah, it's processed. They make shoes out of that—it's terrible."

But back to birds—around the country zoos consecrate the most symbolically powerful of all feathered creatures in the United States, the American bald eagle. Audiences venerate these celebrated birds as well. At City Zoo, two mothers pointed out a bald eagle to their children as they passed its enclosure. "It's really just so beautiful. Stunning. So regal," says one.

"It doesn't even look real, does it?" replied the other.

She makes an interesting point. In *The Elementary Forms of Religious Life*, Durkheim arrives at a revelatory conclusion: among ancient societies the ceremonial *images* of the totemic animal were considered more sacred than the *animal itself*.[24] Although zoos exhibit live animals literally in the flesh, in many ways *real* zoo animals are revered as merely stand-ins for their representational icons—the totemic symbol itself—and not the other way around. We have certainly seen numerous illustrations of this: live clownfish that help children recall an animated cartoon character, giant pandas that resemble juvenile plush toys, and growling Siberian tigers that remind Brandywine visitors and staff of their zoo's own mascot. Zoo visitors commonly express this tension between the iconic and the real, fantasy and authenticity. One summer at City Zoo, a woman pointed out a gorilla in its dedicated pavilion, and remarked, "He is so big. I thought he was fake at first."[25] Meanwhile, the Maryland Zoo exhibits a number of live ravens intended to represent two *other* live ravens, Rise and Conquer, who in turn serve as the official mascots of the Baltimore Ravens NFL football team, handled by Maryland Zoo trainers at home games. In the zoo itself, the raven exhibit's placard compares the natural aggressiveness of the bird species to the Ravens' defense. Other zoos outsource their animals to serve as mascots for professional or collegiate athletic teams. For example, the Cougars of the University of Houston have an official live mascot named Shasta who lives at the Houston Zoo. The wildcat sometimes makes public appearances with a costumed human cougar mascot, also named Shasta.

As arguably *the* animalistic totem of the United States, American bald eagles residing in zoos perform a similarly symbolic role. At its Eagle Canyon exhibit, the Elmwood Park Zoo in Pennsylvania evokes national allegiance and pride:

### The Bald Eagle: A Symbol of Strength and Hope.

In 1782, Congress chose the bald eagle to be our national emblem because of its majestic looks and strength.

Today, the eagle appears on the Great Seal of the United States, on our money, and on many state flags, including Pennsylvania.

The bald eagle has proven to be a worthy symbol. It has prevailed over the threat of extinction, inspiring further efforts in conservation. The eagle has become more than a symbol of our nation; it has become a symbol of hope for our future.

At the Cape May County Zoo the pageantry surrounding its bald eagles is unmistakable. Facing the eagle exhibit and its live birds stands a statute of an eagle on a pedestal with the engraving, "BALD EAGLE HABITAT DEDICATED TO ALL VETERANS." An American flag stands directly behind the statue, while plaques abutting the eagle enclosure itself memorialize two American veterans, Charles A. Burke, a chief yeoman from the U.S. Navy who fought in World War II, and Warren H. Westover, U.S. Coast Guard. At the bald eagle exhibit at the Philadelphia Zoo (where one of the birds is named Glory), a father held his three-year-old daughter and told her, "The eagle is a very special bird in America. There are pictures of it on our money, and at the White House, and it's illegal to kill one."[26]

Perhaps not surprisingly, the Smithsonian National Zoo features an entire American Trail exhibit celebrating mammals and birds of North America: gray wolves, river otters, American beavers, ravens, brown pelicans, hooded merganser ducks, gray seals, California sea lions, and, again, the American bald eagle. Brimming with patriotism and an eco-friendly message, a placard nearby the eagle exhibit reads, "The Pride of a Nation." Once endangered due to pollution, habitat loss and hunting, the bald eagle has recovered, becoming one of very few animals to be removed from the endangered species list. Now found in regions where it was long absent, the bald eagle symbolizes our nation's commitment to protecting wildlife and wilderness." Next to the bald eagle's enclosure sits a collection box adorned with an image of the stately bird, only far more photogenic than the actual live eagle on display. "Be an *American Trail* Blazer! Contribute today to fund the National Zoo's critical science and animal care programs."

In fact, sometimes seeing the mere image of a charismatic animal can feel equivalent to observing the genuine article. One reason why different zoos seem so similar to one another is that many purchase their identifying animal signage from the same set of companies, including Sanctuary Supplies, which outfits Metro Zoo. These placards typically feature flattering stock photographs or illustrations of the exhibited animal in question, just like Chinese-American takeout restaurants display enticing images of General Tso's chicken and Kung Pao shrimp that may bear little resemblance to the actual dishes they serve. At the Cape May County Zoo, a family with a young daughter in tow visually scanned the ocelot exhibit in search of its inhabitant, but the spotted wildcat was nowhere to be seen. The girl then stared wistfully at the generic photograph of an ocelot exhibited on its accompanying placard. "But he's so cute!" Her mother to the rescue, she snapped a picture of her daughter next to the photograph of the unnamed ocelot in lieu of the actual animal. The little girl could not have been happier.[27]

# Ugly Ducklings

If charismatic megafauna and beautiful birds enchant mass audiences, far humbler creatures generate fear, disgust, or repulsion. Anthropologist Mary Douglas observes that the abominations outlined in the book of Leviticus (those animals considered unclean and therefore not kosher by religious Jews, even today) include creatures that defy categorical simplicity. Pigs are cloven-hoofed like cattle, but do not chew their cud. Shrimp and lobster live in the sea but lack fins and scales, unlike fish. Insects fly, but lack feathers.[28] This particular attribution of stigma is by no means universal; after all, beloved penguins are birds that cannot fly, just as dolphins and humpback whales are mammals that swim. Still, just as "ugly ducklings" abound in the wild—consider the star-nosed mole, proboscis monkey, elephant seal, blobfish, and African manatee—even curated zoos exhibit categorically strange or otherwise unpopular animals.[29] Yet even the ugly ducklings of the zoo manage to attract enthusiastic attention from small cross-sections of zoo audiences and staff alike, if not larger fan bases.

Here is one example. According to a March 2001 Gallup poll, Americans fear many things: public speaking, flying on an airplane, thunder and lightning, crowds, small spaces, and great heights. Their biggest fears also include a number of animals, including dogs, spiders and insects, mice, and Americans' most reported fear—snakes.[30] While some readers will be shocked that "man's best friend" makes the list, few will be surprised by the pervasiveness of human trepidation surrounding cobras, pythons, boas, and anacondas. In *The Interpretation of Dreams*, Sigmund Freud observes how "the natural human fear of snakes is monstrously intensified in the neurotic." Among Freud's patients, previously frightening encounters with ordinary backyard snakes provided the symbolic material for even more intense attacks of hallucinatory panic.[31]

At the risk of mixing animal metaphors, snakes are one of the many ugly ducklings of the zoo. Amber, a twenty-two-year-old Metro Zoo keeper, recognizes the generalized fear of snakes shared among zoo visitors: "Everybody's afraid of snakes. I've had guests that won't even go past the exhibit, because they think that somehow the snake is going to jetpack its way through the glass and just wrap around their neck." One summer day at City Zoo, a mother in her early thirties stood behind her two young daughters in the reptile room as they watched a python resting in its enclosure. As the snake began to slither, the mother grew noticeably disturbed, clutching her collarbone and dropping her jaw. "That one is moving! Oh, Lord, I cannot handle this." One of her daughters said, "Mommy, look!" but she just laughed uncomfortably and shook her head back and forth. "Umm, no, I don't think so. I'm going to have to leave soon—I just cannot handle this." At a local street fair I represented Metro Zoo by handling Kaa, a Colombian red-tailed boa constrictor that I kept wrapped around my left arm. One terrified guy said, "You've got a bigger set than me," while a neighbor *raced* away from me, a look of panic in her eyes.

What explains the widespread anxiety surrounding snakes? The biologist Edward O. Wilson argues that the irrational human fear of snakes and its attendant symptoms (nausea, cold sweat, shortness of breath), or *ophidiophobia*, is universal, an innate and evolutionary trait adapted over millennia spent surviving in the wilderness alongside these venomous creatures. The argument is a compelling one. As Wilson observes, throughout human history snakes have caused sickness and death; snakes live on every continent on Earth except Antarctica; and perhaps as a result, representations of the snake and its mythical counterpart, the serpent, can be found in all civilizations as "the animate symbols of power and sex, totems, protagonists of myths, and gods":

> The mind is primed to react emotionally to the sight of snakes, not just to fear them but to be aroused and absorbed in their details, to weave stories about them. . . . For hundreds of thousands of years, time enough for the appropriate genetic changes to occur in the brain, poisonous snakes have been a significant source of injury and death to human beings. . . . It is notable that the phobias are most easily evoked by many of the greatest dangers of mankind's ancient environment, including tight spaces, heights, thunderstorms, running water, snakes, and spiders, but are rarely evoked by the greatest dangers of modern technological society, including guns, knives, automobiles, explosives, and electric sockets.[32]

Of course, while Wilson's evolutionary explanation thesis may be persuasive (if nonfalsifiable), it can hardly explain the crucial *differences* among human societies, or even the sharp discrepancies observed within our own. While a plurality of Americans report being fearful of snakes (again according to Gallup), this majority is quite slim, at 51 percent.[33] In fact, whenever I exhibited boas or other snakes, reptile fans would frequently stop to share stories about own pet snakes, or the garter snake they captured in their backyard that morning. At the aforementioned street fair, heavily muscled and tattooed men approached me all day to divulge loving memories of their pet snakes and iguanas, while back at Metro Zoo two women savored an opportunity to study Hogwarts, the small western hognose snake I held in my hands. One remarked, "It's a *Harry Potter* moment," and her friend smiled, nodding in agreement.

There is also a significant gender gap: according to Gallup 62 percent of women report harboring this reptile phobia, compared to only 38 percent of men.[34] (Obviously, men might be underreporting fear in their responses.) Of course, such statistical probabilities hardly provide a totalizing portrait, especially given that nearly all the zookeepers, educators, and volunteers responsible for handling snakes at Metro Zoo are female. Christina, a twenty-four-year-old keeper and former zoo educator, wears a tattoo of a snake twisted into the shape of an infinity symbol near the nape of her neck. Her fondness for snakes extends to the intimacy she feels toward Sambora, one of Metro Zoo's northern pine snakes, despite the fact that most visitors find snakes, in her words, "slimy and gross and scary." As she explains, "He's just my absolute favorite animal in the zoo. I love Sambora, the pine snake. I love him completely. When I worked in education, I used to love taking him out because he's impressive. He comes out of the bag with that huge hiss, and he's big and just *mean*-looking. I would try to give myself extra time when I would work in the reptile room, because I would just sit up there with him."

"Which one is he?" I asked.

"It's the black-and-white one—*so pretty*. I would just sit up in the reptile room, with him on my lap, and he would just fall asleep in my arms. I adore him. . . . I always felt such a connection with him, because he was just so grumpy, but wonderful at the same time. I like grumpy animals, I don't know why. I guess I just like that they still have that spirit, still have that kind of fight in them, because I think they should. Just because they're in a zoo, it doesn't mean that they can't still be that wild animal."

"Usually when people talk about cuddling with an animal they mean a furry one, not a snake," I pointed out. "But you really seem to experience a connection with him."

"Yeah, like I said, he's probably my favorite animal in the zoo, hands down."

"And what's that connection you feel?" I asked.

"I don't know. I guess for me, it's the fact that he can be super-grumpy and very aggressive, but when I take him out—when *I* handle him—he just calms right down. And you feel special when that happens, when you have an animal that feels so comfortable with you that they don't mind relaxing with you in a way that they normally wouldn't."

Besides snakes, what other "ugly ducklings" do zookeepers defend against public fear or derision? Whenever I interviewed keepers or educators I asked them to choose one animal whose polluted reputation they would rehabilitate if they could. The most popular response was bats. Certainly, the widespread fear of bats is palpable: in fact, while the Maharajah Jungle Trek at Disney's Animal Kingdom exhibits two species of bats, the Rodrigues fruit bat and Malayan flying fox, Disney provides alternative pathways that allow anxious visitors to bypass these nocturnal creatures altogether. ("Those not wishing to see bats, go this way," the exhibit redirects its audiences.) But according to Krista, a twenty-eight-year-old keeper at Metro Zoo, bats deserve to be reconsidered by the public, "because they're really awesome":

People see bats and they're just like, "*Ah! It's a giant flying rat! Eeeew, kill it!*" And no—they're so helpful, and they're really awesome. So our fruit bats, it's really funny: I'll go in to clean their exhibit, and the fruit bats are tiny. They're like the size of your cell phone. And people are like, "*Oh, my God! She's in there! Oh, my God! She's freaking out!*"

But then they ask me why I'm not inside with the jaguar. They're like, "*Did you pet the jaguar?*" And I'm like, "No! *That's* a dangerous animal." But they freak out and get terrified when I'm in with the tiny little bats. And I was telling them that these guys are not the ones you're going to find in your house. These are *fruit bats*. But the ones you find in your house are super-helpful. Do you know why there are so many mosquitoes now? When you were younger, there didn't seem to be that many, but now there are tons. . . . Because there are no more bats!

Some bats are really cute; some bats are really ugly. But some bats are adorable. . . . They get an extremely bad rap, but they are extremely helpful—the most helpful mammal that we have. If you want something cute and fuzzy to save, I'd say, go with bats.

Daphne, who along with Tyler served as Metro Zoo's other education volunteer coordinator, agreed that bats were ecologically vital as well as attractive. She said that although most "people would be happy if they never saw a bat again," they simply "don't realize how important bats are, and that they are dying out":

> Bats pollinate: without bats, the price of produce is going to go sky-high for peaches and mango and nuts. Plus, did you hear about the tiger mosquito? It's an invasive species that just came up from Texas, and they are all over the place. And this was a very bad summer with all the mosquitoes and the West Nile and everything, and that's because the bats are dying. We need the bats. And they are adorable. *They are adorable.* We put up a bat house outside one night and we had a fire, and we saw one swoop over the deck! And my husband and I were so excited. My sister was with me, and she thought that we had lost our minds.

## Bugging Out

One notable development among American zoos has been their growing attention to invertebrates, particularly insects and other arthropods. In an effort to teach visitors about conservation, zoos have increasingly begun exhibiting these helpful creatures, touting their contributions to the Earth's planetary fitness. For example, the San Diego Zoo exhibits Egyptian dung beetles, giant millipedes, honeybees, leaf-cutter ants, bird-eating spiders, jumping sticks, and hissing cockroaches, celebrating their usefulness: "Bountiful and Beneficial: Insects make a world of difference. They don't have skeletons, but insects are the backbone of most ecosystems. They make up about 80 percent of all animal species on Earth! As they buzz, burrow, and hunt, these mighty invertebrates pollinate flowers that yield food for the rest of us; aerate and fertilize soil; control plant pests; recycle dead plant and animal matter. Each of the more than two million insect species plays a vital role in the environment." Other prominent insect-themed zoo exhibits launched in recent years include "It's Tough to Be a Bug," a permanent multimedia installation in the Tree of Life at Disney's Animal Kingdom based on the characters from the 1998 Pixar film *A Bug's Life*, and the 2012 Xtreme BUGS Exhibit at Chicago's Brookfield Zoo. (Perversely, the Brookfield Zoo's insect exhibition's primary sponsor was Terminix, the largest pest control company in the world. At different stations kids could pretend to be professional bug exterminators, and purchase fly-swatters at the accompanying gift shop. Even stranger—visitors were invited to taste foods *prepared* with insects. I reluctantly sampled Cinnamon Bug Crunch with fried caterpillars, Six-legged Salsa with mealworms, and Hoppin' Herb Dip with crickets.)

Admittedly, insects have a somewhat soiled reputation in American society (let's call them *un*charismatic *micro*fauna), perhaps none as severe as cockroaches. While we merely shoo and swat horseflies, we wage chemical warfare on cockroaches with the aim of *exterminating* them. Yet they are a vital part of the biosphere. By feeding on decaying organic matter, including rotting wood, leaf litter, dung, fallen fruit, and dead plants and animals, they maintain the health of the Earth's soil. They are natural recyclers, the world's original composters. As the San Diego Zoo reminds its visitors of African hissing cockroaches, "Our ecosystems could not function without their dirty work!"

Still, these outreach efforts have not been easy, largely because most adults find cockroaches repulsive, and many are surprised to see them displayed at the zoo at all. I observed this audience repulsion on a summer afternoon when I first presented Charlie, a Madagascar hissing cockroach, to visitors at Metro Zoo.

"Gross!"

"Disgusting!"

"You're very brave—I don't think I could touch him."

Frankly, I was surprised I even got Charlie out of his tank without crushing him to death. He lived with about sixty of his friends in one of Metro Zoo's administrative offices, along with their African millipedes and emperor scorpions. (Perhaps I shouldn't refer to them as "friends," since they can be terribly vicious. In fact, Charlie was missing one of his legs because another cockroach ate it.) Of course, Charlie was not his real name. The cockroaches at Metro Zoo didn't have names; there were far too many to keep individual track of each one. Still, I liked to pretend they had names, as way to make visitors feel more comfortable around them—I thought of it as anthropomorphizing with a purpose. This is fairly common in American zoos. Whenever Tyler, the education volunteer coordinator at Metro Zoo, exhibited a cockroach to a family he temporarily named it after one of the Brady Bunch siblings—

Marcia, Greg, Jan, Peter, Cindy, Bobby—and then told visitors that there were six cockroaches who lived at the zoo, each named for different Brady kid.

One week I really gave it my all. My hands cupped around his segmented body and half-chewed limbs, I positioned myself at the front gates and announced to passersby entering and exiting the zoo, "This is Charlie!" At first it worked, since what child can resist a tiny, hand-held creature named Charlie? But eventually their parents caught on.

"I'd have a heart attack if I saw that somewhere else," one mother told me.

Another mom agreed. "If I saw that in my house, it would be up for sale the next day." I persisted, and eventually my entreaties became something of recurring joke for some of the more vocal grown-ups.

"Um, no thanks!"

"Bye, Dave—I don't think so!"

"Oh, jeez, not that cockroach again!"

Even Hilary from the zoo's development office seemed turned off. "Eww . . . cockroaches are not my thing," she sneered.

The repulsion expressed by American zoo visitors encountering *Gromphadorhina portentosa* is perhaps to be expected. As anthropologist Hugh Raffles observes of our alienation from these and other six-legged arthropods, "So much about insects is obscure to us. . . . We simply cannot find ourselves in these creatures. The more we look, the less we know. They are not like us. They do not respond to acts of love or mercy or remorse. It is worse than indifference. It is a deep, dead space without reciprocity, recognition, or redemption."[35] Even among surveyed policy makers from conservation-oriented NGOs, insects inspire less attention than birds and mammals, even very rare species.[36] No wonder American audiences routinely dismiss those displayed in zoos. In the rainforest exhibit in New York's Central Park Zoo, an elementary-school-aged girl and her mother passed by a display featuring leaf-cutter ants. "Is this the queen ant?" the girl asked, pointing to a picture of an ant on the sign.

"I don't know. Does it matter?" her mother responds.

Other visitors seemed just as dismissive. "Why would anyone want to look at ants in a zoo?"

"I don't know, don't all ants look the same?"

"Yuck, I hate bugs."

But the one thing Charlie and his fellow arthropods have going for them is that among preschool and elementary schoolchildren, some have yet to be socialized by their parents to regard insects as pests, even cockroaches. Moreover, American middle-class children spend their early years surrounded by a universe of pop-cultural touchstones that normalize insects as cute and lovable creatures, similar to other beloved animals. Kids absorb these G-rated images of insects in nursery rhymes ("The Itsy-Bitsy Spider," "The Ants Go Marching"), children's books (Eric Carle's *The Very Hungry Caterpillar*, E. B. White's *Charlotte's Web*), animated films (*A Bug's Life*, *Antz*, *Bee Movie*), and even anthropomorphized corporate-brand mascots, like the Honey Nut Cheerios bee.

Of course, this isn't to say that a cockroach crawling across a child's cereal bowl during breakfast would cause anything but hysteria. But zoos keep their own cockroaches safely contained within a sanitized and nonthreatening context of captivity. And so, during my many stints with Charlie and his fellow insects I observed curious children engaging with these six-legged creepy crawlers, despite their parents' dismissiveness. Some seemed thrilled merely by the roach's power to disgust others, as two boys did upon approaching Charlie and me: "Awesome! That's awesome!" (Gross-out kicks do not discriminate on the basis of age, either. One time as a family approached the exhibited cockroach with revulsion, the grown-ups decided to photograph the critter with their accompanying children, instructing them, "Pretend you are going to *eat it*, and we'll send it to your mother and she what she says!")

But cockroaches are not just for generating disgust as fun. On another occasion, a little girl visiting with her summer camp acknowledged of the cockroach, "It's cute!" and her adult chaperone asked, teasingly, "Who said that? I'll put one in your pillowcase!" As another little girl's mother took a photograph of the cockroach, her husband chided her, "Why would you want to take a picture?" but his daughter defended Charlie—"He's cool!"

Insects like cockroaches have an additional leg (or six) up on more normatively charismatic zoo animals insofar as they can be viewed by visitors up close in a way that more popular megafauna cannot. Unlike gorillas, jaguars, or elephants, which must be kept secured at a distance within enclosures made of steel bars, thick glass, or expansive moats, a hissing cockroach can be exhibited at close range in a small two-and-a-half-gallon terrarium, or in a zookeeper or docent's open palm. Young children appreciate these moments of intimacy with tiny zoo creatures, no matter how small.[37]

Admittedly, a small number of arthropods do attract a select group of adults, along with their kids. I found that some curious parents with small children occasionally gravitated toward tarantulas, scorpions, and other arachnids that evoke a sense of danger and foreboding, just as snakes do. Far more frequently, such creatures scared visitors out of their minds. At Metro Zoo I sometimes exhibited Tippy, a Chilean rose-haired tarantula that frightened mothers and fathers alike.

"Wow—that's terrifying. That is huge."

"Oh my gosh, that creeps me out!"

"I'm sorry, sir—I'm not avoiding you, I'm avoiding what's *in your hand.*"

Others responded as if the tarantula was a house pest: "If it's jumping, I'm swatting!" ·

"I vacuum spiders up."

"I appreciate what spiders do for the ecosystem, but I do not want them in my house."

Yet despite their arachnophobia, I found that some of these same parents would engage with me if their children displayed an interest in the spider, which gave me a chance to make a pitch on its behalf. In the case of Tippy, I usually pointed out to the adults that some arachnids take only thirty seconds to mate, during which time the female sometimes *eats* the male. (Upon hearing this, the reactions from women were overwhelmingly positive. As one camp counselor slyly remarked, "I feel like that's typical of female species.")[38]

I have also received mixed signals from audiences when displaying giant African millipedes, as many enthusiastic grown-ups find them both creepy *and* cool. One woman joked, "Can you put him on the ground as I go *squish*," stomping her foot on the ground. "I'm only kidding. . . . Can I touch it?" Although we regularly kill insects in everyday life without guilt or shame, her comment seemed off-putting in the context of the zoo. Imagine a visitor joking about crushing the neck of a baby rabbit or lion cub with their heel.

Not surprisingly, the most popular insects among adults visiting Metro Zoo are the enchanting butterflies and moths that reside in a specially designed greenhouse where they feed on the zoo's in-house "nectar," a sweet concoction made from sugar, water, bananas, and beer. We value butterflies just as we do flowers: not only for their beauty, symmetry, and color, but also for their fragility and brevity of life.[39] The delicateness attributed to butterflies even operates on an organizational level at the zoo. One May morning, Metro Zoo staff accidentally admitted entry to an elementary school class previously banned from the zoo for misbehaving. Of all the zoo's exhibits, volunteers were called upon to protect the butterfly house from these "little monsters." (I personally heeded the call along with Marty, a retiree. Nothing untoward happened on that day, although that summer a boy purposely stomped on a spicebush swallowtail, nearly killing it. His mother gave him a time-out.) Of course, to some parents even the most fragile of pests is still just a pest, enchanting or otherwise. When I presented guests with a mounted Atlas moth specimen—among the largest moths in the world—one mother admitted, "I would freak out if I saw that in my closet. I would be swatting it."

## Where the Wild Things Are

Outgrowths of ecological diversity in American cities—the public plazas where urban dwellers feed street pigeons, the urban waterways where city folk engage in night fishing—remind us how impossible it is to separate human civilization from what we think of as the natural living world.[40] City zoos may be pavilions for captive animal exhibition, but like other urban gardens and parklands they are also ecological habitats in their own right, rife with creatures well adapted to peopled environments. Children pleasantly surprised me whenever they responded more enthusiastically toward the feral animals they encountered at the zoo than more "official" exhibited animals. I first noticed this behavior when observing my son Scott and his friends react to the incidental and somewhat mundane presence of wild animals at the zoo as if they were intended attractions. During one May visit to City Zoo, two wild geese walking along the pathway intrigued Scott and his pal. One of the boys suddenly exclaimed, "Look at the squirrel in the trashcan!" Sure enough, an ordinary squirrel had attacked a nearby receptacle, likely in search of a discarded bit of overpriced lunch. (Among the zoo's more popular concession items are its fried chicken fingers, which the zoo serves to visitors even as it keeps other chickens on exhibit in nearby coops in the children's farmyard.) This happened again at Zoo America in Hershey, Pennsylvania. We approached a black bear exhibit, and Scott acknowledged the giant beast. But in the next moment, he pointed out an ordinary spider spinning a web on the barrier of the exhibit. The tiny arachnid soon took up all of his attention.

I observed other kids engaging in similar public behavior. At City Zoo, a little boy enjoyed watching the chickens penned in their yard, and then followed around a wild pigeon, clucking after it. In this case the pigeon and the chickens must have seemed interchangeable in the boy's mind. As if to emphasize the point, I later spotted a girl chasing after another pigeon, and actually calling out to it—*"Chicken!"*

The sense of wonder that feral animals inspire in children reveals itself during the most surprising moments. On a late August morning at Metro Zoo, Sonora and I relaxed outside beneath the zoo's tall cigar tree, enjoying the last warm days of summer. (Sonora loves the heat since she hails from the American Southwest, and also because she is a chuckwalla, a scaly lizard that normally lives in the desert.) Accompanied by his family, Alfie, a charmingly inquisitive boy, approached me with a wide grin, and immediately began telling me about his pet leopard gecko, when suddenly his sister saw a figure cut across the zoo, and with excitement she blurted out, "A SQUIRREL!" Even Alfie, a self-identified reptile lover, noticed the squirrel's antics. "Cool, a squirrel *and* a lizard—two small animals. Cute!" he gushed, without a touch of irony. The adults all giggled, bewildered that a wild squirrel could upstage an otherwise eye-catching lizard in a zoo teeming with exotic rhinos, jaguars, and boa constrictors.

This is a fairly common occurrence at zoos, as children take endless delight in the wild animals that frolic about their public gardens. Yet parents often protest, insistent that their kids show more appreciation for the zoo's exhibited creatures than the more mundane and possibly diseased animals that linger in between their enclosures, scavenging half-eaten hot dogs and other assorted litter left behind by visitors. At City Zoo, three children watched squirrels climb into a set of trash bins, foraging for snacks. "Take a picture, take a picture," one boy said to his mother.

"I am *not* taking pictures of squirrels eating garbage out of the trash. I go to the zoo to take pictures of *real* animals," she answered.

"Aren't squirrels real animals?" the boy asked, smiling.

"Well, yes," the mother said, "But they are not *zoo* animals."

Is the mother correct? Perhaps to a point, yes: but I am not so sure her children are as foolish as she thinks. After all, unlike most of the zoo's exhibited animals the squirrel is *truly* wild and free, brandishing a kind of beastly unpredictability that caged zoo creatures rarely can.[41] Given that research shows that zoo visitors are more easily enchanted by exhibited animals engaged in physical activity—in fact, viewing times are about twice as long for active animals as compared to when those same creatures are dormant—it should not be all that surprising that feral rodents regularly compete with captive and often sedentary zoo creatures for human attention.[42]

The blurriness of the imaginary boundary separating human civilization from nature comes to life when a feral animal disrupts the zoo's fantasy landscapes with a burst of tooth and tail, feathers and beak. (This is especially the case when wild geese and mallard ducks land in open-water exhibits specifically labeled as belonging to the Humboldt penguin, Caribbean flamingo, or black-necked swan; or when white-tailed deer living in the Rock Creek parklands surrounding the Smithsonian National Zoo occasionally make their way into its Asian elephant exhibit.) Children recognize and appreciate these moments far more than their parents, not only because they have not been socialized to recognize the arbitrary legitimacy of certain animals over others in zoo settings, but also because they have been *specifically conditioned* to value small yet frantically active wild animals as magnetic and exciting creatures. Again, children's literature and film are full of examples of anthropomorphic wild animals with human-like personalities: *The Jungle Book, Winnie the Pooh, Bambi.* Picture books and cartoons treat pedestrian animals of the city no differently. Pigeons are portrayed as wisecracking characters in Mo Willems's popular children's board books featuring titles that scream—*Don't Let the Pigeon Drive the Bus! The Pigeon Finds a Hot Dog! The Pigeon Has Feelings, Too!* Feral rodents scurry through the fields of juvenile mass culture: Disney's Mickey and Minnie Mouse (and chipmunks Chip 'n' Dale), Rocket J. Squirrel of *The Rocky and Bullwinkle Show*, and a whole warren of rabbits: Peter Rabbit and his sisters Flopsy, Mopsy, and Cottontail; Thumper; Bugs Bunny; Roger Rabbit; the Nesquik Bunny; and the Easter Bunny. These cultural touchstones allow kids to appreciate urban pigeons, squirrels, raccoons, and even mice as worthy creatures, rather than filthy nuisance animals. No wonder that during a summer outing to City Zoo, Scott (at age four) and I watched the orangutans at play, when Scott suddenly got very excited. "Daddy, I saw a little mousy crawl through there. It wasn't dead. It was real!"[43]

In comparison, adults can usually get excited only by the types of feral animals that inspire awe outside the zoo grounds as well. For instance, in recent years birds of prey from eagles to red-tailed hawks have gained notice in the American cities where they sometimes hunt, or raise their babies. (An example: for at least three successive spring seasons starting in 2011, the *New York Times* broadcast from its online "Hawk Cam" images of red-tailed hawks nesting in Washington Square Park on a twelve-floor ledge of New York University's Bobst Library, to surprisingly popular fanfare.)[44] At Metro Zoo, grown-ups peer skyward as wild black vultures swarm overhead. Occasionally these marauding birds attempt to steal the meals of horse meat regularly distributed to the zoo's eagles in their open-air display. Zoo audiences often wonder why the vultures aren't kept in enclosures, mistaking the feral scavengers for those birds in the zoo's exhibited collection. (In point of fact, Metro Zoo does keep and train one black vulture, Decker, for its educational programs, but he never flies without a harness.)

At other times the wild vultures put adult visitors on high alert. One May I was handling a chinchilla outside when a father approached with his daughter. He asked me if there was any danger of the black vultures harming the zoo's animals, such as the very chinchilla resting in my arms. In all my time working at Metro Zoo, no vulture ever attacked me or any other animal in my care, although one rainy day in December during my falconry training Decker did both defecate and throw up on me. On the other hand, Atlantic seagulls practically terrorize the guests at the coastal Cape May County Zoo as they swarm the food concession tables in search of loose french fries.

Zoos blur the boundary between managed and wild populations in other ways. Rodents seem to reflect this ambiguity particularly well. During my stint volunteering at City Zoo, staffers spent an early morning meeting debating how to most effectively and humanely trap mice in the zoo's offices and workstations. (Glue traps, which ensnare live mice that may suffer for an extensive period of time before suffocating or starving to death, are considered less humane than snap traps that kill instantly.) The discussion was notable for its emphasis on how to exterminate vermin with compassion. Indeed, it is difficult to imagine many work environments where the pain threshold of mice, rats, or other pests would be discussed with even a whit of concern. But while some mice and rats at City Zoo are treated as carriers of pestilence, other members of the same genus and/or species at the zoo are fed daily and groomed for educational programs and kids' birthday parties. Some rodents, like Bonkers the kangaroo rat, live in enclosures *in the very same rooms* where mousetraps baited with peanut butter line the walls. (Similarly, exotic cockroaches, spiders, and beetles from the zoo's collection reside in the same storage rooms where flypaper and ant traps catch far less-pampered bugs.)

So at zoos, some rodents are trapped and beheaded, while others are considered worthy of being given human names as if they were beloved pets. Of course, not all zoo mice and rats are kept in cages; some are stored in freezers and prepared as food for other zoo animals. I will never forget the first time this reality confronted me as I helped Craig, a college intern, feed a bowl of frozen mice to City Zoo's turkey vultures, burrowing and spectacled owls, red-tailed hawks, and other birds of prey. Zoos purchase feeder mice in bulk from wholesale breeders, and they come packaged frozen in a variety of sizes, from extra-small "pinkies" to midsize "fuzzies" and larger "hoppers," as noted in the last chapter. Other feeder animals distributed to animals at zoos include rats, rabbits, quail, chicks, and guinea pigs.

To sum up, among zoo rodents, some live in captivity and are fed tasty morsels daily; some roam wild, seeking out tasty morsels in trash bins until they are caught and killed as pests; and other rodents *are* the tasty morsel for captive carnivores higher up on the food chain. Little wonder that parents and children argue over the legitimacy of squirrels, pigeons, and other feral animals habituated to urban life, especially since the zoo itself traffics in its own inconsistencies and ambiguities concerning the relative status of these animals. In fact, in recent years some zoos *have* been exhibiting many of these common city animals that have adapted to human environments. The Philadelphia Zoo displays multiple breeds of pigeon (fantail, magpie, capuchine, nun) and two female Harris's antelope squirrels named Itsy and Bitsy. Meanwhile, in 2011 the Phoenix Zoo took in four endangered Mount Graham red squirrels to prevent their extinction during that summer's wildfire season.[45] Of course, across the United States one geographic region's nuisance animal may be another's exotic zoo creature. This explains why armadillos are considered little more than road kill in Texas but used for educational programming at both City Zoo and Metro Zoo.

## Even Better Than the Real Thing

Zoological parks are filled with not only live animals (captive and feral), but still and silent replicas of living creatures as well: warthog statues cast in bronze, playground apparatuses shaped like alligators, sculptures of African elephants cut from granite. Historically, gardens from imperial China to monarchical Europe have featured lion sculptures,

mechanical birds, stuffed taxidermy, and all varieties of animalistic to-
piary artwork as substitutes for living wildlife.[46] Today these remnants
are most readily found in city zoos that carry vestiges of their nineteenth-
century metropolitan heritage and Victorian style into the new millen-
nium. A bronze statute of a tigress and her cubs with a peacock in its
mouth was presented to New York's Central Park Zoo in 1867, and it
remains on the grounds to this day. The Thirty-Fourth and Girard
Street entrance to the Philadelphia Zoo features a statue of a dying lioness stuck with an arrow; she lay beside a male lion and her two cubs.
Sculpted in bronze, this powerful piece first premiered at the 1876 Cen-
tennial in Philadelphia, and today its foreboding image continues to
confront visitors entering the zoo's gates.[47]

These and similar animal statues of bronze and granite may be renowned works of public art, but adults would be forgiven for expecting
children to ignore them on their way to more animated creatures frolicking among the naturalistic habitats scattered throughout modern
zoological gardens and parks. Yet those adults would be wrong. In fact,
these immobile hulks of metal and stone rarely want for attention from
young children who sometimes seem to *prefer* spending time enjoying
these statues instead of with live creatures roaring and whooping
nearby. Inside the entrance to the Philadelphia Zoo stands Heinz
Warneke's giant *Cow Elephant and Calf,* a set of granite pachyderms
built in the early 1960s. Standing tall at eleven feet, this statue drives
children to distraction, especially when their parents insist on redirect-
ing their attention to a set of Colombian black spider monkeys swinging
about in a nearby exhibit, as illustrated by the following observation:

> One September morning a mother and her daughter passed the monkey
> enclosure.
> "Want to see the monkey? See him?"
> "No," the daughter protested, distractingly looking around the zoo.
> "Come here!" the mother said pointedly, taking her daughter by the
> hand and dragging her to the railing. Yet something else caught the girl's
> attention.
> "Look! Elephant statue!" the girl said, eagerly pointing toward *Cow El-
> ephant and Calf,* past the monkey exhibit, expressing her affection for the
> larger and perhaps more celebrated charismatic mammals.
> "Okay, are you done with the monkeys?" her mother asked, helplessly.
> But her daughter was already off, running toward the statue. "Awesome!"
> the girl yelled as her mother hurried to catch up.

This contestation of meaning plays itself out over and over again. Moments later at the Philadelphia Zoo, a boy asked his mother, "Do you see that statue?" pointing to the elephant sculpture. The mother replied with a corrective, "I see that statue—but look at the *real* animal," insisting that he prioritize the live monkey over the elephant replica.

These types of interpretive confrontations—what sociologist Erving Goffman once referred to as *frame disputes*—between parent and child occur with surprising frequency.[48] In recent years zoos around the country such as Zoo Miami, New York's Bronx Zoo, Salt Lake City's Hogle Zoo, and the Philadelphia Zoo have begun featuring animal sculptures made of thousands of colorful Lego bricks. At the Philadelphia Zoo, a mother and her son stopped to admire an enormous Lego penguin. But after a short while the mother suggested, "Come, let's go see if there are real, live penguins to see."

But the son knew what he liked. "No, this is more fun. Take a picture of me with the penguin." The mother took the boy's hand and said, "Come, I'll take a picture of you with the live penguins. We didn't come to the zoo to take pictures of you with Lego penguins."

"What's the difference?" the son protested. "It'll look the same in the picture. This penguin is a lot cooler." Sensing that she would lose this particular argument, the mother capitulated, taking out her iPhone and snapping a picture of the boy with the Lego penguin behind him.

"Okay, now let's go find the *real* penguins," she said. "You'll see how cute they are when they walk and swim." Syracuse sociologist Marjorie DeVault's research on public behavior among families visiting zoos illustrates how common these types of normative disputes can be, especially as parents with small children attempt to orient their kids' attention toward more socially appropriate objects of display. When successful, they socialize their children to prioritize the zoo's exhibited animals over more "mundane features of the environment," such as rocks, squirrels, guardrails and enclosures, maintenance areas and equipment, or unlabeled ponds or enclosures. DeVault writes, "The child learns, for example, to see the animal, but not the cage."[49]

But given the importance of curiosity and imaginative play among preschoolers and early-elementary school kids, we should *expect* that young zoo visitors would be especially attracted to bronze or brass statues and other makeshift apparatuses such as handrails and stairs in public settings like zoos. Like their teenage counterparts who engage in risky street play such as skateboarding, fixed-gear biking, and parkour, children at the zoo often foreground the materiality of the built environment over its more culturally appropriate uses, usually to the consternation of their frustrated parents. At City Zoo, a mother with two toddlers in tow stood at the lemur exhibit. Meanwhile, her son ran up and down the exhibit's railing, running his hands along the bars, while her daughter attempted to climb *through* the bars of the railing. "Are you guys even *watching* the lemurs?" the mother asked.[50]

Stone statues of a certain height not only are accessible playthings for young kids to climb and explore, but also verbally and physically engage as make-believe companions. Children can meaningfully participate in imaginative and playful encounters with statues and other animal representations as interactional surrogates, just as humans can enjoy forms of sociable play with *actual* animals such as their pets even though genuine interspecies understanding may not be possible.[51] For these reasons, zoos have made great strides in the past several decades by providing the public with interactive play areas conducive to early childhood development and sociability—call it behavioral enrichment for people. In 1985, Philadelphia Zoo's designers renovated an antiquated antelope house originally built in 1877 into the Treehouse, a creative play space where children climb dinosaurs and larger-than-life honeycombs, crouch inside giant flowers and flamingo eggs, and play hide-and-seek in a multistory tree with outlook perches spread throughout its enormous trunk. Across the country, local zoo patrons who can afford annual membership subscriptions allow their children to spend lots of free time exploring play areas during zoo visits, whereas out-of-town tourists often discourage such diversions. Given the enormous time demands experienced by vacationing families, they prioritize encounters with living animals over seemingly frivolous attractions that children might actually prefer.

The popularity of brass monkeys and other zoo statues among children also draws from the same rivers of commercialized popular culture in which contemporary American children so eagerly swim. As every parent knows only too well, young children keep large collections of animal replicas in their bedrooms—only they are plush and furry, rather than bronze and granite. (They are also sold in zoo and aquarium gift shops around the world: white polar bears, gray elephants, blue whales, green frogs, orange clownfish, and countless others.) Children recognize that like their stuffed animals, zoo animal statues offer a far more idealized representation of their particular species than the genuine article ever could. Given children's familiarity with sanitized and perfected images of animals from television, film, and popular picture books, they gravitate toward lifelike statues, stuffed replicas, and artworks less capable of generating disappointment (or revulsion) than a napping lion, caged jaguar, urinating goat, or molting bird. This is especially the case for Lego animal statues, which are quite literally built out of children's toys.

Moreover, compared to the live creatures at the zoo, animal statues are far less threatening when placed in proximity to young children, as kids occasionally point out to their parents. On a sunny but chilly November afternoon at the Philadelphia Zoo, a young boy climbed onto the snake statue in front of the Reptile and Amphibian House. The boy called out to his mother, "Take a picture of me! Take a picture of me!"

His mother answered dismissively, "Why? I know what you look like, and the statue isn't real!" Spotting a peacock walking along a nearby path, she offered, "Get down and stand near the peacock—then I'll take a picture."

The boy got off the statue and cautiously followed the peacock down the path, but he wasn't so sure about his mother's plan.

"Will it bite me?" he asked his mother.

The mother assured him, "No, it's fine, get closer." But her boy seemed reluctant, and the mother once again said to her son, "It's *fine*, get *closer*."

But the boy froze up, insisting, "No, I'm afraid!" After clambering back up the statue, he pled, "Take a picture here. *I don't care that it's not real*." The mother just shook her head and took the picture. Nature is culturally constructed, and so it is no wonder that children must be socialized to not merely recognize but actually *value* the genuine article over its simulated facsimile at the zoo.

## Children as Animals

Every American family knows that zoo visits provide opportunities to photograph not only exotic fauna, but one's own children or grandchildren as well. In many ways, human children represent our final breed of categorically unusual animals found at the zoo. Like pets and other companion animals, children are both dominated by adults and simultaneously showered with affection, treated with equal parts care and condescension. As suggested by the cultural geographer Yi-Fu Tuan, "Whatever views a mother may have toward her infant, in the actual practice of mothering she has to treat it as an incontinent young animal and even as a thing. . . . The small child is a piece of wild nature that must be subdued and then played with—transformed into cute, cuddly beings or miniature adults as the mother or the surrogate mother sees fit."[52] According to Erving Goffman, adults often regard children as nonpersons in everyday interactions among themselves, just as they might with animals. We expect children and animals to lack self-control, and so we allow both to devolve into all sorts of otherwise private behaviors in public, from self-indulgently playing with toys to cackling and even urinating.[53] Adults enjoy children and nonhuman animals, even if they belong to other families, as creatures to be displayed, admired, and asked to perform tricks most adults would find personally debasing.

One day at City Zoo a man and woman stood to the side of the polar bear exhibit. The woman pointed to a toddler-aged child (not her own) and turned to her husband, "Honey, look how small she is." The man replied, "Babe, we are here to look at animals, not kids." Many grandparents would disagree, of course. One June morning at Metro Zoo I greeted visitors with a woma snake, a large python from the dry deserts of central Australia. Two grandparents began arguing when the grandfather started snapping a photo of the snake.

"Why are you taking a picture of the snake?"

"Don't you want a picture of the snake?"

"No, I don't want pictures of the snake. I want pictures of *Olivia*," their granddaughter.

As it happens, children themselves are not immune to their own charms, and can always be found checking out one another on the zoo grounds. The morning I stood at Metro Zoo's entranceway holding Hero, the spiny-tailed lizard, a child walked right past us, staring at all the children around him. "He's more excited about the kids," his mother pointed out, especially in comparison to the nearly comatose lizard in my hands. That spring I also stood at the Metro Zoo gates holding Hogwarts, the western hognose snake. While a mother tried to get her daughter to focus her attention on the snake, it was clear that the toddler was distracted by all the older children flooding into the zoo, who as a group generated much more excitement than a small greenish-brown snake ever could. Eventually the mother physically lifted the girl up out of her stroller and pointed her head toward Hogwarts's beady eyes, but her gaze just drifted up toward my own (admittedly entertaining) face, and I smiled. *"No, don't look at him. Look at the snake!!"* the mother shrieked in frustration.

Children and animals sometimes *literally* resemble one another. The diets and eating schedules of kids and zoo animals alike are both carefully planned and enforced by their supervising guardians, just as both require careful management of their respective digestive wastes. Zookeepers and parents also provide their charges with toys, puzzles, and opportunities for supervised play and other constructive activities designed as enrichment.[54] Moreover, contemporary middle-class parents allow their children a great deal of expressive and creative autonomy when in public—especially with regard to their appearance—which is why zoos bring in so much revenue from painting children's faces to look like tigers, lions, butterflies, and bunnies.[55] Many children also arrive at the zoo wearing costumes depicting real and imagined animals from dragons to dinosaurs to bumblebees. If some parents encourage their kids to dress up like animals, others compare their children's everyday behavior to that of zoo animals. At City Zoo, a mother carried a curly haired girl dressed in pink over to a monkey exhibit. "He's swinging like you. Say *'Hi, monkeys!'*"

In the river pavilion, another mother told her son, "The otter has water toys, just like when *you* swim."

At Metro Zoo, a dad compared his son to a fruit bat. "Hey, it's eating bananas! Hey, buddy—he's just like you!"

(Children themselves made these comparisons as well. Upon spotting a plush monkey and toy octopus in the porcupine house, a boy called out to his father, "Daddy, that's a stuffy! Look, they have stuffies!")

In an era when adults sacralize their children and thus love and protect them as both precious and priceless beings, parents treat their kids as zoo animals in one other way as well.[56] While middle-class parents may allow their children a great deal of *expressive* and *creative* autonomy, they give them very little *physical* autonomy or freedom of mobility in public places, thus confining them as one might a captive animal.[57] Parents corral their children around zoos together in protective bubbles of interpersonal space, emphasizing the social distance that families construct between themselves and other spatially proximate groups of strangers when in public.[58] In an era of exaggerated and overhyped fears of pedophiles, predators, and overall stranger danger, some parents literally keep their children on leashes while at the zoo by making them wear backpacks with stuffed monkeys, kangaroos, or other animals with extra long tails that function as tethers.[59] (Thus we have children *dressed up* and *face-painted* like captive animals wearing backpacks that *look* like captive animals so their parents can walk them around on leashes as if *they* were captive animals, all so they can traverse the zoo looking at *other* captive animals that they can then compare to themselves.)

Just like conservation biologists tag endangered species with GPS-tracking microchips in the wild, parents dress their children in matching T-shirts so they can easily be spotted from afar, both for their own safety as well as for the parents' convenience. One April morning at the Smithsonian National Zoo, Scott and I encountered a group of kids participating in their friend Emily's seventh birthday party all wearing identical tops, each one a screaming neon shade of bright yellow and green. I mentioned to Emily's mother, "Wow, I could see you guys from space." She responded, "That's the idea! We're outnumbered today." On a sunny day in June at City Zoo, a worried mother with long blonde hair and a furrowed brow called out for her son, a six-year-old boy, who had raced out of her reach. Upon his retrieval, she admonished him for needing to "know better," warning him, *"Bad things happen to people every day. Someone could kidnap you."*

These fears are only augmented by the fact that many American city zoos originally built in the nineteenth century happen to be located in areas that today have lost much of their luster. Trips to the zoo therefore invite suburbanites and other out-of-towners to negatively judge the surrounding urban environment—and, by extension, the entire city or even urban life itself—as stereotypically dangerous and unsafe. For example, the Philadelphia Zoo is bordered by railroad tracks and West Philadelphia neighborhoods of poverty and urban decay. When parking at a zoo-owned lot on nearby West Girard Avenue, out-of-town visitors cannot help but notice the boarded-up buildings lining the street, a symptom of decades of abandonment and neglect. One afternoon as Scott and I crossed West Girard, we overheard a family from nearby Delaware walking behind us. Their little girl had observed three cars run a red light, and she asked her parents, "Why aren't those cars stopping?" Her mother replied, *"Because this is Philadelphia."* The zoo attracts suburbanites and tourists from outside the city, and so although lousy traffic behavior knows no bounds, the mother used this as an opportunity to depict the city as a place of incivility and danger. Through these kinds of interactions, middle-class parents affirm to their kids and to one another that the city is patently unsafe, especially for young children.[60]

"*Bad things happen to people every day. Someone could kidnap you.*" Are these safety concerns warranted? In point of fact, children are far more likely to be severely injured or killed by members *of their own family* than by strangers.[61] With regard to incidents of kidnapping, Hanna Rosin reports in a cover story on "The Overprotected Kid" in the *Atlantic* that "the kind of abduction that's likely to make the news, during which the victim disappears overnight, or is taken more than 50 miles away, or is killed," remains "exceedingly rare." Meanwhile, *family* abductions *are* on the rise, due to the high prevalence of divorce and their resultant protracted custody battles. Nevertheless, Rosin observes that "overall, crimes against children have been declining, in keeping with the general crime drop since the 1990s."[62]

Still, fears abound, even in the relative safety of zoos that employ security guards, watchful docents, guest services personnel, and zookeepers with firearms training. These kinds of panics are only augmented by paternalistic campaigns like the Masonic Child Identification Program (CHIP), which encourages parents to have their kids photographed and videotaped, their fingerprints inked, and their DNA collected from hair samples, nail clippings, and an oral swab, thus "providing parents with identifying information they would need in the event their child ever goes missing." I mention the Masonic CHIP program because I first learned of it when a local Masonic Grand Lodge organized a special station at a free children's fair held on the grounds of Metro Zoo. Leave it to the Masons to needlessly raise the anxieties of parents while surrounded by enclosures filled with *actually* threatened species and their endangered young.

## Making Meaning at the Zoo

American city zoos are cultural worlds, urban landscapes where human primates generate collective meaning and sensibility, distinction and disgust. Full of pine snakes, penguins, and fruit bats, zoos illustrate how families and other audiences attach shared sentiment and symbolic significance to animals. Like nature itself, the culturally derived meanings visitors attribute to pandas and tigers and elephants may be the product of human imagination and myth, yet such fictions are no less powerful to those who hold fast to them. In the Anthropocene we will need to reckon with these myths, the product of collective prejudice and mass desire, if we are to stave off what scientists refer to as the sixth major extinction event in the history of life on Earth.

At the same time, zoos fulfill real human needs. Just as fragile ecological habitats around North America have been usurped for oil drilling, hydraulic fracturing, real estate development, and suburban sprawl, we have seen a similarly devastating loss of truly public spaces in the urban metropolis where people can freely and unabashedly socialize in one another's presence. While many zoos may be privatized, themed, consumption-driven, and charge admission, they can nevertheless serve as oases of relative accessibility and inclusion in the contemporary city, especially when compared to heavily policed downtown corporate plazas, high-income residential developments, luxury shopping malls, upscale coffee bars, and well-financed business improvement districts. This is especially the case for municipal or otherwise publicly funded zoos that do *not* charge admission to visitors, such as the Saint Louis Zoo, Chicago's Lincoln Park Zoo, the Cape May County Zoo, and the Smithsonian National Zoo.[63]

Yet if the consistent differences between adult anxiety and childhood desire discussed in this chapter reveal anything else about contemporary zoos, it is that different audiences—what literary critic Stanley Fish refers to as *interpretive communities*—rely on zoos to satisfy cultural agendas that sometimes align with one other but occasionally diverge.[64] Nearly all zoo visitors share a desire for *recreational* experiences full of opportunities for entertaining diversions, whether enjoying penguins waddle around their polar exhibits, buying stuffed toys resembling them, admiring Lego statues of them, or simply watching an IMAX movie about them.[65] But in addition, parents approach the zoo with a set of *didactic* aims, both to teach their children about animals and the natural world and also to socialize them to recognize the normative rules and social expectations governing modern public life. Children's cultural agendas at the zoo are far more *affective* as they find pleasure in not only watching pandas and polar bears, but also investigating the wild lives of squirrels and spiders, the hardy exoskeletons of cockroaches, and the railings, staircases, statues, and other material elements of the zoo's built environment. Some adults participate in *documentary* projects, whether by video-recording the antics of giant pandas or snapping endless series of photographs of their own grandchildren.

Then again, many zoo audiences enjoy zoological parks and gardens as protected habitats of *sociability*, whether generations within families enjoying each other's company, toddlers observing other children, or even romantic couples strolling arm in arm among the zoo's pastoral attractions. (Perhaps the latter are immune to the zoo's pungent aromas and crowds of whining preschoolers.) Local zoo members with annual subscriptions have the luxury of permitting their children extra time to fully explore the play spaces of zoos, while out-of-town tourists generally seem less interested in such pursuits, given the often rushed nature of their visits.

These varied audience dispositions draw on the dual promise of zoos as public arenas designed for recreation and education, fun and enrichment, entertainment and enlightenment. Later in the book, we will explore how zoos try to engage visitors by mobilizing public desires for learning and cultivation, followed by a closer look at how zoos employ the pleasure and postmodern buzz of popular culture and mass spectacle to entertain their audiences, sometimes to the detriment of more educational goals. But first, let's take our safari beyond the crowds and further into the backstage veldt of the zoo to meet the largely female workforce responsible for caring, feeding, and cleaning up after its wildcats, monkeys, alligators, macaws, and many other fauna, charismatic or otherwise. These zookeepers provide a model for the dedicated stewardship needed to tend to our warming planet and its biodiversity of endangered and vulnerable species in the Anthropocene. They also represent the lengths to which people can construct emotionally meaningful relationships to nonhuman animals, and a shared sense of affinity and commitment to one another.

# Chapter 3

# Birds of a Feather

## Zookeepers and the Call of the Wild

**ADORNED IN MY STANDARD ISSUE** zoo worker's uniform—green shirt, name badge, work boots, and obligatory khaki shorts—I began my first full day of work as a volunteer at City Zoo by signing in at the employee's security entrance and heading down to the children's area, which featured a duck pond, rabbit hutches, and barnyards where an assortment of farm animals resided: cattle, chickens, pigs, goats, sheep, guinea fowl, and a horse and donkey. After finding the drab break room and depositing my lunch in a refrigerator covered with magnetic poetry about monkeys, I looked around at the diverse group of high school and college interns, one of whom asked me if I was a retiree, even though I was only in my thirties. (Zoos often bring on young unpaid interns to assist zookeepers in cleaning exhibits and enclosures in addition to handling a variety of other animal care-related tasks. They also typically require a trial period of internship before hiring a new keeper, even one with prior experience at another zoo.) Throughout the day the zoo was thick with adolescent gossip. Elizabeth, who was soon off to begin her freshman year at a local college, complained aloud that when she dumped her high school boyfriend Justin, he practically sobbed non-stop, for days. She thought he was being super lame.

Under the blazing August sun, our morning responsibilities mostly involved maintenance work. First we all assisted in unloading two palettes of hay into a wooden shed; gathered wheelbarrows, rakes, and shovels; and then we really got to work. In ninety-three-degree heat we shoveled the refuse hay, straw, and various piles of fresh, fly-infested animal dung from the dirty barnyards into our wheelbarrows and emptied them in an industrial-sized dumpster brimming with manure. As I cleared the pungent excrement from under one of the cows, it turned and came toward me. I backed away, slowly but steady. I didn't want to make it angry. It may or may not have mooed at me.

As we continued shoveling I noticed out of the corner of my eye a family looking on with equal parts fascination and revulsion. Suddenly I heard the mother tease her son, in reference to us, "Hey Oliver, I think they need some help!" as if trying to gross him out by the mere *thought* of cleaning up cow dung, ignoring the fact that at least three of us actually were, in fact, doing just that, and within earshot.

Many of the tasks required of zookeepers can be physically disgusting as well as socially degrading and morally ambiguous, a kind of labor the University of Chicago sociologist Everett C. Hughes referred to in the early 1950s as *dirty work*.[1] Yet zookeepers nevertheless take great amounts of pride in their careers and the dirty work they perform on the job, both literally and figuratively. On this leg of our safari we go behind the scenes to explore the backstage culture of zookeeping, all with an eye toward understanding how keepers attribute value and significance to their work. Like zoo audiences, keepers assign shared meaning to the captive animals that surround them, and in doing so, they too collectively construct the natural world. Zookeepers also activate moral and symbolic boundaries to emphasize their affection and commitment to what they see as the central mission of zoos: to provide health, comfort, safety, and care to the zoo's resident creatures, fish or fowl.

At the same time, the ties that bind among zookeepers extend not only to the prehensile-tailed reptiles and furry mammals under their care, but also to each other. As a predominantly female workforce, zookeepers collectively participate in what the sociologist Erving Goffman once referred to as a shared *community of fate,* as colleagues who maintain similar professional roles, gender norms, and concerns over social status and the significance of their work.[2] Like birds of a feather, zookeepers develop intimate relationships with one another—tight-knit flocks bound by a shared sense of camaraderie, collective identity, and a moral commitment to zookeeping as far more than simply a job. Their selfless dedication to the care and protection of animals provides a worthy model for the stewardship efforts required to preserve vulnerable and threatened species amid the growing environmental crisis.

# Dirty Work

In Peggy Rathmann's children's book *Good Night, Gorilla*, a lovable but absent-minded zookeeper strolls the grounds in the dark, bidding good night to all his four-footed friends. Unbeknownst to him, a gorilla steals his multicolored keys and frees his fellow zoo animals, who quietly follow the keeper home to wish him and his sleeping wife a hearty good night of their own. Naturally, hilarity ensues. Actual keepers themselves report that the laypeople they meet outside of work similarly imagine zookeeping to be a bit of a leisurely romp. Ashley, a keeper at Metro Zoo, recalls once being asked, *"Oh, do you get to pet the lions?"* Of course, the daily rounds of a zookeeper look like anything but a romp, leisurely or otherwise. The zookeeper's workday at Metro Zoo begins at seven o'clock in the morning, rain or shine. Lauren, an auburn-haired zookeeper in her late twenties, explained the tedious and often backbreaking labor required while tending to her assigned area as part of a four-keeper shift:

> We get in, and we check to make sure if anyone is on vet care, or needs to take meds. We go down to each exhibit to get a visual on every single animal: the falcons, the bison, the elk, the ferrets, the burrowing owl. We also have to check on the breeder mice [live mice that keepers must kill immediately before serving to certain zoo carnivores], the squirrel monkeys, and the bats. After we check on the animals, we all clean the barn together. I take the truck and go straight down to the bison, clean the bison yard of hay and poop, which takes about a half hour or forty minutes, and after we put out the bison we clean the bison holder, which is another half hour or forty minutes. Then we put all that poop in the truck, and drive that entire pickup truck full of poop down to the dump truck, and we load all the poop onto *that* truck, which is another twenty-five minutes or so. Then we go back down and clean the elk holding—more raking poop and changing out the water and picking up the hay. And then we clean the prairie dog enclosure, rinsing out the old hay and scattering new hay. Then we go clean the ferret exhibits, which are these five-hundred-pound concrete blocks that we

have to roll out on wheels, change their straw, wipe down, and roll back in. (And if we don't place it correctly, we have to pull it back out and try again.)

And then the falcon exhibit: we scrub the branches with water and a scrub brush. In the summer you have to weed-whack the keeper area between the bison holding and the bison yard, the falcon exhibit, and the prairie dog exhibit. And then once we are done with all of that work, we've got the feeder mice: three days a week they get full changes. We have eight mouse tanks right now; we have to take all of the mice out, take out all of the substrates, scrub down the tank, fill it with an inch layer of bedding, put all of the wheels and beds and houses back in, and do that for eight tanks. Then we have the bats, which we clean, take out their food, scrub down their branches, pull the newspaper out, and put in new newspaper, and new water. And then the squirrel monkeys have two holdings. Lots of poop everywhere, all over the place—on the walls, on the branches. We put feed in one holding, and then clean and put out fresh enrichment, and then clean the other holding with lots of scrubbing and hosing.

And then at the end of the day, we run around and bring everyone else in. If we don't have diet help [volunteers and interns often assist with animal diet preparations], then we dedicate an hour to making fruit and vegetable diets in addition to all that work. On a four-keeper day, that's pretty basic.

After an arduous eleven-hour workday (minus an unpaid lunch break), Metro Zoo keepers finally leave the zoo at about six in the evening; they repeat this exhausting routine for four days in a row, including weekends and holidays. The job inflicts a chronic and painful toll on its workers. At twenty-seven, Lauren had already worked at a number of zoos around the country, and worried how long she could keep up the pace and physical demands of the profession:

I don't know that my body could handle another thirty years of this. I have been doing the work off and on for eight years now, and I already experience quite a lot of pain. My shoulders don't like going high over my head. I am seeing a chiropractor for my lower back and my hips, and at the end of the day, when I have to walk the whole zoo and check everyone's locks? Some days that's kind of uncomfortable for me: I'm sore, or I'm tired, or my knees go out because we're always jumping fences. Our elbows—because the way we shovel and rake the poop and stuff—it's like a really intense tennis elbow. I know a few people in [a midwestern AZA-accredited zoo] who had to quit. One was a primate keeper who was drinking two martinis a night to get through the pain in her elbows, and she eventually just had to stop.

Krista, a twenty-eight-year-old keeper who always wore hip glasses at work with stylishly translucent plastic frames, agreed. "It's just hard work and it hurts my back, and I don't really like doing it. It's drudgery and you have to do it every day, and it's just like, 'Ugh, I have to do this *again.*' And then by the end of the week your back is killing you, but you do it because you have to."

If this laborious work demands both physical strength (some of the keepers I met at Metro Zoo played competitive sports as NCAA athletes) and tireless patience and endurance, it can also be potentially dangerous. To restrain large animals during veterinary procedures, Metro Zoo keepers use tranquilizer darts filled with ten cubic centimeters of the drug Carfentanil, an animal anesthetic ten thousand times more potent than morphine; at such dosages it can kill humans instantly. Lead keepers must maintain proficiency in handling firearms, as they would serve as the primary shooter if one of the zoo's ten deadliest animals escaped its enclosure: American alligator, bighorn sheep, American bison, cougar, elk, jaguar, Chacoan peccary, pronghorn, rhinoceros, and gray wolf. Zoos consider this type of (mercifully rare) emergency a "Code Red," one so calamitous that keepers obsessively perform rituals of double- and triple-checking enclosures to assure themselves they have been properly locked. Christina, the twenty-four-year-old keeper with the snake tattoo, puts her finger to her nose before touching each lock. Heather, a twenty-eight-year-old keeper and veterinary technician with brown hair and funky purple nails, says, "Oh, I look like a moron because I shake [the lock]. Like, I have to make sure I shake it, and then I leave, and then I'm like, '*Did I shake it?*' And then I shake it again. 'Oh, wait. Did I shake it?' Especially in the jaguar or the cougar area, I just shake the locks a million times before I leave, or feel comfortable leaving." (Fortunately the only animal escape that ever occurred on my watch at either zoo happened when a younger City Zoo volunteer left one of the chicken coops open, and I wound up chasing a gigantic rooster named Eisenhower around a nearby pony corral for ten minutes.)

Of course, zookeeping is not only labor-intensive and dangerous but also filthy, sometimes nightmarishly so. In their daily zoo rounds, keepers regularly engage in dirty work normatively considered repugnant, from cleaning out hippopotamus pools to scrubbing gorilla cages to picking up elephant dung. According to Krista, parents rarely suppress themselves when remarking to their children within earshot of zoo staff, "'You wouldn't want to do that job, huh?' I hear them talking to their kids, 'Eeew, that's gross.' And it is—it's gross."

My own experience as a zoo volunteer reflected this reality as well, albeit on a diminished scale.[3] At City Zoo I spent mornings outdoors cleaning and disinfecting animal enclosures, and shoveling cow manure, rabbit and goat pellets, and all manner of just plain shit—horseshit, donkey shit, chicken shit, macaw shit, duck shit—in extreme weather conditions ranging from summer swelter to wintry wind-chill temperatures in the frigid teens. (Whenever I shoveled hay and dung at the zoo, I could taste it in my mouth for days afterward. It also turned out that it made my wife allergic to me.) Over the course of my research countless animals urinated, excreted, or vomited on me.

At both City Zoo and Metro Zoo I also prepared diets for the zoos' resident animals, work that alternated between dreary and gruesome. At Metro Zoo my routine normally began in a drab dry-goods stockroom, where I would measure out heavy buckets of grain from alfalfa hay pellets to bison feed.[4] Next I would prepare carnivore diets for cougars, jaguars, bobcats, bald eagles, and peregrine falcons, among other flesh-eating creatures. For the zoo's eagles, hawks, and other birds of prey, I carefully weighed out portions of raw horse meat ground up with organs and bone, and prepared beef diets soaked in buckets of cow's blood for our wildcats. After dishing out the ground meat and innards, I used a soup ladle to pour additional blood over the food as a kind of au jus gravy for the ravenous beasts. The awful stench stayed with me for hours, as did the juicy blood spatter. In addition to their ground equine, I prepared bowls of defrosted mice, rats, quail, and baby chicks for our birds of prey. At City Zoo I was also frequently tasked with splitting open the body cavities of the thawed rodents with a paring knife in order to insert additional portions of raw meat into their tiny carcasses. It was grisly work.

According to Louise, an experienced volunteer in her sixties who supervised interns working in Metro Zoo's kitchen, one of her young charges quit on her first day after she vomited from the look and smell of the meat alone. This is not uncommon. According to Ryan, a twenty-eight-year-old former "animal experience specialist" at a prominent aquarium who continues to work with live animals as an educator for a natural history museum, "I've seen this with people who come in as volunteers—they just want to work with the animals. They want to see this cute parrot, or this cute bunny, and after a week of picking up shit and scraping up dead mouse guts, they're like, *"I'm done."*

Such distasteful work is not only physically defiling but *socially* polluting as well. Lay people associate the handling of bodily fluids and other offending contaminants with low-status occupations (hospital orderlies, school custodians, sewer workers) performed by marginalized people at the bottom rungs of society. Parents regularly emphasize the janitorial duties of zookeepers to their children as terrifying cautionary tales, and multiple keepers noted to me how frequently they overheard zoo visitors remark to their children or grandchildren that the purpose of a college education was to avoid stigmatized labor such as shoveling animal dung—even though 82 percent of today's zookeepers actually *have* a college degree, a requirement for most animal care jobs in AZA-accredited U.S. zoos.[5] For example, the keepers at Metro Zoo had all earned four-year undergraduate diplomas in biology, wildlife management and conservation, zoological science, or another related field. Again according to Lauren, "I've had parents see me cleaning the elephant yard [at a midwestern zoo], point to me, and say to their kid, '*See, honey? This is why you get a college degree.*' And I wanted to say, 'Excuse me—I actually spent a lot of money on a college degree to be able to have this job.' So I hate that stereotype that we are just grunt workers." Lily, a well-traveled twenty-nine-year-old keeper with raven hair, attended a large public university and has received grief from her college classmates about her low-status career choice:

My friends, especially my college friends whom I've known for a really long time, gave me a lot of crap for it. They were just waiting for me to get over this "stage," this thing that I was trying out "temporarily." They really didn't think I was doing this long term, and they really didn't think that it was a real job—and they said as much. I did pretty well among those in my group of friends in college, and I think they were expecting me—in their eyes—to "do something" with my degree. But this is totally what my degree is! I studied biology, and I'm now working in a field that is based around animals! But my college friends were not supportive at all, and they still aren't, especially because I live with my mom for financial reasons because I can't afford to live on my own.[6]

Also, my extended family has not been very nice about my work. . . . My aunt's parents were around at some holiday and they were, like, "Oh, so when are you going to get a *real* job? I mean, you can't possibly think that *this* is what you're going to do."

I experienced this kind of shaming from time to time myself. One day while working at Metro Zoo, a coworker directed me to the jaguar enclosure where one of the wildcats had just vomited. When I arrived, a little girl who had been observing the jaguar pointed out the vomit and told me, "I feel bad for you." I asked why. "Because you have to clean up throw-up," she explained. In fact, people find zoo work so objectionable that the state that governs Metro Zoo allows first-time DUI offenders to perform community service hours there as part of its Accelerated Rehabilitative Disposition program as an alternative to a traditional court trial and sentencing. (Yet interestingly enough, regulations actually forbid these scofflaws from shoveling manure at the zoo, as they lack the academic credentials to do so. This is not a joke.)[7]

Moreover, keepers must routinely perform certain tasks that those outside the zoo world may find *morally* repugnant in addition to simply degrading. Among such on-the-job duties, the zookeepers I interviewed from City Zoo and Metro Zoo reported having neutered sheep, killed feeder mice with their bare hands and a brick, and gassed live and previously exhibited chicks two weeks after they had been hatched. (In this last instance, the chicks were sent to the zoo's commissary, where they would be distributed to carnivores as fresh meals.) Keepers also take part in euthanizing old and terminally ill animals, following up such operations by performing necropsies on the deceased. As a rather extreme illustration of this necessary yet macabre practice of handling dead animal remains, Heather cried as she recalled using an electric saw to decapitate a recently departed two-year-old jaguar named Seamus in order to examine his brains immediately after his death.[8] While these practices may be justified as necessary for the management of live animals in captivity within the zoo world, it is easy to imagine how they might give outsiders pause.[9]

Ultimately, the persistent stigma of zookeeping is best reflected by its low economic market value. According to the U.S. Bureau of Labor Statistics, the mean annual wage of a nonfarm animal caretaker in May 2012 was $22,370, with a mean hourly wage of $10.75 per hour. To put this in perspective, the mean annual wage for *all* occupations in that year was $45,790, more than *double* what many zookeepers earn.[10] These national averages reflect full-time zookeepers' pay at Metro Zoo, which during the time of my research ranged from $10 to $12 an hour. (At City Zoo, I met many keepers who made only $10 an hour, as well as seasonal keepers who earned even less.) These low wages persist in a socioeconomic context that demeans women as workers and devalues their labor. In 2000, women represented 75 percent of all zookeepers nationwide, just as they made up a majority of the keepers with whom I interacted at City Zoo, and the *entire* zookeeping staff at Metro Zoo, with the exception of its head animal curator.[11] Once a male-dominated occupation, zookeeping has become increasingly feminized over the past several decades, just as have other animal care professions such as veterinary medicine.[12] Female-dominated occupations typically pay less than those filled predominately by men, and it has been shown that at least in some fields (such as publishing and bartending) decreased or stagnant wages have actually led to their feminization by discouraging male participation.[13] Moreover, American women are also far more likely than men to work in occupations that, like zookeeping and animal training, involve caregiving, and such jobs actually carry a relative wage *penalty*. Given its cultural associations with nurturing and motherhood, the delivery of care has historically been considered women's work that females are assumed naturally predisposed to perform out of love, empathy, and virtue, rather than for earned remuneration. For these reasons zookeeper wages have remained stagnant even as job requirements demand increasingly higher levels of education, work experience, and skill.[14]

Even with college degrees, keepers at both City Zoo and Metro Zoo struggled to make ends meet. Many lived with their parents to save money, and some picked up additional hours working part-time retail jobs to cover their routine expenses. Yet even when she worked at a big-box pet store on top of her already taxing full-time zoo job, Krista felt like she was "just barely scraping by." As she explained, "For the amount of education I have, plus the experience I've had, what I get paid now is not at all—what are the words I'm looking for? It doesn't really *add up*. . . . My family sees it as, 'I know you like your job, your job is really awesome, but you're tired all the time, and you can't pay your bills. You should maybe look into becoming a nurse.'" For most keepers, monthly bills include not only living expenses but also inflexible student loan repayments that begin soon after graduation.

The sad irony is that in some ways Krista is lucky to be getting paid at all. As I noted above, before keepers can land even minimum-wage zoo jobs they must first gain experience by working as *unpaid* interns working alongside paid full-time keepers, sometimes for years. Eventually, after a lengthy period of what some might consider indentured servitude, some interns do in fact get hired by their zoos, but often only for part-time seasonal work, a purgatory that promises low wages without health care benefits or paid vacation time. According to Heather, one zoo hired her to work on a part-time basis, and whenever she managed to max out her allotted hours, the zoo would actually lay her off and immediately hire her back, just to avoid paying her benefits.[15]

In contrast, the chief executives of accredited AZA zoos have enjoyed fabulously lucrative incomes in both good times and bad—especially relative to their lowest paid workers, just as in the corporate world more generally.[16] According to tax returns submitted to the U.S. Internal Revenue Service, president and CEO Deborah Cannon of the Houston Zoo earned $316,309 in annual base salary plus bonus and incentive compensation in 2011. In that same year, Douglas Myers of the San Diego Zoo earned $321,948; Raymond King of Zoo Atlanta, $370,616; and Stuart Strahl of Chicago's Brookfield Zoo, $409,821. The highest paid zoo executive in the country that year was Steven Sanderson of the Wildlife Conservation Society, which operates the Central Park Zoo, Bronx Zoo, Queens Zoo, Prospect Park Zoo, and New York Aquarium, all in New York City—he took home over a million dollars in compensation in 2011.[17]

Even *among* zoo executives, some seemingly nonessential administrative posts often pay more than the highest ranking jobs involving the management of animal collections. For example, according to tax records the chief *marketing* officer of the Philadelphia Zoo received higher compensation in FY 2013 than the zoo's chief operating officer responsible for actually running the day-to-day operations of the institution and ensuring the health and safety of its entire animal population—a zookeeper among executives, one might say. As in other occupational realms, intraprofessional status in the zoo world is, perhaps ironically, predicated on one's *distance* from the messy realities of actual zoo work.[18]

# The Call of the Wild

Zookeeping is tedious, physically taxing work that most people would consider revolting, not to mention low paying, socially disreputable, and potentially dangerous. Of course, plenty of jobs involve performing stigmatized labor, each unpleasant in its own way: janitors, garbage collectors, rat exterminators. But zookeepers are different, because new hires require a bachelor's degree and an internship period of unpaid work experience. Zookeeping therefore better resembles the dirty work of physicians and nurse practitioners—occupations that educated college graduates choose for themselves, rather than working-class jobs that the unemployed accept under conditions of necessity and constraint. This begs some questions: As recently minted college graduates, why do zookeepers *elect* to work under such physically and socially abject conditions? What attracts them to the zoo, and why do they stay?[19]

First, we must consider the contextual basis of disgust. Although zoo personnel may handle blood, feces, and dead animals, this particularly colorful aspect of their jobs does not radically differ from the dirty work performed by medical examiners, morticians, and other professionals who handle similarly polluted contaminants: over time they grow accustomed to working with bodily waste and other biological refuse, and come to experience its constant presence as routine and thus mundane, rather than loathsome.[20] Whether slicing up freshly expired rats or clearing urine-soaked newspaper from a birdcage, keepers learn to handle tainted impurities with detachment, at least within the professionally defined context of zoo work. According to Amber, a tall twenty-two-year-old keeper with straight blonde hair,

> It's part of the care, so you have to do it. I don't mind it as much as a lot of people would. . . . I know a lot of people that will pick up poop with their hands, and they'll take part in necropsies and things like that, but seeing a person with a big open gash will freak them out. So it's just funny how accustomed you get to certain things, and then other things you just don't want to be around. I've heard of older keepers who've picked up all kinds of poop for years, but when it comes to changing a diaper they get skeeved out. . . . It's funny what you can get used to.[21]

Some zookeepers actually develop professional sets of best practices centered around the most seemingly mundane and unpleasant chores: Tracy, a seasoned keeper volunteering at City Zoo, even referred to raking hay and animal feces as an "art and a science," rather than the tedious and reviled tasks of a stigmatized occupation. As Hughes reminds us, "To the layman the technique of the occupation should be pure instrument, pure means to an end, while to the people who practice it, every occupation tends to become an art."[22] In many ways, the immense amounts of dirty work required for zookeeping in turn generate strong and reactive workplace subcultures among keepers—supportive social worlds powerful enough to symbolically transform that dirty work into meaningful labor.[23]

Still, while the performance of dirty work around animals eventually loses its ability to shock or disgust among the initiated, it hardly explains why college-educated zookeepers would do it for such low wages, or why interns would do it for *free*—after all, there may be honor in shoveling cow dung, but it is still *work*.[24] Yet in a way the question answers itself: zookeepers and volunteer workers demonstrate such a strong affinity for animals that it compensates for the poor work conditions and low (or nonexistent) pay associated with their jobs. In fact, many keepers claim a long-standing and exclusive desire to work with animals from a very young age. For these keepers, their love of animals feels so inborn and natural to them that they glorify zoo work as more than simply a job, but as a calling: a sense of self-knowledge that they were *meant* to provide care and comfort to animals, and without material consideration. As Lauren explained, this calling—the call of the wild, if you will—comes from "knowing that in your career you have the ability to make the lives of those animals so much better."[25]

As keepers explain it, their calling has deep roots that reach back into their biographical memories of adolescence and even childhood. As Ashley, an athletic twenty-four-year-old keeper with a pierced lip and tongue, recalls, "I've loved animals since I was able to walk, basically. . . . I definitely knew at a young age. As far as my memory stretches, I knew I liked animals."

Krista similarly recollects, "I always liked working with animals. That was originally my goal the whole time growing up, to work with animals," and in her youth her parents always surrounded her with house pets—cats, fish, parakeets, canaries, hamsters. Amber fondly remembers being the "Dr. Doolittle" of her family:

It's kind of weird because I grew up on Long Island in a suburban area. Like, there weren't a lot of animals. I didn't live on a farm or anything. There was really no reason for me to grow up having this strong affection or desire to work with animals, but I did, ever since I was young. One of my earlier memories is getting the first pet I remember, which was a rabbit. And my dad and his grandfather used to raise pigeons. Not many people know this about me, but it was a really cool hobby. And my grandfather used to have a coop in his backyard, and when I was five or six, sometimes they would be looking for me, and I would be sitting in the coop, just watching the birds. So, I guess, that's how I realized that I really liked animals. And then I got into the *National Geographic* books and always surrounded myself with animal books. Going to school, anytime animals were brought up, it would always pique my attention. My whole family would refer to me as Dr. Doolittle. And anytime there was a problem with anything—the dog was sick, or they found an injured bird—they would always call me up, as if I had vet skills to handle it. And then, actually, as I was growing up, my first job was at a pet store. And I worked there for about four years, and I really, really enjoyed taking care of the animals. I still go in there and work the cash register sometimes.[26]

This kind of affinity for animals became apparent very quickly when I observed zoo staff at work. Many zookeepers and volunteers employed baby talk when "speaking" to the animals in their care. At City Zoo, one female volunteer made a habit of using baby talk when simply talking *about* particular zoo animals, especially when saying their names aloud—"Oh, *Flipper.*" At Metro Zoo, a young volunteer routinely kissed one of the Russian blue Dumbo rats every time she held him, explaining that she used to have a similarly affectionate relationship with her pet guinea pig. As Daphne, fifty-seven, an education volunteer coordinator, told a class of new recruits during their training, *all* of the zoo's animals are, "for all intents and purposes, pets." (Daphne herself enjoyed a special relationship with Beaker, an African gray parrot who used to perch at her desk all day long.) Meanwhile, many keepers blurred their home and work lives by caring for quarantined zoo animals in their bedrooms, basements, and backyards, or by adopting zoo animals before the veterinary staff euthanized them. At City Zoo, Jess, an animal keeper and curator, sheltered a hen and rooster at her house for three weeks before they could be tested for avian influenza and cleared for their zoo residencies. Over the years Sharon, a herpetology keeper, had adopted (and thus saved from euthanasia) many zoo creatures, including a tiger rat snake, a prehensile-tailed skink, and a parrot that used to bite people. In fact, plenty of zookeepers maintain stocks of pets at home, even though they already care for ample numbers of exotic animals for a living. Heather grew up with frogs, snakes, rabbits, and a dog, and today she also raises three guinea pigs and two lizards at home, on top of her keeper job. Zookeepers and related staff at both City Zoo and Metro Zoo felt so strongly about animal welfare that in their spare time they raised funds for rhinoceros and habitat conservation as part of the American Association of Zoo Keepers' Bowling for Rhinos campaign.[27]

Many zookeepers internalize the call of the wild so strongly that it makes even dirty work feel ennobling.[28] Krista explained that while her siblings work at "jobs that pay them well, I have a job that I absolutely love." When I asked Lauren what made zookeeping worthwhile, particularly given the low pay, she pointed out that she breaks up the less satisfying moments of the workday by training a number of her favorite animals:

> I am finding time to train the squirrel monkeys and the wolves. You have to find the time, at least ten minutes with the squirrel monkeys, training them to station. The wolves I'm training to target and accept a pressure on their shoulders; we are trying to get them injection trained. It's probably ten or fifteen minutes of training, and ten or fifteen minutes of prep time. So between the two animals, it's probably forty minutes to an hour, time that you have to squeeze in somewhere. That *twenty* minutes that I spend training the animals, ten minutes with the squirrel monkeys and ten minutes with the wolves, might be enough to make the other *nine hours and forty minutes* worthwhile. . . . I have a relationship with the animals that I work with, and that relationship makes it worthwhile.[29]

Keepers at Metro Zoo expressed those human-animal relationships in numerous ways. One summer morning a lead keeper prepared a birthday "cake" for Ayo, one of the zoo's two giraffes. She made the dessert by grinding up his usual Wild Herbivore grain diet along with chopped carrots for the base, molding that into layers, adding water and freezing it, scattering more grains on top (as if they were sprinkles), and topping it all off with a carrot instead of a candle. On the other side of the zoo, Ashley maintained a personally fulfilling relationship with a pair of cougars, Fred and Ginger, despite the fact that she would never know whether her feelings toward them were reciprocated.

She explained, "I invest a lot of time in the cougars, so I feel like we have a pretty good relationship. You know, whether or not they would remember me when I left is a whole different story."

"Well, I can interview you; I can't interview *them*," I pointed out. "So just tell me how *you* feel about them."

"I love the cougars . . . I feel like they kind of know who I am. That's my feeling. I'm not 100 percent sure, obviously. I can't interview them either.[30] But Ginger does this little—I wouldn't say they are *meows*, but noises. And I mimic them back to her. So I kind of, I guess, talk to her, so to speak. I have no idea, obviously, what I'm saying to her. But I kind of talk to her, and she seems to follow me. Both of them—they know that usually I bring treats. So, that's probably the best explanation as to why they follow me around. But I feel like there is an established relationship there."

"Do you remember a particular interaction that you've had with them?"

"Ginger does, kind of, follow me around. [The training consultant] mentioned that she might have been limping, so one day we were trying to see if she was. So we had one keeper on the one end, and she's like, 'Can you get her to walk towards you?' And I'm making the noises and walking down. . . ."

"What kinds of noises?"

"It's like a *meow*, but she kind of just does a half-*meow*."

"Like a cougar noise?" I ask, unhelpfully.

"I don't know if cougars out in the wild make that noise. Obviously I probably wouldn't want to find out. Neither would you, I'm sure. But it's almost like a half-meow, which I feel is definitely something that you would find in captivity rather than in the wild. I definitely don't think that they make that call to each other. But even Fred, who is even a little less vocal, will start to make some noises. They definitely purr. So when you're in there, they love attention, and it doesn't matter if you're just, you know, talking to them, or anything like that. They love the attention, so you can hear the purring when you're talking to them in the holding area."

"So I'm at one end of the exhibit, and I'm trying to get her to walk towards me. So I'm like calling her name—*meow*—and walking down, and she walks down. And we walk back, and we found out that she was limping. We're not quite sure why. She could just be a big faker, we don't know," Ashley laughs. "It's just fun to see that they kind of recognize people, even if it is just because they know you bring extra food."[31]

Ashley illustrates how emotionally rewarding she finds interacting with Fred and Ginger, regardless of whether they base their entire relationship with her on cat food. In light of these nonmonetary rewards, when asked to envision alternative career paths keepers inevitably select some other occupation in which they would continue to train animals, or else work in field conservation or zoo management more generally. In the latter case, Lily points out that many would likely transfer their need to directly interact with their favorite zoo animals to caring for pets dwelling in and around their homes. "People who have gone from zookeeping into management or another field all of a sudden develop a little zoo at home," she explains. "It's when they get away from zookeeping everyday, as a part of their life, that they start accumulating their own extensive animal collection at home. I would hate not working with animals, but if I was able to get into something else more conservation-based, like working at a wetlands or with wild animals, I could see myself getting more animals at home."

## Symbolic Distinctions and Moral Boundaries in Zookeeping

For zookeepers the call of the wild may emerge from a lifelong affinity for animals, but they maintain it through the performance of boundary maintenance on the job. Working alongside keepers at City Zoo and Metro Zoo, I began to understand how they forge shared work identities built around a culturally constructed yet nevertheless powerful set of symbolic distinctions between nature and civilization. First and foremost, keepers enjoy the sensation of working outdoors rather than in a sterile office environment protected from the elements, and they prefer the exertion of manual labor to the bureaucratic humdrum suggested by administrative busywork. Ashley explains that while as a child she dreamed of a career as a veterinarian, ultimately she realized how much happier she would be working "hands-on, more directly in the field" as a zookeeper, rather than at a desk. "I don't mind the physical work at all. I have played sports since I could walk, so physical work actually suits me better than sitting down in an office. It's definitely better for me to be outside doing something all day."

Lauren similarly prefers working in the woodsy outdoors, inhaling the same fresh air as many of the animals in her care. "I'm not the type of person who would be happy sitting at a desk all day. Sometimes I really like my assigned area because I'm *outside*, I'm doing *physical* labor, and *it just feels good* (when I'm not dying because it is too hot or too cold). Sometimes I just really like being *out to do work*."[32]

In addition to these largely aesthetic boundaries, keepers also activate and reinforce a set of *moral* boundaries to differentiate themselves from other zoo workers, as they see their professional role as closer to nature (and thus more central to the mission of contemporary zoos) than those working in administrative departments such as accounting, development, or guest services.[33] I first discovered this when I asked Lindsey, a thirty-year-old keeper, if she knew anyone from City Zoo's media relations department. She responded in an uncharacteristically aggressive way: "Are those the women who I always see wearing *high heels* to work? At a *zoo*?" Given that zookeepers typically wear heavy work boots to navigate muddy exhibits and soggy backstage areas all day, they disparage the less functional and more stylish clothing worn by administrative zoo workers as incompatible with the rugged work of zookeeping in naturalistic spaces—especially, it seems, with regard to footwear. Nicole, a former development assistant at City Zoo, explained,

> They called us the "click-clack girls" because of the high heels. I personally thought it was stupid to wear high heels at the zoo when you are walking around all day, but [the others] would dress to the nines, in fancy shoes. There was one woman in marketing who, until she was eight and a half months pregnant, wore her fancy Jimmy Choo high heels all the time. They didn't like the animals any less. But it definitely contributed to this sort of feeling [from zookeepers]: *"They're different. They don't get it. They don't understand."* Part of it is also that the girls that go into corporate sponsorship are really interested in the networking and the business side of things and they have business degrees, whereas the people that go into being keepers are more interested in day-to-day animal care, and more willing to get their hands dirty. . . . And it was hard, because to [the animal staff] we all looked the same. They saw us going to galas and cocktail parties, having receptions, and taking people to dinners, while they were really sustaining the zoo by taking care of what needed to be done in order for the zoo to even function.

It is interesting that animal keepers emphasized their distinctive work roles in terms of footwear. In a quite literal sense, work boots allow zookeepers to walk directly upon the sodden terrain of the Earth's surface, whereas high-heeled shoes elevate administrative workers a few inches *above* the ground, thus detaching them from the soiled zoo grounds in a physical as well as a professional sense. Likewise, animal keepers at City Zoo regularly joked about the formal attire the zoo director wore on a regular basis. To them, his black suits, silk ties, and shiny wingtips seemed inconsistent with the zoo's organizational identity as a center of wildlife care and conservation.

While the keepers at Metro Zoo largely maintained friendly relations with the zoo's office staff, they also pointed out differences in work attire to distinguish themselves from these office workers. One zoo-keeper referred to the mostly female office staff as "the girls up front, in their dressy clothes, their hair, their makeup, and their cute little out-fits," suggesting that such gender performances on the job were incompatible with legitimate zoo work.[34] More importantly, keepers derided the zoo's office staff for their relatively low level of personal commitment and shared sense of stewardship over the zoo's animals. According to Lily,

> They are very removed. Sometimes they just don't have a clue as to what is going on in the zoo. I remember one girl: she had been working here for months, and she said, "I have never actually walked down to see the prairie dogs," and I asked her, "What do you mean? *Come down and see the prairie dogs.*" That's just frustrating because these are our babies, and we're taking care of these animals, and you're not coming down to see all of the animals? It's a little frustrating. Not caring about what type of animals we have. . . . I hear people talk about the cheetah and I'm like, "You work here—you should know it's a jaguar." . . . We definitely have our little group, and I think it's a good thing that we're separate from them.[35]

Keepers did not limit their activation of moral boundaries to distinguishing themselves from administrative office staff, but also targeted top zoo executives and board members who they criticized for siphoning zoo resources away from the needs of their animals. According to Lauren,

> How money is spent can be really frustrating. When you see offices being renovated and money being spent on cabinets when we could instead replace a fence on an exhibit that's falling apart. To the keepers it seems unsafe, but to the management it's good enough. . . . Why is a brand new office renovation (or new TVs for the zoo, or a new sound system) a priority over an exhibit renovation to make it better for an animal, or to make it better for the keepers? We can't have a little bit of money spent toward a better entrance to that exhibit? It can be really frustrating.
>
> Sometimes, you can't judge everything: for instance, Metro Zoo just had some offices renovated, but [a furniture store] donated some of that stuff. So obviously Metro isn't spending the money inappropriately there, because that was a donation. But when money comes in from the board

sometimes, or donors will donate money and want it to go to something specific, nobody sits and talks to them and explains that, actually, "You are spending all this money on x, y, z thing for this animal that you really care about, but did you know that this animal would be better off with something else?" For example, you are spending all this money to cut down the trees in an animal's exhibit because you don't think that they look nice, but that animal loves those trees. Can we do something *else* to make the exhibit look nicer? . . .

If they would just take the time, we would have an ambassador, somebody who is on the keeper staff who could be present at these meetings when these things are discussed, at least just to give our perspective. . . . So, it's just that disconnect. And we don't really have that voice in any of these sorts of meetings. And that is true at every zoo that I've ever worked at.[36]

Keepers also direct moral outrage toward zoo guests whose public behavior betrays a lack of respect for the zoo's animals. According to Ashley, Metro Zoo visitors taunt the animals constantly, especially the peacocks that range freely over the zoo grounds:

We had a woman, she was with her children. . . . She was chasing around Stanley, one of the peacocks, trying to get him to display his feathers. Apparently she had told her children, "I know how they display their feathers. They have to be agitated." As if they have to be angry to display their feathers, to tell you to back off or something. So Amber went outside and said, "Please don't chase the peacock. Obviously it stresses our animals out when you do stuff like that." And she said, "Oh, really? I was just trying to display his feathers." And Amber explained, "Actually, the reason why he'll display his feathers is because he's usually looking for a mate. And unless you are a lady peacock—which, no offense, you're not—he's not really interested." And the lady's like, "Oh, well, I thought he did it because he was agitated." And Amber just looked at her. "Either way, please do not chase my peacock!" And of course, right after that, we saw people trying to *feed* Stanley. And we were like, *"Please don't feed Stanley."*

As in many other professions carried out in the presence of paying clients, such as teaching, medicine, dentistry, entertainment, and the arts, zookeepers employ moral boundaries to distinguish their professional status as animal experts from what they consider the poor judgment of lay audiences.[37] These boundaries help keepers maintain shared work identities that emphasize their authority to protect the zoo's creatures from the excesses of irresponsible visitor behavior—just as over time I myself grew more protective of the animals I routinely presented to the public, including Kaa, Hero, Hogwarts, and all of the giant cockroaches.

For instance, during our interview Heather happened to mention in passing, "There was a group of school kids that went around putting Slim Jims in everyone's exhibit, and then the wolves had diarrhea, the cougars had diarrhea. . . . People do crazy shit." This seemed like an instance of moral boundary making, and so I asked her to elaborate.

"A lot of zookeepers come off as nasty or mean because people are so ignorant and do stupid things," she explained. "Like, for instance, last week. You were there Friday? . . . I must have seen like twenty chaperones just on their phones while their kids were running around like idiots and going crazy. They broke, like, four split rails; they ripped signs off of exhibits. . . . Meanwhile, I was standing there, yelling at an adult showing a child how to feed the elk standing on the split rail, next to the one that was already broken. So I said, 'Excuse me, sir. Please do not feed the animals.'"

"What was he feeding them?" I asked.

"Grass. Like, they just picked the grass and then tormented them."

I wasn't quite sure why a person would try to torment a thousand-pound animal with antlers, one of the largest land mammals in North America. Honestly, I wasn't even sure *how* one would go about tormenting an elk. "What do you mean, torment them? They were offering them grass and then pulling it away?"

"Or teasing them . . . sticking it up their nose."

Huh. I didn't see that one coming.

"But the person I was yelling at decided to ignore me, and didn't even turn around. And then a guy behind me chimed in, '*Oh, you mean we're not allowed to feed the elk their natural food?*' And I was like—I just started yelling."

"What did you say?"

"I was like, "NO! Because it's assholes like you that teach their children that it's okay to do whatever they want to a zoo animal. It is a federal offense to abuse animals in a public setting. You cannot abuse a zoo animal—this is ridiculous."

Confident that zoo workers were not allowed to curse out visitors, I double-checked. "Did you actually call him an asshole?"

"Yeah, I did. . . . That dude didn't even acknowledge that I was standing right next to him, asking him politely to not feed the elk. And then that guy had to be a dick. *'Oh, we're not allowed to feed them their natural food?'* No. You're not. I don't see your uniform."[38]

# Birds of a Feather

Keepers cope with the difficulties and hazards of zoo work in numerous ways. Having aspired to professionally care for animals since childhood or adolescence, they experience their career paths as honorable and worthwhile. Although zoo work can be painfully arduous and at times socially stigmatizing, keepers take pleasure in their encounters with animals during training sessions and in moments of everyday caregiving, rendering such work deeply meaningful. They mobilize moral boundaries that emphasize their kinship with zoo animals while defining themselves in contradistinction to dolled-up dandies in dress shoes, white-collar desk jockeys, and irreverent guests. While tense encounters with visitors obviously add to the unpleasantness of zoo work, they also allow keepers to maintain a firm and confident sense of self-respect and shared identity as the most capable stewards of the wildlife collections they have grown to care for with both undying devotion and hard-won expertise.

Yet zookeepers experience this sense of virtuousness and professionalism not merely as solitary individuals, but collectively as well, as illustrated by the camaraderie they generate when working together in collaborative teams throughout the workday. Although it has become common to assume that zoo workers relate better to animals than other humans, keepers laugh this off. As Amber points out, "I think that is definitely a stereotype, because everyone says it: *'Oh, I like animals more than people. I don't want to work with people.'* They don't realize that in the zoo world, you *are* working with a lot of people."

Krista explains, "You become pretty close to the people you work with. You have to trust them to get their stuff done, you have to trust them if you need help with animal restraint, to know what they're doing, to trust them with animal care." Already bound by their shared commitment to Fred, Ginger, Stanley, the other Stanley, and Beaker, Metro Zoo keepers thrive on the experience of teamwork and the sociability and trust it engenders. Like birds of a feather, keepers maintain a collective identity rooted in a mutual sense of dignity and purpose. Even when individual keepers decamp for other zoos, their coworkers present them with a personalized scrapbook of photographs and memories of their time spent together in Metro Zoo's cages and corrals.[39]

Of course, just because one shares a dignifying professional identity with others, it does not necessarily follow that one must always act dignified. Shoveling the bison yard makes for tedious work, and keepers liven things up by singing made-up songs and telling jokes about—well, by now you can probably guess. As Christina explained to me, "We're a very joking group. Not the most P.C. of all-time. But it's just constantly having that joke and making fun of something, not taking everything so seriously. You've got to just kind of let things go."

I braced myself. "What kind of jokes do you guys make?"

"There are so many," said Christina, laughing. "Honestly, everybody has a really weird sense of humor. And I feel like this is probably across the board with keepers. The amount of poop jokes that you will hear . . . lots of poop jokes. But also, our team is kind of weird because we like to sing a lot. So a lot of times we just jokingly take whatever we're doing and put it to the words of some song."

"Give me an example?"

"Heather got—here we go with poop jokes again—a bunch of fecal loops. They're little loops that you use to get fecal samples from an animal. So, I don't know if you've ever heard of the song "U Got the Look"?

"By Prince?"

"Yep. 'She's got the loop! She's got the loop!' Yeah, and it'll go on."

According to Krista, "We make up our own words to songs that are animal related. When I first heard the line 'I'm gonna pop some tags,' from 'Thrift Shop' [by Macklemore & Ryan Lewis featuring Wanz], I thought it was 'I'm gonna pop some *jags* [short for jaguars],' because I didn't really know what he was saying. So we made up a whole song with that . . . and it became just a whole weird song about animals."

But as Krista would go on to explain, keepers also filled their workdays with "lots of singing and dancing, lots of inside jokes, and lots of just being loud and crazy" to help them take their minds off some of the more unpleasant aspects of zoo work and animal care:

We deal with a lot of death and things that are morbid. Like with the [carnivore] food that we feed out, and then when animals die. When animals get sick there's lots of blood and vomit and poop and stuff involved, so it makes our conversations open to anything. We don't really have a line anymore. So in our conversations, we have to kind of balance that by just being fun and crazy.

We'll leave the music on in the truck and we'll sing. There are already radios in every enclosure and holding area. It's for the animals, but it's also for us when we're in there, just to sing and dance. A lot of times I feel like I'm living in a musical, a real-life musical, because every once in a while you just burst out singing and dancing.

Heather agrees. "A lot of times, if you don't poke fun at it, you're just going to sit there and cry all day. You know what I mean? Because all you can do is laugh at some of the shit that happens." Tony, a older City Zoo keeper in his mid-sixties, would similarly spend his workdays amusing himself by telling one-liners and singing oldies hits: "I Can Help" by Billy Swan, "Cuban Pete" by Desi Arnaz, "We Are the World" by USA for Africa. During my first few weeks working at City Zoo, Tony and I had developed a friendly, joking rapport. He had taken to calling to me "The Professor" (one of the seven castaways on the 1960s TV sitcom *Gilligan's Island*), and asking me questions about my teaching, my family, and even my Jewish background. One day Tony entered the zoo kitchen belting out "Over the Rainbow" from *The Wizard of Oz*, and then announced, "This one's for you, Professor!" and began chanting, "*Hava nagila, hava nagila . . . .*"[40]

The fellowship keepers enjoyed when working alongside one another—sharing jokes about fecal loops, rapping about jaguars, laughing and dancing to get through an emotionally difficult workday—extended to their nonworking hours as well. At a small but lively Halloween party held after hours on Metro Zoo's grounds, keepers munched on hoagies, gossiped, and played *Dance Dance Revolution* on a Nintendo Wii system. Heather dressed as a bison, replete with a pink, numbered ear tag; Krista, a raven; Lily, a "kitty-cat" princess. (Another keeper described her own costume as someone "being attacked by outbreak monkeys.") Perhaps not surprisingly, zookeepers often take vacations together, and when they travel they visit—wait for it—other zoos. (Talk about a busman's holiday.) During one such jaunt, a bunch of keepers and zoo educators piled into a single car to road-trip across the Midwest to visit the Columbus Zoo and the Saint Louis Zoo. Of course, such trips allow keepers to relax and enjoy the zoo from a visitor's perspective, without job responsibilities or khaki and green work clothes to muss up. But keepers also use these zoo visits as opportunities to reflect on professional issues related to their work: how to design a better tiger exhibit, how to prepare more effective behavioral enrichment, how to best distribute balanced diets to a gorilla troop. In doing so, keepers reinforce their own collective sense of themselves as zoo and animal experts, and as members of a larger professional community. Krista recalled one recent zoo trip: "I think the last zoo I visited with the other keepers was the Cape May County Zoo. And we were just like, 'Aw, look what they did with their animals!' 'Our exhibit is way better than that!' Or if we can find a keeper, we'll talk to them, just kind of exchanging ideas. The zoo community is really small, but there are new ideas, and people are always trying to do new things with their animals, and it's kind of the best way to learn from each other—go see other zoos and what they're doing." Indeed, just like the obsessive sociologist who is forever jotting down field notes at his own family's dinner table, sometimes zookeepers just can't help themselves. Ashley recalls a trip to a local safari park one summer with some of the other keepers: "So we went through the safari: we had giraffes licking the windows of our car, and we had a heron checking himself out in someone's side-view mirror. We got to see all the animals they had there."

"But when we were looking around the safari, we were definitely asking each other, '*How do they clean all this?*'"

Keepers assign shared significance and sentiment to the animals they regularly care for, and in doing so they collectively construct the animal kingdom just as much as zoo audiences do on the other side of the exhibit fence. They express their intimacy with nonhuman animals by treating them as they might people—they speak to their cougars, prepare birthday cakes for their giraffes, and kiss Dumbo rats. In many ways zookeepers face the same challenges as those working in the arts, or the culture industries: they must endure periods of unpaid internship work after paying for expensive college educations, they work for low wages under precarious conditions of employment, and they experience extreme levels of income inequality relative to their bosses. Yet they readily accept these costs because, as in the arts, the potential emotional rewards and possibilities for self-fulfillment are so great.[41] Birds of a feather, zookeepers develop close relationships with not only their animals but also each other, their community bound by a shared sense of collective dignity and moral purpose. Their selfless dedication and collaborative work ethic provide a worthy model for the stewardship required to protect the Earth's biodiversity in the anthropogenic age of environmental catastrophe.

For the next leg of our safari we will remain behind the scenes, but turn our attention toward another set of human primates native to the zoo: the educators and volunteers responsible for interpreting the zoo's landscape and its resident creatures to visitors and surrounding communities. In public presentations, live animal shows, and one-on-one animal encounters, these teachers culturally construct the natural world by artfully translating the empirical realities of the Earth's environment into accessible narratives and nuanced arguments about the animal kingdom and its habitats. Their commitment and drive illustrate the crucial need for diffusing established knowledge about the life sciences and an appreciation for the planet's biodiversity in the wake of the climate change crisis, and they will share with us both the enormous pleasures and difficult challenges involved in educating the general public about these ecological issues.

# Chapter 4

# Life Lessons

## The Zoo as a Classroom

**"WHAT IS SPECIAL ABOUT** a bird's bones, and why it is so?"

This one was easy—bird bones are hollow, which makes them lighter than they otherwise would be, and thus more capable of flight.

"*What is a Jacobson's organ?*"

What is a *what?* Is that a musical instrument? The pancreas of a goat?

"*Compare and contrast precocial and altricial chicks.*"

Oh, brother.

While midterms and finals have been a regular part of my life for as long as I can remember, during the last fifteen years or so I have been the one designing and grading the exams, not taking them. Yet there I was, crouched over my desk, gnawed pencil in shaky hand, sweating through my training exam, the last step on my way to becoming an official education volunteer, or docent, at Metro Zoo. Since training began, my life had been devoted to cramming my head with animal facts—zoological taxonomy, the difference between genotypes and phenotypes, the reproductive systems of marsupials—you name it. Our training began on a Saturday morning in a crowded conference room filled with the year's docent class: seven retirement-aged adults, thirteen high school students, and me. (In volunteer training, we were all equally in the dark about the Jacobson's organ, or at least the grown-ups were. Not the students, though—those kids were *really* smart. At any rate, the Jacobson's organ is an olfactory sensor set at the base of the nasal cavity of certain animals. For instance, snakes and lizards use the organ to smell by delivering to it odor particles captured in the air with their tongues.)[1]

In addition to the edifying content of exhibits discussed in earlier chapters, zoos also formalize their educational programming through departments that develop curricular materials and recruit educators for animal encounters and other instructional events and activities. At City Zoo and Metro Zoo, both dedicated staff members and unpaid volunteer docents delivered this programming. Paid educational staff worked closely with keepers, who were generally about the same age and shared similar college backgrounds, while volunteers included both teenagers and older adults of retirement age. On this next stop on our zoo safari, we explore how educators construct nature by distilling the unruliness of the surrounding environment into what are quite literally *life lessons*—accessible cultural narratives and refined understandings about the many marvels of the Earth's biosphere, including the animal kingdom and the biodiversity of species. We will also catch glimpses of the curiosity and wonder occasionally experienced by children and grown-ups alike at zoos, and the challenges of not just broadcasting to a truly mass audience of spectators but actually reaching them, captivating hearts and minds alike. But first, let's meet some of the zoo educators and volunteers responsible for engaging with these visitors, as their lives are as colorful as those of the zookeepers we just recently met on our safari trek.

## Making a Living, Making a Life

A twenty-three-year-old education manager at Metro Zoo, Kelsey probably enjoys talking about her job more than any person you have ever met. When I met up with her on campus for the first time, I couldn't help but notice that she was wearing her usual Metro Zoo sweatshirt, and I asked her offhandedly if she wore it much outside of work, unprepared for the elaborate explanation that was to follow:

This is my zoo sweatshirt jacket, and I'll wear this out on days that call for a sweatshirt. I am very proud of where I work, and I absolutely love my job. I want people to talk to me about it, as an advertisement for the zoo. I've had people come up to me and be like, "Metro Zoo—I really love it but I haven't been there in a couple of years." Then I get to tell them about all the great changes we've made. Even if I'm *not* wearing my uniform, usually the first thing out of my mouth is, *"Hey, I work at the zoo. What do you do?"* And I know that keeps the conversation going.

I am very passionate about what I do. I'm very passionate about animals; that's my private life, and my work life. Like, when I'm not at the zoo I'm probably reading about hyenas, because that's where I want to go next. I've done several different presentations at different colleges and different places on hyenas, just because I love hyenas. *That has nothing to do with the zoo—I just love this animal.* I actually did a radio talk show at a local university a couple of weeks ago on how they help intersex humans.

I'm so passionate about my job that it is *who I am*, so I want to talk to people about it. Whether I tell them funny stories, or I get the point across that bats need help because bats are in a huge conservation crisis now with white-nose syndrome, or anything like that, I love talking about the zoo, so I'll wear this sweatshirt. Now, I will not wear my *uniform*, because that smells really bad. Even though I do laundry—when a porcupine pees on you, you can never get that smell out. So that sweatshirt has been washed fifty thousand times, and my roommate tells me that it still smells like urine. So, that's only for work, whereas I will happily wear out a clean sweatshirt.

This is what a normal conversation with Kelsey is like—a dizzying surge of animal facts and personal commentary that quickly jumps from spotted hyenas to bats to porcupines. In fact, Kelsey proceeded to spend the next ten minutes explaining the unusual anatomy and habits of the hyena, its reputation in the zoo world, the scientists who study them, and Kelsey's feelings about it all, a sure sign of her commitment to zoo and animal education, whatever the forum. ("The female spotted hyena is a natural animal model born with an elongated clitoris." "I had previously bought a hyena shirt from Kay E. Holekamp, who is my ultimate, favorite spotted hyena scientist." "There are only a small number of zoos with spotted hyenas in captivity. I've been to four of them now.") In fact, Metro Zoo hired Kelsey not only because of her vast zoological knowledge, but also for her extraordinarily abilities as a quick-thinking public educator.

I interned at Metro Zoo, and then when my internship finished and I went back to college to finish my senior year, I pretty much went to Metro every other weekend, just to keep me in their minds. But every time I went in, there was always, well, not a problem, but something that I helped them with, unexpectedly. So, when they had the American crocodile in the exhibit with the American alligator, Scout, they had taped Scout's mouth, and they were trying to introduce them to be in that exhibit together, but Scout was attacking the crocodile, and they needed all the keepers to intervene and save the crocodile, and then decide what to do with it after that. The problem was that they had scheduled an alligator keeper talk at the time that Scout was attacking the crocodile.

I happened to be around that day, and I told the keepers, "Look, you go handle the alligator-crocodile situation, and I'll go do a keeper talk at the bobcat exhibit instead," because all the people were gathered at this alligator exhibit. So there I am, in street clothes, doing a keeper talk on the bobcats, keeping the public's attention from this weird scene.

Another time I went in randomly and the keepers were short-staffed or short on time or something like that, and so they asked me if I would do the *alligator* keeper talk. And again, I was just there, in street clothes, doing the keeper talk. No big deal. And then I was asked to help with some zoo programs held at a local science museum, and that was the day I was offered the education position.

Not surprisingly, Kelsey's enthusiasm for zoo education comes from an enormous affinity for animals. As a child she grew up with dogs, fish, parrots, cockatoos, cockatiels, rabbits, and hedgehogs as pets. As if to emphasize the point, she described to me a complicated scheme to deliver a screech owl, two opossums, and a woodchuck from the New York City suburbs to Metro Zoo shortly after she began working there. Her recollection of transporting the owl alone illustrates how zoo educators like Kelsey blur the boundaries between home and work, just as keepers do.

"So I went to the suburbs over Easter break to pick up a screech owl named Hooter, and he stayed the whole weekend with me in the suburbs. At that time we had three dogs, seven birds, a couple of fish, an iguana, a hedgehog, and a rabbit. I had to feed him mice, so I had to ask my mom, 'Hey, do you know how we are going to have Easter ham and all that good stuff in the fridge? Can I put these dead mice in there? Do you have a problem with that?' So she allowed me to have a collection of dead mice in the fridge with all the Easter food.

"Then I drove Hooter back to my apartment, and he stayed the night in my bedroom with me. My roommate, bless his soul—he has dealt with a lot from me. I had to ask him, 'Hey, do you mind if I have dead mice in the fridge?' And he didn't, so I fed the owl, it's flapping around trying to eat things. . . . It's not easy to have nocturnal animals staying in your apartment because they stay awake at night while you are trying to sleep, so the little owl is flapping around all over the place and hooting at everything. . . . At the time I had three mice, so I had to angle the owl away from the mice so he wasn't looking at them like, 'Hey, I want to go eat your mice,' and my mice weren't looking at my him, like, 'Oh my god, this thing is going to eat me.'"

I did a double take. "Wait a minute—these mice were *pet mice*?"

"Yes, I had pet mice." Of course she did. "I actually had a lot of pet mice, but at this time I only had Fudge, Misty, and Omar."

"So in your apartment, you had mice in the fridge, and then pet mice . . . and those were in your room also?" I verified, just to be sure.

"Yes, because mice are nocturnal as well, so I didn't want to move my mice and wake up my roommate. So I left them all in my room."[2]

Like the zookeepers discussed in the last chapter, zoo educators love animals, and their homes often resemble the zoos where they work. But the education staff members I met at both City Zoo and Metro Zoo were so much more than that—they were incredibly knowledgeable experts who *needed* to perform that expertise to others, almost obsessively so. As Kelsey admitted to me, "I only go to zoos with my roommate, actually, because he just lets me tell him things, because he doesn't know anything about animals. And I love educating, so even on my days off, I'm like, 'Johnny, let me tell you all about this animal.' And I just vomit words onto him."

In addition to paid educators like Kelsey, Metro Zoo audiences also interact with a team of unpaid education volunteers, or docents. Again, docents typically hail from two generational groups—high school- and college-aged adolescents (more on them later) and older adults of retirement age. In accordance with recent sociological research on volunteer work, most adult zoo docents are women.[3] Daphne, one of Metro Zoo's education volunteer coordinators, explained that these women typically share both a strong affinity for animals and a desire to "make a difference," especially those who find themselves experiencing a family or work transition:

Most of the zoo's docents are older women who have raised their children, and they have their weekends free now. And they all love animals, that's what it is. The zoo wouldn't have them if it weren't for the animals—they could go be a docent in a museum. They want to be around the animals. And the fact that they can come and actually touch animals that they wouldn't have at home, that is the carrot on the stick to get them through the door, and then you hook them with the conservation angle. You show them that they are a small piece of a larger concern, that they have a chance to make a change. And that is how you keep them, with the idea that they make a difference. Everyone wants to make a difference. . . . But your average woman who works in an office, who is married to the same guy—what difference is she going to make? So you give her a chance to come in and make a difference by talking to people, and that is irresistible. Especially in this day and age, with social media and the computerization of everything? Direct contact—connecting with other human beings—is irresistible.

Leslie, sixty-five, a retired social worker and vocational consultant who turned to the docent program at Metro Zoo after her husband passed away, explained why she chose to volunteer at this transitional stage of her life:

This is a time when people have some time to do things, when they can turn to what is most important to them in life and devote a little extra time and care to it. . . . It's not like I have a million dollars to give to somebody. What I have to give is my time, so I do that. I've done the best that I can do for something that is so important to me—animals and wildlife, and the Earth in general—so in the end, I feel like I can be happy at what I devoted my life to. I don't have kids, and so much of my life was all about my career, and that was fine. I just didn't want that to be my only statement about my life.

Louise, a fervent animal enthusiast who is also in her sixties, began volunteering at Metro Zoo fifteen years ago because she "had an unfulfilling job" and "needed something else" in her life. Today, she volunteers at the zoo twenty or more hours per week:

I was an administrative assistant; I did it because they paid me well, but it was like a dead-end job. I needed something else in my life, but I didn't know what. . . . Fortunately, I saw this little ad in the newspaper—they were looking for docents. *"Come to this meeting at Metro Zoo, and see if you are interested."* So I did, I went. And that was it. I found it and never looked back.

I like it . . . I finally found an outlet for my message—about conservation, animal rights, the welfare of endangered species, the animals' place in the world. It's a love of animals—a general love of animals. You know, I watch a lot of nature shows on television, and I always end up crying if I see that some species is in danger. It *always* comes with big cats. Anything about cats can make me cry, bring me to tears . . . I can tell you, my husband does not want to go see *The Lion King* on stage with me ever again, because all I do is cry from the beginning to the end. It got so bad one day that people were actually turning around and looking at me, because I was sobbing.

Kelsey calls me a dreamer, an idealist. But I like to escape from reality a lot of times because I know it hurts. I know reality is where all the endangered species live. They live in reality. If someone doesn't take up their cause, then I'm afraid that they will all go away. For the past twenty-five years, I've gotten on the animal bandwagon. It just gets more intense as I go along. It doesn't wane. It gets stronger.

I belong to the World Wildlife Fund; the National Wildlife Federation; the Friends of the Florida Panther; the Mountain Lion Foundation . . . I have a list THIS long of organizations that I support. My husband and I don't have children, so when we make out our will, everything that we have goes to animal organizations—zoos, refuges, sanctuaries, our zoo—because to me, they are our children. They are what we have to look out for.

While women made up the vast majority of Metro Zoo's cast of volunteers, a small handful of men participated as well. Richard, sixty-three, attended the docent training sessions with me, and also volunteered on projects with the zoo's maintenance crew a few times a week. A former veterinary microbiologist, Richard helped build or repair exhibits for the zoo's cougars, coati, and red-handed tamarins. As he explained to me over lunch at a pizza shop just off the zoo grounds, "At my age, sitting down at a desk all day is not good. I'm a hands-on person, and I like putting my skills to use to build things—simple carpentry, electrical, plumbing. It's a craft; you build stuff from scratch. . . . The reward is that you see the result, and how people use it—it's not abstract."[4]

At the same time, Richard, like Leslie and Louise, also expressed a deep love for animals, and a social ethic of volunteerism. As a child he kept many pets, including gerbils, ducks, chickens, rabbits, hamsters, and dogs, and brought all sorts of wild animals into his yard, from lizards to small snakes. ("My parents would say, 'I bring the zoo back into the house!'") In his professional life he worked with laboratory animals, and upon retiring he wanted to stay busy in an environment surrounded by wildlife. So Richard works six-hour shifts at the zoo, two or three times a week, for free. He maintains much credence in what he calls his "family doctrine": "You have to volunteer sometime in your life," whether during your career, or when you retire. Having learned this social ethic from his family in Hong Kong, he imparts it to his own children. (He proudly reported that his daughter, an attorney, performs pro bono work assisting homeowners threatened by foreclosure.) Richard occasionally gets grief from the blue-collar guys on the zoo's maintenance staff, given his willingness to perform manual labor without pay. One guy would tease him, "Come on, Rich, why are you *really* doing this?" But Richard harbors no hidden agenda: "It's what they call *society*—you're either in it, or you're out."

Retirement-age docents may choose to volunteer in order to better the world around them, but along the way they inevitably discover a new world of sociability and fellowship made up of similarly minded people experiencing the same stage of life. According to Leslie, "I didn't really think at first about the friendships with people that I have so much in common with. . . . But that is definitely a big plus because I immediately connected with people, and I really enjoy doing stuff when we go on little trips here and there, like to a big reserve or something together. And that's fabulous. I love it. So that is a big plus. . . . We really care about animals and go out of our way to do things that foster their care and conservation." During my time preparing animal diets at Metro Zoo I worked side by side with Leslie and Louise, and their spirited conversations often jumped from their strong feelings about safari hunters ("Disgusting! I put them in the same category as pedophiles!" "Give me a gun and I'll take care of them! Or better yet, I'll attack them from behind, like a lion, and gouge their eyes out!") to whether George Clooney or Antonio Banderas was the sexier actor. (Leslie preferred Clooney, while Louise was more partial to Banderas.)

Leslie and Louise circulated among a larger group of female docents who not only volunteered together but also exchanged animal-themed gifts and greeting cards during the holidays and socialized outside the zoo, whether visiting a nearby wolf reserve or attending animal-related movies such as *Water for Elephants, Big Miracle,* and *African Cats.* (In December 2011 my family and I joined Leslie and a group of about a dozen other Metro Zoo docents at a nearby multiplex to watch the Cameron Crowe film *We Bought a Zoo.* It was quite a scene: during the movie's more bombastic moments some of the ladies shouted at the screen, *"That would never happen!"*) Like Metro Zoo's keepers, the docents occasionally travel together, often to zoo conferences where they too could visit zoos and animal reserves in the area, as well as enjoy general sightseeing. Leslie recalled their trip to Los Angeles for the 2012 annual conference of the Association of Zoo and Aquarium Docents:

> There were about eight or nine of us, so we had at least four of us in each room. We just had a good time. It was just a laugh. We had tours of historic homes and stuff—we saw where Lucille Ball used to live, Jack Benny had his home, and we also saw where Jennifer Aniston's home was, John Stamos, where Jennifer Lopez and her last husband had their place. . . . Then some of us went to the Santa Barbara Zoo, while the others went to an aquarium, and they went whale watching and stuff. And then we went to Stearns Wharf; they had this great display of jellyfish. Unbelievable. It was really beautiful. And I went to the Living Desert, in Palm Springs . . . and that's a very rich, rich area, but the docents were wonderful. And you could tell that some of them were very wealthy, but they were so committed. We also went to the La Brea tar pits, and the L.A. Zoo. It's fun—we know a lot more about the animals, so we just fill each other in on stuff. It's like going to the zoo, but going with people who really, really care, and are pretty knowledgeable about it.

Louise compared the camaraderie of these trips to the peer-group intimacy experienced in college or even the military, except in their case among seniors who love animals:

> The most fun we have are at those conferences, because we all room together. It's like a college dorm. We are kindred spirits. Half of us are retired and half of us are not. But because you have common interests, it brings you together, and then you realize, *"Gee, we really are a lot alike."* If I had friends at work, I don't even see them anymore. I don't care to see them. I don't have anything in common with them.
>
> But this is different. We are bonded by something deeper than just social events. We have beliefs that we share, and common goals. We are all one

big family. I've never been in the Army, but we are all in the same brigade or platoon, or whatever you want to call it. We all get to know each other and trade stories, and we have a bond. My zoo friends are probably my best friends, really, because like I said, they are kindred spirits. I really put a lot into those words. We were born to be together.

If these retirement-age docents share common bonds that draw them close as "kindred spirits," younger high school- and college-age education volunteers similarly herd together. While working at Metro Zoo I often took lunch breaks with these adolescent students, and occasionally presented animals with them as part of the zoo's educational programming. In many ways they socialize just like regular teens: watching YouTube videos, discussing the newest *Harry Potter* movie, stealing french fries from each other's lunches, playing harmless pranks on one another. But they quite obviously took their role at the zoo very seriously. Although a volunteer coordinator in a leadership role, Tyler nevertheless conveyed a youthful demeanor of his own, especially as an alternatively dressed college student who kept his hair dyed dark red (and sometimes green). Perhaps as a result, he skillfully connected with Metro Zoo's young volunteers, or *junior* docents, and understood their specific needs. He explained,

> For these kids, they see it as not just a job, but for a lot of them a very core piece of their identity. . . . They are *zoo people*. Most of them, on their Facebook profiles they list "Docent at Metro Zoo." It's not just something that they *happen* to do. It's not like, *"Yeah, I volunteer at the zoo, at the homeless shelter, and I'm on the volleyball team, and I do Spanish club."* It ranks above those sorts of extracurricular activities in importance, given how they identify themselves. I think because of that they are apt to take it a little bit more seriously than Spanish club.
>
> The biggest perk is definitely the animals, but I think that for a majority of them, they also really love being able to feel like they are helping the zoo, and that they are making a difference. I think it makes them feel like they are contributing something to the world. It's a little more than, *"I go to school everyday."* I'm going to use this Spanish club analogy again now, but it's like, what do you accomplish by going to Spanish club besides putting it on your transcript? Whereas this feels a little more purposeful—you are helping this community institution that makes the people of Metro happy, and you are helping it to thrive and enrich the experience of the people who come there, and teach people to love animals.

While the opportunity to work with animals and participate in a be-loved local institution clearly motivated these youth to join Metro Zoo's volunteer program, the friendships they too developed among kindred spirits encouraged them to stay, just as for retirement-age docents like Leslie and Louise. Again according to Tyler,

> I don't think its necessarily what's going to draw people in, but I think once they start meeting each other and bonding and developing friendships with each other, they find a lot of like-minded people. Daphne and I have always seen the program as a haven for people who don't fit in. A lot of kids who are bullied, who are generally loners who don't always fit in at their schools, then get to come and be a part of the docent program, and find like-minded people with whom they fit in. So that becomes a part of it, too—it's having a sense of place within the zoo and feeling a part of it. I know a lot of the kids do feel that way, like the zoo is their second home.

Daphne agreed, pointing out the degree to which these teens looked out for one another, regardless of the personal challenges they faced: "During the last five or six years, we were getting more kids who had things like Asperger syndrome, Tourette syndrome, and most of our kids had ADHD. We had people with all kinds of things that we had to work through. And we found that being in this group—and it was a loving group—these people really grew to care about each other, these kids, and they would support each other, help each other." Over time I noticed how strongly these teens identified with the social world they had created for themselves inside the zoo—some fraternized to-gether outside the zoo, others paired up into romantic couples, and all began inflecting their conversations with zoo talk. One summer after-noon during lunchtime the teens began comparing hand gestures named for animals—awkward jellyfish and starfish, the snail fist-bump—and later the girls joked about bringing the zoo's ferrets to their proms as their dates. They also bonded over shared experiences, both in the zoo and outside of it. In fact, during a training session more than one high school volunteer revealed that after they suffered bloody scratches on their wrists and forearms from a bearded dragon named Smaug (a common occupational hazard, as I too learned the hard way), teachers or guidance counselors from their school had pulled them aside, concerned that they might be cutting or otherwise mutilating themselves.

# Animal (and Human) Tricks of the Trade

Again, zoo educators do not merely love teaching—they *need* to teach. As Hannah, thirty-three, a former education program manager at Metro Zoo, explains, "It's such an overwhelming feeling when you teach someone something new, whether it's an adult or a child." Of course, engaging with the public in a crowded zoo poses plenty of challenges. One concerns the wide generational diversity of zoo audiences. Daphne pointed out that as a zoo interpreter, "You might be talking to a group that might have a two-year-old, a five-year-old, a ten-year-old, up to a sixty-five-year-old, and you have to get your message across to *all* of them." Sometimes ideological or lifestyle differences between zoo visitors and docents cause discomfort and hinder conversation. Louise offered an example: "I had one incident where I was standing by the cougars, and I heard a little kid say, "Hey, Daddy, isn't that just like the one you shot?" And I got so upset that I had to walk away because I was just, like, *'If I go near that person, I will scream at them.'*"

Zoo educators and docents also contend with audiences less enthusiastic about learning than experiencing the zoo as a place of entertainment and leisure. Ryan, the former "animal experience specialist" at a prominent AZA aquarium who now works as an educator at a natural history museum with a live animal collection, recalled, "The reality of it is, I think most people who come in are just looking for that shiny stuff and a factoid or two. They're looking to see a bunch of sharks and other cool animals, and touch a bunch of stuff—not necessarily walk away with a greater knowledge about the animals." A number of zoo workers mentioned how much this lack of audience interest personally disappointed them. Rosemary, twenty-eight, who worked in City Zoo's education department before moving to another zoo, explained,

> I had a feeling that whenever I worked at City Zoo, and I was interpreting for people, they didn't want to listen to me. They wanted to see the animals and move on. Even though what I had to say—I'm telling them [the animals'] names, I'm telling them how old they are, I'm telling them when their birthdays are, where they came from, if they had any babies—stuff that people want to know, they're like, *"That's a lion; I gotta go."* People from the outside, they come to see animals because they think their kids want to see them, and then they just want to move on. . . .

Kids are fine. It's just the adults—unless it's a big group of school or camp kids. Then they're crazy; they're just nuts, and you don't want to deal with them. Groups with just one chaperone for like twenty kids, and they're just like, "*Whatever, go do stuff.*" And they're chasing chickens, they're trying to catch pigeons, which is gross, and they're all trying to sneak food into the petting area, they're pushing little kids off of stuff—that's when kids become an issue.[5]

Rosemary found this behavior especially disturbing in light of how little some visitors appeared to know about zoo animals in the first place:

Most of my thing with guests is just their ignorance about what they're looking at. I don't expect you to know the difference between a tamarin and a squirrel monkey. But when you go to the zoo and you are looking at a lion, and you ask me if a female lion is called a *tiger*, then I'm a little confused. Or take the hippos at City Zoo: they get those big, round, red and blue jolly balls to play with. [Visitors] will be like, "Oh, my goodness, it laid an egg!" They'd be serious. These are grown-ups! Hippos laying eggs. . . . When the tiger cubs were born, they asked me when they were hatched.

[There is] just a lack of knowing fundamental differences among the animals, like a mammal, a bird, a reptile. Birds especially. People don't understand the differences among birds. They think that, like, a chicken and a crane are not at all the same. Well, they've got feathers! They're birds. They've got beaks. They've got wings. They're *birds*. When I was an education intern, just standing there, I felt like I was on repeat—I would just stand on a box and yell at people, because that was the only way to get through to them.

To ease the challenge of interacting and engaging with a wide range of audiences varying in age, degree of interest, and basic knowledge about animals, zoo educators and docents employed certain tricks of the trade, routines on which they could confidently rely. For example, when preparing for presentations involving multiple animals, educators selected from an established menu of themes that had successfully intrigued zoo audiences in the past, much like television studios rely on *evergreens*, conventional topics and narrative plot lines that can be endlessly recycled.[6] At Metro Zoo, recurring programs included *Animal Adaptations*, in which staff demonstrated how wild species develop adaptations to their environments, such as the webbed feet of alligators, the prehensile tails of Solomon Islands skinks, or the carapaces of leopard tortoises. *Suitcase for Survival* taught zoo visitors about the harms of animal diseases such as the chytrid fungus, which threatens frogs and other amphibians throughout Central America, and white-nose syndrome, which compromises the health of North American bats. Always eager to share new knowledge, Kelsey herself contributed a segment on Australian marsupials and the spread of Tasmanian devil facial tumor disease to the *Suitcase for Survival* program. Kelsey also ran *Animal Amore*, an age-restricted presentation on animal sex habits, every year on Valentine's Day. Brooke, twenty-seven, a former Metro Zoo educator who now works at the same natural history museum as Ryan, described another popular program, *Animals with Bad Reputations*:

> One of my favorite classes to teach is about reputations. We bring snakes, we bring alligators, we bring vultures, we bring bugs, tarantulas—and we knock all of the bad reputations out. A lot depends on how [the audience] drives the lesson. Okay, we bring the animal out, and we figure out what it is. Now you tell me why you don't like it. . . . *"Okay, well, snakes are slimy."* Well, snakes aren't slimy—we can talk about that. Ninety-nine percent of the time they say, "That thing is going to bite me." Well, let's talk about why it would bite things in the first place: maybe because it's food. Am I food to this? No, I'm not going to fit in this snake's belly; it can tell I'm not food to it. *"So okay, it's not going to bite me, this thing's going to attack me."* Well, why are these things in the wild? They want to stay far away from us, unless we go near their young or anything like that. So a lot of that class is getting them to think about all of those bad things—*"I hate snakes, I hate reptiles," "I hate this, I hate that"*—and maybe changing their mind. They don't have to love them, but maybe change their mind as to why they thought they hated them. I love that class because it's a challenge for them to come up with all different reasons why we don't like things, or why animals are portrayed in certain ways in movies, like sharks, alligators, and wolves, and then we can think about them in real life.

Dangerous animals such as barracuda and crocodiles mesmerize children and adults alike, which is why when I performed the animal show at Metro Zoo for my son Scott's sixth birthday party, I selected (with his guidance) the favorite standby *Deadly Creatures* for a theme, and brought out Chompers, an American alligator, and Kaa, the Colombian red-tailed boa constrictor, both of which I managed to handle without incident. (I performed before a room full of kindergarteners from Scott's class, their parents, and at least one terrified father with a deep-seated fear of snakes. The kids were not permitted to touch the live animals, but they seemed to enjoy inspecting an extra-long snakeskin and a model of an alligator skull. Afterward, there was cake.)

Finally, both Metro Zoo and City Zoo educators lead a recurring program named *Best/Worst Pets*, largely out of a sense of public responsibility. According to Kelsey,

> When I do a program, it is [often about] doing your research before you get a pet, because I have twenty pets in my parent's house because *no one* did their research, and I don't want that to happen to other animals. You know, maybe somebody took a turtle from the wild; I don't want you to do that. Leave the turtle in the wild. Turtles live to be over *eighty years old*, and I don't think it's a good pet anyway, so don't do it. So my message for any age group is, "Do your research before you get a pet. Don't take animals from the wild."
>
> I had one person tell me, "*Oh, I want to get an otter as a pet.*" And I was like, "Well, actually, that's a really bad idea, because otters have extremely sharp teeth, extremely sharp claws, you know? They need to be in the wild, and if you get one as a pet that means you took it from the wild." And they're like, "Well, I *still* want it as a pet." And I'm like, "Did you not hear anything I just said? Really?"

The wild and exotic pet trade causes enormous harm to animals as well as people, and many zoo educators view programming on the topic as a public necessity.[7] In fact, my very first educational program at City Zoo involved running an activity table on *Best/Worst Pets*, where I conducted interactive surveys with visitors and passed out pet ownership guides. (Besides common domesticated house pets like dogs and goldfish, the zoo's Best Pets recommendations included ball pythons, corn snakes, Honduran milk snakes, fat-tailed geckos, and fire-bellied toads. Not surprisingly, its Worst Pets included alligators, crocodiles, bats, venomous snakes, coyotes, wolves, bears, and tarantulas, as well as less obviously disastrous choices: diamondback terrapins, box turtles, hermit crabs, and wild-caught fish and coral.)

Educators also employ this theme in part because the owners of unwanted pets frequently try to pawn them off on local zoos, especially when they prove too large or laborious to handle. According to Sharon, the herpetology keeper at City Zoo, people abandon rejected pets at the zoo all the time, from rabbits and iguanas to even dogs and cats. Three additional City Zoo keepers confirmed this. One recalled how "at City Zoo, people threw chickens over the side of the fence," while another noted that visitors brought their unwanted Easter chicks to the zoo after they outlived their cuteness as babies. After months of hearing about this strange kind of animal trafficking, apparently a consistent problem among zoos, I asked Daphne how often Metro Zoo received requests from the public to adopt their rejected pets. "A lot, at least once a week," she explained. "Sometimes people would just dump them, or they'd show up at the front gate. . . . Exotic birds, alligators, those are the biggies. Turtles."

"They would actually dump them at the zoo?"

Daphne nodded. "The terrapins, just last year. Someone had called— this person must have been so stupid. They emailed us about taking the terrapins. They said they would give us money and the tank, and we didn't have a chance to get back to them. So they brought a bucket, and they put it at the front gate. I think it had fifty dollars in an unmarked envelope, as if we weren't going to know where they came from."

"And they were their pets?"

"Right. They were diamond-backed terrapins—a threatened species. You aren't supposed to have them as pets. And we knew who it was [that abandoned them], but we could use them—so we kept them, but you're really not supposed to."

I asked Daphne if she could recall any other cases of unwanted pets abandoned by their owners at the zoo.

"Yes, groundhogs," she replied. "Somebody left them at the service gate, which was bad. It was lucky that I found them because they put them right at the bottom of the surface gate. And they were baby ground-hogs, the kind that fit in the palm of your hand. They were in a knit hat. I saw them when I got to the gate, so we brought them in, after they first went through a local wildlife center. But I never want anybody to say on outreach or on program that one of our animals came [from a pet owner], because then everybody will come to our door with their pets that they don't want. So I always stressed that we got them from other zoos or rehab facilities, although we did take some pets."

"But you don't want people to know about it?"

"Oh, god, no, because they will all be at your gate—alligators, be-cause people buy them and they grow so fast, and turtles, too, because people get tired of their turtles. You know, they like them for a bit and then they grow. Red-eared sliders [a species of semiaquatic turtle], oh my goodness. Everybody has a red-eared slider. They either dump them in a pond, or they want others to take them. . . . They would go, 'Oh well, the zoo would take it.' Well, no, it doesn't work that way.

"Look at all of our parrots that were donated. . . . The macaws? They were all donated. They were pets; they lived for so long, and then people didn't want them anymore. They're messy and they're loud. In Beaker's [the African gray parrot] case, her owners retired and moved to Florida."[8]

Recurring programs provide zoo staff with a common and shared repertoire of familiar themes that both ease collaboration among educators and have proven successful among audiences in the past.[9] Zoo interpreters also devise specific strategies for engaging visitors of different ages. Holly, thirty-one, a former Metro Zoo educator who also works at the same natural history museum as Ryan and Brooke, drew on film references to instruct differently aged audiences about particular species:

> Sometimes we used movies as a reference. With the burrowing owl, if they've seen the movie *Hoot*, you can give that reference. For children, the burrowing owl is this animal that they've never seen before; they don't really know where it lives. But then you ask, "Have you seen the movie *Rango*?" which a lot of kids have—the mariachi band are burrowing owls. Then the kids are like, "Oh, okay, it's a desert animal. Okay, so they live where those lizards live, they live where that orange toad lives."
>
> The barn owl that we used is also the [same species as the] lead character in *Legend of the Guardians*, which is an owl movie. A lot of people think it's Harry Potter's owl, but his owl is a *snowy* owl, and so when I use a barn owl I will specifically mention that Harry Potter's owl is a snowy owl, while the owl in *Legend of the Guardians* is a barn owl.
>
> If I specifically saw someone of my age group [early thirties], then I'd reference *Labyrinth* [released in 1986] with David Bowie, because the bird that shows up at the beginning is a barn owl.

Of course, engaging diverse audiences is as much a matter of delivery as content. According to Hannah, preschool-aged children require lots of activity-based learning pitched at a most elementary level:

> We would have them do something basic. You would have to work with kids that were just at the age of coloring, but make it fun. They're coloring the turtle shell this color, and they'll color the other kinds of softer scales a different color, so they know that the softer scales are on the arm and the

head area, but the harder scales are on the shell, because it's a harder surface. And then we bring out a turtle, so that they can feel the difference. We found anything that was interactive to be the best way of getting messages across for education.

Live animals fascinate children at this age, and they always assaulted me with a barrage of questions whenever I handled an unfamiliar creature in public—*"What's its name?" "How old is it?" "What does it eat?" "Can it fly away?"* (In these moments kids also occasionally betrayed their lack of familiarity with the normative rules of the zoo, as when they would they ask me of the animals I handled during such presentations, *"Where did you get it?" "Is it yours?"*) Attention spans at this age also tend toward the lower end of spectrum, just as young children can always be counted on to respond with hysterics to anything evocative of toilet humor. As Brooke observed, "Sometimes [when presented with an animal] they're like, '*Whoa, that's so cool!*' and other times if an animal goes to the bathroom, you've just lost them—they are more intrigued with how they just went to the bathroom, and they've lost focus on the animal. A lot of the time with the birds, when they go to the bathroom you'll lose them for about two minutes."

As for older grade-school kids, Brooke noted that "if they're interested, they start asking you at third, fourth, fifth grade—they start coming up with these really interesting questions, more than just '*What does it eat?*'" She offered some examples of the kinds of questions asked by children in this more advanced age group:

A lot of it has to do with it eating other things—*"If you had an owl and a bunny out, what would happen?"* Although an owl could eat a rabbit out in the wild, we would explain to them that, obviously, our owl has not killed a live thing in years, so it would probably just stare at the rabbit, or the rabbit might try to run away. They can come up with interesting scenarios about that. Some lizards can drop their tail—*"So if it drops its tail once, can it drop it again?"* Since it drops its tail at certain parts—*"If it drops it here, can it drop it there? Can a bigger part drop off there and still grow back?"* It's a lot of in-depth questions about these animals.

On the other hand, for middle and high school kids Brooke recommends, well, getting dirty. "Sometimes for them you have to go with the gross factor. You have to go with that thing that will make them go, '*Did she just say that?*'" As an example, she suggested that she might present a turkey vulture to a middle school class visiting the zoo, and explain that the bird urinates and defecates on its own legs to cool itself off on hot days, and vomits up road kill as a defense mechanism. Similarly, one morning I chatted up a high school class visiting Metro Zoo's animal kitchen by showing them the giant white rats I had prepared for the falcons' and eagles' lunch. The students seemed equally fascinated and repulsed. One young man even approached me to photograph them.

Meanwhile, educators agree that adults—even the parents of small children—are generally the most difficult visitors to impress. I learned this firsthand as a zoo docent when mothers or fathers would abandon their children at one of my animal presentations in order to sneak away to make cell phone calls—although, as Holly pointed out, absentee parenting provides teaching opportunities for educators to exploit. "If mom or dad is fiddling on their phone, you can take a moment and talk to the kid." Once at Metro Zoo two mothers dumped off their kids at my kiosk to smoke their Marlboro Lights under (perhaps fittingly) the zoo's cigar tree. According to Tyler, his greatest challenge as a zoo educator was always connecting with adults:

> It's the hardest. On some level you feel like you've been given implicit permission to approach children, but to approach adults it's almost like [they'll be thinking], "*You'd better have something that's worth my time. I want to hear something that I haven't heard before.*" They are definitely the hardest group to connect with.
>
> You also get a lot of couples that go to the zoo because it's a date location and their interest is mostly in having a place where they can go together, walk around, and hold hands. Most of the [childless] adult visitors you see are couples using it as a date environment. I find, as an interpreter, that those are the most difficult ones to approach, because they are really not interested in an educational experience.

A final strategy employed by zoo educators and volunteer docents concerns the performance of expertise in public, particularly in cases when visitors ask questions that the zoo interpreter cannot readily answer. This happened to me on a number of occasions at both City Zoo and Metro Zoo when I was thrust into situations with little background knowledge of the animal I was tasked with exhibiting. While I generally confessed to lack specialized knowledge about the creature in question, these moments were often tinged with embarrassment. One time an adult guest asked me about the regional habitat of the spotted Atlas moth I was displaying outside Metro Zoo's butterfly house, and when I pled ignorance he immediately whipped out his smartphone to look up the answer online and announce it to nearby zoo visitors. (One of the largest moths in the world, the Atlas moth resides in the tropical forests of Southeast Asia.)

This was not just a problem I faced, either. City Zoo and Metro Zoo often gave their volunteers limited information about their assigned program animals based on the predictable set of questions most frequently asked by visitors (such as the animal's name, sex, age, and diet). While this tactic usually worked, it was hardly foolproof.[10] (Perhaps it is more surprising that it ever worked at all.) Brooke recalled overhearing docents and other volunteers at Metro Zoo (as well as noneducation personnel such as maintenance and guest services staff) give incorrect answers to visitors' questions all the time. In fact, by their own admission docents frequently gave out erroneous information, albeit unwittingly. One docent confused Senegal with Somalia while another told an audience that the Nelson's milk snake was indigenous to the American Midwest instead of Mexico. Of course, sometimes volunteers quite consciously dissembled in front of guests, as when a Metro Zoo docent, unaware of the given name of the spiny-tailed lizard she was handling, simply told audiences its name was Grover. (One of the zoo's keepers told me that the guest services staff behaved even less responsibly when running zoo tours and birthday parties, claiming "they just make up shit as they go.")

Like volunteers, experienced educators are hardly immune from making errors in public, although they generally try to stop themselves first. As Ryan admits, "I'm sure I've given out wrong information, but never purposefully. If I don't know, I'll fess up to it right away." As for Kelsey, the performance of expertise can sometimes seem like an obsession. "I try to keep up on all the animals that we have in our collection, just because I don't like feeling that I don't know something. If someone asks me a question, and I don't know the answer, I will tell them '*I don't know*,' and then I'll research that question—and if they give me their email address, I'll email them the answer. *It's so important to me to know it.* My home life is my work life"—in fact, Kelsey lives only three minutes from the zoo—"so I'm constantly reading, watching Nat Geo Wild, all that stuff."

Then again, interpreters often rely on learned guesswork when responding to zoo guests' queries. Ryan explains,

> I've worked with animals long enough where I can take educated guesses. [A visitor might ask], "How many eggs does a turkey vulture lay?" I'm really unsure, I don't remember, but my guess is that they probably have fairly small clutches, because they are big birds. I'd say they probably lay maybe two to four eggs. I don't know for sure, but that's my educated guess. [The correct answer is two eggs.] I like to give visitors sort of a hint. If I'm totally clueless, I won't just pick it out of thin air, but a lot of times when I give the educated guesses, it's fairly close, because either I subconsciously remember, or else I go by process of elimination.

Tyler also admits that under certain circumstances he will employ guesswork when interacting with zoo audiences.

> We always tell people that you want to be honest with guests in terms of your own knowledge limits. If you don't actually know the answer to a question, don't make one up—*unless* it is one where we can break the rules: names, ages, and occasionally life spans. Kelsey and I were talking about this: you're generally going to pick an age range which is going to be somewhere between like, "Oh, this is an animal that lives five years, this one lives ten years, this one lives twenty years." Just based on approximating, and then picking the age of the animal itself, it's going to be three or seven, almost always. If you don't know how old the animal is, just go with three or seven and that's a good enough answer.

Given the guesswork on which educators occasionally rely, and the knowledge gaps experienced by volunteers and professional zoo interpreters alike, how do staff maintain their authority as experts when addressing guests? In an impressive show of teamwork, docents and educators collectively adhere to a code of mutual and reciprocal dependence by never correcting or challenging one another in public.[11] According to Daphne, "You never correct somebody in front of a guest. It's demeaning, it lowers their confidence, it makes us look bad to the guest, and it's just not nice. We support and help each other, and that's how we work." She further explained, "That's why it is a community, and we build each other up. The idea is not to make someone feel like they are stupid. It's hard—it's hard to get the confidence to get up in front of people and tell them things. And anything that shifts away from that confidence is bad for morale, bad for the group, and bad for the person who had to summon up the balls to come in to do this in the first place." Of course, this collaborative agreement among educators protects not only the status and personhood of the individual interpreter but also the legitimacy of the entire zoo staff's collective expertise as a whole. As Ryan observes, "You don't want to devalue the reputation" of the entire institution by revealing even one erroneously delivered piece of trivia. "If 99 percent of what you say is accurate and interesting, and people are getting excited and learning about it, and you make *one* mistake, you don't want them to forget the whole thing, and say, '*This guy doesn't know what he's talking about.*'" At the same time, Ryan knows the dangers inherent in allowing the delivery of unauthorized nonsense to the public continue unabated: "If someone is going out and they're just talking crap, talking about all of the wrong stuff, you have a serious discussion with that person afterward and say, 'You have to get this stuff straight—you can't just go out there spouting off all of this wrong information.'"

Tyler wholeheartedly agrees with Daphne's sympathetic approach to protecting the legitimacy and egos of her colleagues, all while acknowledging its implications for the overall educational mission of the zoo:

> Obviously, there is a certain type of etiquette to interpreting that we go over with people. I think the idea is that we don't want to embarrass people. It doesn't build good camaraderie, and it doesn't make people feel like they can improve. It's embarrassing and it doesn't make people want to rectify the situation. If you get called out in front of a whole crowd of people for having your information wrong, it just makes you feel like, "*I quit. I'm done.*" Whereas afterward, it's easier to [explain to volunteers], "This is a critique—this is something I'm telling you so you can improve and not make this mistake in the future." Yeah, on some level it is about making sure there is camaraderie among volunteers, and that people understand that we are here to help each other grow.
>
> Sometimes it can be the hardest thing in the world to do, to stand there. I had a junior docent this summer named Jack. It was his first summer. When [volunteers] are up front greeting, I will occasionally listen in on their interpretations, just to get an idea of where they are. And he's telling them, "This is Kaa and he's a boa constrictor, and he's fully grown, *except they can grow to be about thirty feet long, and they can eat crocodiles.*" I wanted to jump in so badly and just scream "Abort! Abort!" After the visitors all walked away, I was just like "Jack, nothing you said was right! Abso-

lutely none of it." If you don't tell them that, they are going to keep on giving out wrong information.

At the same time, Tyler recognizes that the cost of protecting Jack from certain public embarrassment—even just once—may be measured by the number of visitors leaving Metro Zoo "thinking that boa constrictors are school-bus-sized snakes that crawl around South America eating crocodiles."

## The Rewards of Life Lessons

As I spent more time around zoo educators and interpreters at both City Zoo and Metro Zoo, I gained a greater appreciation for the pure pleasure they received from connecting with the public. One rainy weekday in September I assisted Metro Zoo's education department with a lunchtime program for a group of twelve senior citizens residing in a nearby assisted-living facility. The seniors slowly walked into the education hall with their walkers, and sat at tables with place mats made from construction paper while awaiting the first four program animals before lunch: Pepé, the skunk; Trix, the giant Flemish rabbit; Sol, the prehensile-tailed skink, and Beaker, the African gray parrot. For me, the program was a bit of a personal disaster. After Christina presented the skunk, I tried picking up Trix but Christina and Kelsey said I was not holding its paws correctly—"No, *that's how you would hold a baby*." I tried a few more attempts before they abruptly took back my rabbit duties. Meanwhile, an elderly gentleman explained that he used to trap rabbits as a child, skin them, and sell them to Sears and Roebuck.

Next I hoped to regain some credibility by nudging Sol, the prehensile-tailed skink, onto a stick. This proved fairly easy compared to handling Beaker, who bit me on my forefinger. Christina managed to coax Beaker out of her cage right before the parrot defecated all over the floor, mercifully missing us as well as the seniors. As I went to go clean up the mess with paper towels, Christina offered Beaker a cookie in return for speaking to the crowd—when out of the corner of my eye I spotted Gladys, an elderly woman wearing a white cardigan sweater and purple pants, leaning toward the parrot with something in her hand, ready to feed her. I looked down at her hand, and gasped—it was her medication, a small pill she tried to give Beaker as a tasty treat.

Quickly I asked Gladys—loudly—"*Is that your pill?*" so that an attendant could intervene, which she thankfully did.

Among the groups that regularly visit Metro Zoo, Kelsey rates working with senior citizens among her most satisfying experiences as an educator, as she described to me at length:

> It's working with senior citizens who either are just really old, or have Alzheimer's, or dementia. . . . What we want to do for these people is make them feel appreciated, give them something that might spark a memory. If they have Alzheimer's, animals can do wonders. They can make them remember things that they haven't remembered in a while; they can make you alive. They can bring you that light that you've been missing.
>
> We want to hear their stories. Like when we took out the skunk, Pepé, those senior citizens that might have had Alzheimer's, or dementia, were telling us, "When I was a little girl I used to have a pet skunk, and I would take it to church with me," and you're just like, "Oh, my goodness." And the nurses and attendants come up to you afterward, and they say, "We haven't heard stories in so long. You don't know what you've just done for us."
>
> Once they know that they can tell stories, they're going to tell you stories. And then they ask you for your own stories, and since this is a different type of program [compared to one directed toward children], we will give them our stories. (Whereas if a first grader asks, "Where did you work before?" I'd say, "I worked at a museum before, but we have to stick to the [lesson]! Your teacher is paying me to talk about why bats have fur and wings! So we have to stick to that.")
>
> It's so heartwarming for me, and then sometimes it's just funny. We had one senior come to the zoo who had Alzheimer's or dementia, I don't remember which one he had. I actually thought he was pretty with it, until we took him to the peccaries and he said, "Oh, these are the porcupines!" And we were like, "Oh, that's a really good guess, but they actually are peccaries." And he read the sign, pointed at it, and said, "Yep, 1972. That's when I rediscovered the peccary," and I reminded him, "You thought they were porcupines!"
>
> But you know what? We got a funny story out of it, and he had a great time. That's fine! Another time a senior came in, and she told us, "It's my birthday!" and we were like, "Happy Birthday, let's sing to you!" And another senior goes, "It's her birthday every day, because she can't remember! If we went by that, she'd be nine hundred years old." And you're just like,

"You know what? If she wants it to be her birthday today, it can be her birthday today! We'll sing 'Happy Birthday' to you, and that's that!"

We'll let the seniors pet our animals—it's another different way for them to connect, and spark a memory or a story. Maybe their story is, "I used to have a chinchilla coat." Well, we don't really want to hear that when you're petting our chinchilla, but hey! That's what happened back in the 1930s. People had chinchilla coats!

We also have different programs for them. It's not just, "Hey, here's an animal. What do you remember about this animal?" It's talking about zoos in the 1930s versus zoos now. It's talking about movies from the 1950s, and movies that had animals in them, like *Harvey*. That's a 1950s movie with Jimmy Stewart about a giant white rabbit that was his invisible friend. When we bring out our thirteen-pound white rabbit and say, "Hey, can you guys think of a movie that would have had a big white rabbit?" They light up: "Oh, my gosh, *Harvey*! I remember seeing that when I was a little girl." Or, "I saw it in a drive-in movie theater!" . . . And they love that someone that's not from their generation is so interested and able to keep up—to a certain degree—with what they are talking about.

Metro Zoo educators use animals like skunks and rabbits to connect with not only seniors, but also young children, including those with special needs. Of course, working with groups of children with disabilities poses a particular set of challenges, especially given the extraordinary diversity of needs typically represented within such groups. As Holly explained,

The challenges were in the variety of needs in one room. So you'd have someone that was almost nonverbal, and then you'd have someone who was very verbal, but couldn't sit still. It was hard as a teacher, since a lot of the children had their own unique thing. So they'd say, "Okay, so today we have Sally, and Sally is scared of birds. So don't bring the birds near her, but Johnny loves the birds—so bring them near him." So you'd try to remember all of this, and then you'd present a fur, and we'd be like, "Okay, all you guys can touch this," because we'd brought a lot of items that they could touch, since the animals could only get so close. So the challenge was then, as we were approaching Johnny, "Oh, no, Johnny's scared of skunks, take it to so-and-so." You're trying to teach as a group, but you're also individually trying to make that connection for each individual student where one kid is very tactile and loves to touch, while another kid is perhaps blind and *needs* to touch something—but has a fur allergy. Several were on the autism spectrum. Some of the kids were the type that would fidget a lot, with ticks, and then we had a few that were in wheelchairs, which was challenging for physical activities. You're really presented with a full scale of things, and trying to create lessons that fit everyone.

Still, despite these challenges educators found that in most cases children benefited from exposure to zoo animals, even those who responded with quiet observation:

The goal of their lessons was to add variety to what they were doing . . . so for us, it was bringing in something different, something animal-related. A lot of the kids *loved* animals . . . but for some kids, you could take out an animal, and completely captivate them, and they would just be quiet. So, unlike what we would normally do, we would let there be silence. [Then we would quietly guide the children.] *"Okay, just watch [the animal]. Watch as he moves. Watch as he flaps his wings. Watch the snake moving and slithering. Watch what he's doing. Oh, did you see his tongue come out?"*[12]

In some cases children's encounters with live animals generate truly unforgettable experiences, and not just for those children. Ryan recounted to me a moving interaction he had with a young girl. He explained,

There's just one little anecdote I'd like to share, because this was the highlight of my time at the aquarium. I met a mother and a daughter, and the daughter had Asperger's—she was clearly autistic on some level. She spoke very loudly, but was very excited, and had an incredible knowledge of what was going on. I've always found I've connected really well to children with Asperger's and autism. . . . As we moved around [the aquarium] I talked to the mom, and this eleven-year-old girl had gone through all sorts of health problems and crazy surgeries. Asperger's is her most noticeable issue, but it was the least of their worries—the family had gone through so much. But every once in a while they would come to the aquarium—just this mom and her daughter—as a trip to connect. This kid remembered everything about all these animals from our aquarium. She was very protective of the animals, too. When other kids weren't following directions, she would go up and get right in their face, and be like, "No, no, touch it like this," or, "This is a green surf anemone—those are the tentacles; that's where it eats!

You can't touch where it eats!" This girl loved stingrays, so I ended up going to the stingray touch pool area.

This is where Ryan tensed up.

> Sometimes we'd select a few people on certain days to wade in with the stingrays and feed them. There was this family visiting from out of state and it was a teenage girl's birthday or something, and her mom or aunt was one of those hyper-aggressive people who think anything can happen for the right price if you push hard enough and ask the right people. She was trying to get me to bring the teenage girl in to wade with the stingrays, but I told her, "We are done with the wades for today. I'm sorry, we can't just bring one more person in—it's unfair to everyone else. I know you're from out of state, but we have people here visiting from other countries, and we can't do this for everyone." She demanded to know, *"Who is your boss?"* She talked to my boss, who probably talked to his boss, and he comes back again and says, *"You're taking her in for the wade."* . . . I was kind of annoyed about it, but I smiled and went along with it, because first I went to find the mom with the daughter with Asperger's, and I told them come to the stingray touch pool at a certain time—the doors would be closed, but that they should just come in. After the teenage girl waded, I took the girl with Asperger's in for a wade.

"Did you ask permission from your boss first?" I asked.

"No. I was so annoyed with the job I didn't care if I got fired. To me, to let that teenager go, and *not* this girl with Asperger's, with this genuine love and interest, who only came to the aquarium once in a while? It would mean the world to her, so I just did it, and the girl's face just lit up, she was so happy. She said, 'Thank you, Ryan! Thank you!' as she was leaving, and that moment made my entire time at the aquarium worthwhile—it was just absolutely fantastic to me. It turned out not being a problem with my boss, either."

"Still, not that many people would have done that," I told him. "That was very nice of you."

But Ryan didn't believe his display of kindness would have been all that unusual among other aquarium or zoo educators. "I think a lot of people would have tried meeting this girl. My nephew has Asperger's, too, and like I said, I really like talking to kids with autism—especially when they have a focus on animals or natural science."

Like Ryan, docents and other interpreters at metropolitan zoos experience terrific satisfaction working with children and adolescents. Dylan, a twenty-six-year-old zoo educator and self-identified herpetology fanatic, fondly recalled leading a scout group through the reptile room at City Zoo:

> I was doing an overnight program, probably the most fun that I've had at City Zoo. It was a scout group, and they were just going through our basic class that we teach. And when we got to the reptiles I said, "There are four different types of reptiles. There are lizards, snakes, turtles, and crocodilians." Now, there are actually five—there are tuataras, but almost no one has ever heard of tuataras, unless they are from the zoological field. [Also, they are native only to the islands of New Zealand.] The tuatara looks very much like a lizard, and people would say it *was* a lizard, if they didn't know it was a tuatara. But *this kid*. . . . He was a little bit older, about ten or so, and he said, "Wait a minute! *What about tuataras?*" And I just stood there [in shock]. "You know about those?" And he answered, "Well, yeah. It's another type of reptile. They look like lizards, but they're not." And I said, "Give me a high-five, kid—you are awesome."

Dylan, who has the demeanor of a perennially enthusiastic camp counselor, also worked with high school students involved in a City Zoo program designed to provide skills, work experience, and encouragement to disadvantaged youth residing in nearby urban neighborhoods.[13]

> [The program] is designed to give these high school students an opportunity that they might not otherwise have. They get to come to the zoo; they do public education; they do animal care; they do outreaches. When they did public education, that's when I worked with them. I was asked once to go on a field trip with them. For four days we drove to [a nearby city] and we visited six different colleges. We wanted to support them going to college and graduating with a degree. It was awesome. The zoo paid for it, for four boys, four girls, and two chaperones.
>
> They asked me to go because I was a guy, so I could be a male chaperone for the four boys that went. And after this trip, I came back from it knowing that *this is what I have to do with my life*. I have to do youth leadership. That's what I want to do. And I should be more on the ball with looking into graduate school for environmental education or some kind of

youth leadership program, but when I went on that trip, I was, like, *this* is what I have to do with my life.

It's fun. It's fun to be with the kids, doing outdoor environmental stuff. And what I like about it is that when you get to go on the field trips with them, they get to see a little more of you as an individual, and you get to relate to the kids more—you get to joke around with them a little more, and they get to have fun, and you get to have fun, and it creates a better connection with them.

Having worked with Dylan since his first day sweeping goat poop as an intern at City Zoo before he was eventually hired as a full-time educator, I knew of his talents working with children of all ages and backgrounds. It was wonderful to see him evolve from being a young zoo volunteer with a quirky reptile obsession into a mature leader with a sense of purpose and commitment to teaching and mentoring young people. I asked him if any of the kids with whom he had worked stood out in his mind, just as he had made a strong impression on me. He thought about it, and then told me about Julia:

There is one particular kid that sticks out in my mind. Julia came in as a high school intern because she wanted to do a project on polar bears. . . . She was from a very underprivileged family in [a nearby urban area], and in the interview I wondered if she would be a good interpreter—not only because she didn't know a lot, but also because she was not very outgoing. She was very shy and quiet. Boy, was I wrong.

I wouldn't have initially given her the internship, but she ended up getting it because she showed up anyway—showing up is ninety percent of it, sometimes. She turned out to be one of the best interpreters for little kids that we had. I walked into the birdhouse one day, and Julia was doing the bird cart with a bunch of artifacts and things. And one of the games on the bird cart [consisted of] six plastic eggs, all shaped and colored differently for the different types of birds, and there was a little clue that went with each egg, and you had to match the eggs up with the clues. So, you know, the little blue egg went with the robin, and the really big, round one went with the owl, and so forth.

The carts stand about three feet high—they are about as tall as the table here. And if you are a little tiny kid, you can't see over the cart. Well, I didn't tell her to do this, but she took everything and put it right on the floor. She sat down on the floor of the birdhouse, the kids sat around her, and she was playing the game with them. I never taught her to do that—she just did it herself. And I walked in and saw her doing this, and I loved it. And I took a picture of it, and I sent it to the entire zoo, saying *that* is interpretation at its best, right there.

# Don't Know Much Biology

Given these moving recollections, it is a sad irony that the rich diversity among zoo audiences sometimes *hinders* staff from educating the public about basic animal facts and environmental science topics. One October workday at City Zoo, Tony, the older keeper mentioned in earlier chapters, called me over.

"Hey, Professor, is it some Jewish holiday?" He nodded toward the petting yard, pointing out a large gathering of ultra-Orthodox Hasidic Jews in front of the sheep and pygmy goats. I explained that it was the week of Succoth, the Jewish autumn harvest festival in which families dine and pray outside in homemade huts for eight days and nights. Tony nodded in recognition. "Oh, is that the one where they eat and sleep out in the backyard?"

Like other Jewish festivals such as Passover, the religiously devout take off from their jobs and school during the eight-day Succoth holiday, and refrain from engaging in any work at all during first two days. This leaves them with the remaining duration of the festival free to take leisurely excursions with family—often to zoos, as it turns out. In fact, during the later days of Succoth the Bronx Zoo and Central Park Zoo are so crowded with religious Jews from the boroughs of New York City (the New York metropolitan area is home to one of the largest ultra-Orthodox Jewish communities in the world) that a significant number of families decamp for zoos located elsewhere, including City Zoo. After introducing myself to a number of these Hasidic families around the petting yard, I discovered that many of them had in fact traveled all the way from the Williamsburg, Borough Park, and Flatbush neighborhoods of Brooklyn to enjoy City Zoo's autumnal splendor.[14]

After that day I wondered what exactly it was about zoos that attracted ultra-Orthodox Jewish families, sometimes from several miles away. It was easy to see why traditional Hasidim wearing black broad-rimmed hats and fully grown side curls, or *payot*, might feel *comfortable* in public urban spaces like metropolitan zoos. After all, the anonymity and pluralism of the city invites a certain kind of expressive liberty for visitors of dramatically different backgrounds to enjoy the zoo's pleasures on their own terms. This is especially the case for zoos located in cities known for their racial diversity and progressive political landscape, such as my own adopted city of Philadelphia. One beautiful May morning, a father-son play date to the Philadelphia Zoo revealed a striking display of public life as Indian women wearing traditional saris, African American couples holding hands, Goth lesbian couples adorned in black, interracial families with children, and thousands of other guests strolled the grounds, enjoying the day. (In addition to families with children, so many romantic couples can be seen along the paths of the Philadelphia Zoo that it is no wonder Rocky Balboa proposed to his girlfriend Adrian in front of its tiger exhibit in *Rocky II*.) Given that such a wide range of visitors of different races, ethnicities, nationalities, social classes, religious faiths, and sexual preferences take pleasure in city zoos' wildly inclusive environments, they serve as terrific examples of what Yale sociologist Elijah Anderson refers to as *cosmopolitan canopies*—welcoming urban spaces where diverse groups can savor the civility of public life in the company of strangers. Perhaps this should not be surprising: cities are well known for their ever-flowering subcultural worlds, and the visitors to metropolitan zoos reflect the immense plurality of the residential neighborhoods that surround them.[15]

But even if the members of religious sects such as the Hasidim might feel at ease under City Zoo's cosmopolitan canopy, it hardly explains why they would travel during their holiday to look at sheep and goats (to say nothing of the nonkosher pigs wallowing nearby in mud-drenched enclosures). I therefore began approaching Hasidic Jewish visitors whenever I observed them at City Zoo or Metro Zoo in search of answers. One summer day in August at Metro Zoo I met a lovely ultra-Orthodox Jewish family, the Greenbergs, from Flatbush, with three kids in tow. Offering them a hearty *"Shalom Aleichem!"* from behind a table covered with ocelot fur and a jaguar skull, I told the Greenbergs about my experience the previous Succoth at City Zoo. Both the mother and father smiled knowingly, explaining that they considered all zoo animals to be truly "God's creations," part of the spiritual tapestry of everyday life.

Yet the zoo attracted them for more pragmatic reasons as well. According to the Greenbergs, zoos provided them and other religious families with a nonthreatening oasis of wholesomeness in a desert of entertainment options for the traditionally devout. The mother explained to me how difficult it would be to bring her kids to a water park where guests wear swimsuits and little else—hardly an attractive choice for pious people of sexual and cultural modesty. Meanwhile, she humorously pointed out how zoo visitors usually dress fully clothed: "Here, the only ones not wearing clothes are the animals!" Zoos also tend to avoid some of the other trappings of contemporary American popular culture that deeply traditional communities may associate with rank materialism and unwholesome violence. For these culturally conservative groups—Hasidic Jews as well as Black Muslims, Evangelical Christians, Mennonites, the Amish, and others—urban zoos and their animal exhibits serve as family-friendly refuges providing safe haven from the city's more risqué amusements.[16] This is perhaps especially the case for older metropolitan zoos that, like the Philadelphia Zoo, maintain distinctive architectural and aesthetic continuities with their more traditional nineteenth-century pasts, including twirling carousels, train rides, swan paddleboats, hot-air balloons, and other old-fashioned Victorian attractions.

Of course, culturally conservative communities make up just a fraction of the larger mosaic of social life at the zoo, its cosmopolitan canopy. Yet their zoo patronage—along with that of myriad other communities of faith, ethnicity, and identity—emphasizes the extent to which zoos must pitch their educational content to a mass audience spanning vast ideological divides and political persuasions. Zoos commonly try to avoid upsetting their audiences by placing severe limits on the kinds of life lessons their educators can promote to the general public.

A prime example concerns how zoo policies silently constrain educators from teaching the most basic Darwinian principles of human evolution and natural selection to their audiences. Although central to the dominant paradigms that organize the biological sciences, the teaching of evolution is actually deemphasized at Metro Zoo. According to Daphne, "We stay away from evolution. I always tell docents to talk about the process of adaptation instead, because it is just easier, and you can still get your point across without being controversial. We would always say: 'We believe in science.' If someone asks you [about it], you can answer them, but we are not going to put up a big booth about evolution."

As a volunteer coordinator Daphne herself felt entitled to teach visitors about evolution in certain circumstances, in defiance of zoo norms—indeed, her outspokenness on the need for raising public awareness around many controversial issues of scientific and environmental importance was one of the things that made her beloved by the zoo's docents. But not all zoo educators felt so empowered. Tyler considered public discussions of evolution unequivocally verboten at Metro Zoo, as I learned when I asked him whether the zoo had rules about even broaching the subject in public:

> Yes—do not talk about evolution in front of guests. You can't say alligators *evolved* sixty million years ago. You can say, "Scientists *think* that alligators have been the way they are for the last sixty million years." You almost have to qualify it with, "It's the *belief* of *some* scientists; this is not something I am going to teach you as a fact." I really hate that.
>
> I think it is unfortunate that it is seen as a political position that the zoo isn't allowed to endorse. . . . You have to find ways to explain facts about evolution to people without *talking about evolution*.

In some ways, the general code of silence observed by American zoos surrounding not only human evolution but the evolution of *all* living things can be explained by the results of recent public opinion polling. Clearly, the strength of creationism in the United States as a viable alternative framework for understanding the origin of species stems in part from the vigilance of truly committed and well-organized groups of religious people, particularly members of Christian denominations. They include faith-based camps from conservative biblical literalists to theistic evolutionists to more progressive creationists, such as advocates of intelligent design theory.[17] Yet the popularity of both creationism and intelligent design is far more widespread than most believe. According to a 2012 Gallup poll, 46 percent of Americans believe in creationism, the idea that "God created human beings pretty much in their present form at one time within the last 10,000 years or so." Slightly fewer respondents, or 32 percent, adhere to a belief in intelligent design, agreeing that "human beings have developed over millions of years from less advanced forms of life, but God guided this process." Meanwhile, a feeble *15 percent* of Americans believe in evolution as understood by most scientists in the modern world.[18]

Given the confusion surrounding the scientific basis of evolution among wide swaths of the general public, some might argue that zoos ought to assume a leadership role concerning this vital topic. Zoos instead dance around such questions out of fears of alienating the very families that make up their mass audience. According to Hannah, at Metro Zoo "we especially had limits on evolution. We wouldn't be allowed to talk about that because it was so controversial, and they didn't want the zoo's image to be [involved] in any kind of controversy."[19] Educators at City Zoo similarly found themselves hemmed in by these constraints. According to Dylan, "We don't talk about evolution. We actually try to avoid the word *evolution*," explaining how "it avoids awkward situations and scenarios."

Of course, there are some zoos that have been quite progressive about presenting evolutionary theories of human natural history to the public, although they tend to be located in heavily urbanized cities north of the nation's Bible Belt. At the Lincoln Park Zoo in Chicago, the Regenstein Center for African Apes includes an exhibit with interactive stations, each featuring the bust of a different primate—gorilla, gibbon, orangutan, chimpanzee, and bonobo, or pygmy chimpanzee. Children are invited to lay their hands over the bronzed handprints of each of these hominoids until they reach the final station, labeled "HUMAN," with an oval-shaped hole just large enough for a small child to poke his or her head through. There are simple but important life lessons to be taught here: the organized taxonomy of animals and their relatives; the physical similarities between apes and humans; and, of course, the evolution of humankind.

Yet these kinds of educational opportunities are sometimes met with resistance from zoo visitors. In 2014 Megan Fox, a Chicago-based conservative blogger, accused the local Brookfield Zoo of trafficking in "propaganda" and "anti-human prejudice" in one of its primate exhibits: "They're just pushing a Darwinist theory onto everyone who walks through here. They're equating human beings with monkeys."[20] I asked Dylan if he had ever experienced such criticisms at City Zoo when discussing the descent of man from apes.

"Well, humans *are* apes," he explained. "I would teach that humans are a type of ape. I would tell audiences, '*There are only five apes on the planet. There are hundreds of monkeys and several dozen species of lemur, but there are only five apes in the primate family. What are they?*' They would look around and go, 'Gorilla, orangutan . . .' because they were right there in front of their exhibits, and I would say, 'Well, have you ever heard of a gibbon? Then there is also the chimpanzee. . . .' It would take them *forever* to figure out what the last one was, and I would give them clues like, 'You see them everyday, you go to school with them, you see them when you get home, you go on the bus with them . . .' and people, mostly kids, would say, 'I don't know, what is it?' 'You!' I would tell them. 'You're an ape! You don't have a tail, you have opposable thumbs—you know, you're an ape.'"

"How would visitors react to you telling them that humans are apes?" I asked.

"Most of the time they were just like, 'Oh, okay. . . .' But every once in a while, people protested: 'Really? We are? *I don't know about that.*' I would just leave it alone, saying, 'Oh well, everyone is entitled to their own opinion.' I wouldn't argue with them. If you can avoid an argument, do so. If you are going to argue with someone about evolution or something, don't do it when you are trying to put a good image of yourself and the zoo out there, because you don't know what could ensue from it. They might get angry and sue the zoo—crazier things have happened."

This reaction is understandable from the precarious position of low-wage zoo employees, although more sustained zoo programming on human evolution might go far in educating the public on the scientific origins of species as well as the cultural construction of nature more generally. The field primatologist Jane Goodall observes how close human beings are to chimpanzees in terms of brain structure, blood proteins, immune responses, and DNA.[21] According to the biogeographer Jared Diamond, humans share 98.4 percent of our genetic makeup with chimpanzees:

The traditional distinction between "apes" (defined as chimps, gorillas, etc.) and humans misrepresent the facts. The genetic distance (1.6 percent) separating us from pygmy or common chimps is barely double that separating pygmy from common chimps (0.7 percent). It's less than that between two species of gibbons (2.2 percent), or between such closely related North American bird species as red-eyed vireos and white-eyed vireos (2.9 percent). The remaining 98.4 percent of our DNA is just normal chimp DNA. For example, our principal hemoglobin, the oxygen-carrying protein that gives blood its red color, is identical in all of its 287 units with chimp hemoglobin. In this respect as in most others, we are just a third species of chimpanzee, and what's good enough for common and pygmy chimps is good enough for us. Our important visible distinctions from the other chimps—our upright posture, large brains, ability to speak, sparse body hair, and peculiar sexual lives—must be concentrated in a mere 1.6 percent of our genetic program.[22]

Here is another way to think about this: *humans and chimpanzees have more in common with each other on a genetic level than chimpanzees do with gorillas.*[23] As a society we have collectively chosen to categorize humans and animals (or culture and nature) as separate and distinct entities on our modern Great Chain of Being, thereby classifying chimps as closer to otters, salamanders, and beetles than to us. But this has little basis in biology, and far more to do with how we regularly exaggerate the cognitive distances between mental categories, much more so than their actual differences might suggest.[24]

But like the endangered orangutan in its native rainforest habitat, such discussions between staff and the public are extremely rare in American zoos. This presents immeasurable problems for zoo educators attempting to teach audiences about conservation, species extinction, and the current environmental crisis. As Catherine Brinkley, the aforementioned international zoo consultant, pointed out to me on our walk through the Philadelphia Zoo, "If you want to send a conservation message" to the mass audiences that zoos attract, it will hardly move the public "if American zoo visitors can't even handle the fact that we might be in the primate family."

As it happens, even zoos whose exhibits explicitly discuss evolution sometimes discourage staff members from broaching the topic when they personally interact with guests. The Earth Sciences Center of the Arizona-Sonora Desert Museum (ASDM) provides vivid illustrations depicting the 4.5-billion-year-old origins of the Earth and includes a deep timeline outlining the succession of life on the planet in increments of hundreds of millions of years through the Mesozoic age of dinosaurs, and then from the eventual emergence of ancient humans to the present. A garden featuring varietals of the cactus family of plants explains how "flowers are the evolutionary results of extremely intricate interrelationships with the animals that pollinate them." Yet when I asked a senior docent about the teaching of basic Darwinian science at her zoo, she referred to evolution as the "e-word," a target of controversy that ASDM volunteers and staff members apparently avoid discussing in front of visitors, and for the same reasons many other zoos across the country do.

In their live presentations and personal animal encounters, zoo educators and volunteers construct the natural world by translating and distilling empirical knowledge about animals and other living species in the biosphere into colorful narratives and creative performances for zoo audiences. In the wake of the climate change crisis and the Earth's sixth major species extinction event, their tireless (even obsessive) commitment to diffusing established knowledge to the public about the natural sciences and an appreciation for the Earth's biodiversity has perhaps never been more important. But at the same time, zoos must increasingly pander to a diverse array of audiences in an era of skepticism surrounding established scientific principles, including the origin of species, natural selection, and the descent of mankind. Insofar as zoos evade topics deemed controversial by large numbers of Americans, their promise as contemporary centers of conservation and public education fades from view despite the best intentions of zoo educators like Kelsey, Daphne, Tyler, Dylan, Leslie, and so many others.

In keeping with such trends, our sociological safari now takes us back to the zoo's public grounds for a scenic view of its high-flying worlds of entertainment and amusement, where audiences revel in culturally constructed images of wildlife and demand big-top circus performances of zoo animals. On our journey to zoos and aquariums located in some of the most visited tourist destinations on Earth, including Orlando's Disney World and SeaWorld San Diego, brace yourself for a tidal wave of wet-and-wild dolphin and orca shows, amusement park rides, and plentiful gift shops. Even the country's most established zoological institutions from Baltimore's National Aquarium to New York's Bronx Zoo traffic in themed environments, branded diversions, and fanciful myths about the Global South, arguably the societies most immediately threatened by habitat deforestation and the punishing effects of climate change in the Anthropocene. In a consumerist age of nonstop entertainment and simulated reality, American zoos and their audiences collaboratively draw on prejudice and myth to culturally construct nature as an imaginary fantasyland inhabited by fake dinosaur robots and real Las Vegas sharks.

Chapter 5

# Bring on the Dancing Horses

## American Zoos in the Entertainment Age

**ONE OCTOBER AFTERNOON**, a mother with two sons riding in a double stroller stopped to watch the giraffes at City Zoo. The mother appeared exhausted and annoyed.

"Where's the lion?" one of the boys asked.

"They're giraffes," the mother explained.

"What are they doing, Mommy?"

"Resting," the mother said.

"Why?" The mother didn't answer. "Why are they just standing there?"

"I don't know," said the mother.

"When are they going to do something?"

Again, the mother didn't answer. After a few minutes, she said, "I think they just don't move around that much. They just stand around resting, so let's move on."

According to a research report on the Smithsonian National Zoo published in the annual *International Zoo Yearbook*, the official journal of the Zoological Society of London, "Once they were there, many visitors expected to be entertained. They wanted the animals to perform, even when it was unnatural or not possible for them to do so. Popular animals were those that interacted with the visitor or other animals. Patrons gravitated to such animals and were disappointed with sleeping or placid specimens, remarking that 'where there is no movement, there is no fun.'"[1] On this leg of our safari we explore how visitors seek out recreational experiences and amusing diversions at zoos, whether by enjoying cute animals "perform" for them or watching their tail-wagging likenesses portrayed in animated cartoons. We will investigate how audiences experience the zoo in an age of branded entertainment and digital media, particularly by observing how metropolitan zoos and aquariums envelop their visitors in a synthetic aura of enchantment and manufactured authenticity. By examining the ceremonial symbols, silly signs, and whiz-bang simulations that saturate the built environments and cultural landscapes of American zoos and aquariums, we will further our understanding of how these themed attractions sell themselves to audiences as beacons of entertainment and popular culture, all the while generating collective impressions and imagined myths surrounding the animal kingdom and the cultural construction of the natural world.

## The Greatest Show on Earth

While zoos may emphasize education, conservation, and animal care in their mission statements, research shows that zoo audiences tend to desire recreational and entertaining experiences far more than instructive opportunities to learn obscure biological facts about zebras or to be coached on how to best recycle cell phone batteries, or even their own trash, for that matter. At City Zoo, well-meaning patrons regularly leave candy and gum wrappers, soda can pop-tops, and other litter strewn about the zoo grounds despite the obvious choking hazards they present to zoo animals living in open outdoor enclosures. Perhaps ironically, one of the more common varieties of refuse left behind on zoo grounds by visitors includes the blue-and-white wrappers for Purell instant hand-sanitizer wipes. (Presumably some parents think enough about hygiene to sterilize their children's hands while at the zoo, but not enough about their surroundings to find a trash can afterward.)

Despite plentiful signage urging them to do otherwise, some visitors relax during their reveries at the zoo by smoking cigarettes near endangered animals and in plain view of small children, and leave the dirty butts on the ground for peacocks to nibble up.[2]

Perhaps the association of zoos with carefree recreation and entertainment should not terribly surprise us, especially given the strange and often demented history of zoos and animal attractions throughout the ages. In Hellenistic Egypt, King Ptolemy II paraded his enormous collection of zoo animals past Alexandria's stadium in celebration of the Feast of Dionysus. In addition to twenty-four hundred hounds and hundreds of sheep, the procession included ninety-six elephants, twenty-four lions, fourteen leopards, sixteen cheetahs, eight pairs of ostriches, twenty-six white Indian oxen, eight Ethiopian oxen, a dozen one-humped camels, four lynxes, a giraffe, and a rhinoceros. Most of these exotic animals had been captured and poached in the wild from subjugated lands across the African and Asian continents, and publicly displayed as symbols of imperial conquest.[3]

In the Roman circuses of classical antiquity, rulers subjected their animal collections (including humans) to bloody spectacles of butchery and death in staged hunts, free-for-alls among imported predatory animals, and less-than-fair fights between animals and people, including convicted criminals, prisoners of war, and professional gladiators. In A.D. 80, Emperor Titus oversaw the public massacre of nine thousand wild animals. In fact, Titus inaugurated the great Colosseum itself with popular games that included the slaughter of five thousand wild beasts. (Roman emperors also occasionally flooded the Colosseum with water, allowing for gladiators in boats to hunt and kill amphibious hippos, seals, and crocodiles for public sport.) Just a few decades later, Trajan held games of combat during a four-month celebration of his military victory over Dacia in A.D. 106 in which *eleven thousand* animals savagely died, or about one hundred animals *per day*. In Rome's Circus Maximus, Emperor Hadrian would frequently execute a hundred lions in a single show to cheering crowds. Commodus personally slew one hundred bears, six hippopotamuses, three elephants, three rhinos, a tiger, a giraffe, and untold numbers of ostriches, lions, and leopards. The sheer tonnage of gore and flesh boggles the mind.[4]

During the Tudor period in England (1485–1603), the howling royal menagerie locked away in the Tower of London provided animal combatants for dogfights, cockfights, staged bouts between blinded bears and lions and tigers, and contests between bulls and dogs, all held for the entertainment of court guests.[5] Later, in the United States, nineteenth-century circuses, museums, and menageries presented far less violent public spectacles than their European predecessors, but spectacles nonetheless. Phineas T. Barnum's American Museum in New York displayed sharks, seahorses, porpoises, and white whales captured from the wild, as well as the first rhinoceros ever to be seen on our shores. The most notorious circus promoter of his day, Barnum's traveling Hippodrome—"The Greatest Show on Earth"—featured trained elephants, camels, and horses, along with "sixteen ostriches, ten elands, ten zebras, a team of reindeer with Laplaud drivers, a troupe of performing ponies, monkeys, dogs, goats, etcetera." He also exhibited live humans in his circuses and freak shows—including, in Barnum's own published words, "gypsies, Albinos, fat boys, giants, dwarfs," and "American Indians, who enacted their warlike and religious ceremonies on the stage."[6]

In contrast to the nineteenth-century entertainments and amusements produced by Barnum and his revenue-generating circus attractions, the founders of the first American zoos—many of whom would have been Barnum's contemporaries—expressed far nobler intentions. Indeed, the earliest American zoos strived to emphasize public education, scientific research, and wildlife conservation as well as recreation, as reflected in these institutions' founding documents. In 1859 the Zoological Society of Philadelphia established the first chartered zoo in the United States: "The object of this corporation shall be the purchase and collection of living wild and other animals, for the purpose of public exhibition at some suitable place in the City of Philadelphia, for the instruction and recreation of the people." According to an early report to its board of directors, it was "the aim of the Managers, not only to afford the public an agreeable resort for rational recreation, but by the extent of their collection, to furnish the greatest facilities for scientific observation."[7] In 1889 Congress authorized the establishment of what would become the Smithsonian National Zoo for the purposes of "the advancement of science and the instruction and recreation of the people."[8]

Before he was elected to the U.S. presidency, Theodore Roosevelt helped found the New York Zoological Park (better known today as the Bronx Zoo) during the 1890s, largely in the interests of saving the American bison from extinction. According to the historian Douglas

Brinkley, Roosevelt also thought the Bronx Zoo could provide a more enlightened alternative to the "disappointing" zoos of Europe:

> Little educational information was disseminated to visitors about species variation or habitat, and most zoological parks emphasized the freakishness and oddity of their collections. Such come-ons as a six-legged deer in Berlin and a two-headed turtle in London sickened Roosevelt. Worse yet, the animals in European zoos paced back and forth in tiny cages, like prisoners waiting for the end of a lifetime sentence. This kind of backward zookeeping had to end. As Roosevelt envisioned it, their modern New York zoo would be built "on lines entirely divergent from the Old World zoological gardens." The animals would have more room, in open-air exhibits where possible, and broadsheets would be created specifically for schoolchildren explaining the principles behind wildlife preservation and Darwinian evolution.[9]

The New York Zoological Society was therefore incorporated in 1895 "to establish and maintain in [New York] a zoological garden for the purpose of encouraging and advancing the study of zoology, original researches in the same and kindred subjects, and of furnishing instruction and recreation to the people." Its first annual report announced its goals, which included "the systematic encouragement and interest in animal life, or zoology, amongst all classes of the people, and the promotion of zoological science in general," and "cooperation with other organizations in the preservation of the native animals of North America, and the encouragement of the growing sentiment against their wanton destruction."[10] These august institutions were all created during the same intellectual period that saw the publication of Charles Darwin's *The Origin of Species* (1859), the founding of New York's American Museum of Natural History (1869), and the first stirrings of the American conservation movement.

Still, American zoos hardly stood as models of perfection. Just a few years after its founding the Bronx Zoo stooped so low as to exhibit a human being—a Congolese man named Ota Benga—in its Monkey House, where he shared a cage with an orangutan in the fall of 1906. (Prior to this, the orangutan performed for visitors by strutting around in Western-style clothes, wielding silverware at mealtime, and pedaling a tricycle inside his cage.)[11] Meanwhile, zoo audiences themselves could be a rowdy bunch. In the first few weeks following its opening,

Philadelphia Zoo visitors poked a South American sloth to death with their umbrellas and canes.[12] As for Ota Benga, when Bronx zookeepers released him from his cage, crowds continued to pursue him on the zoo grounds, as they did on Sunday, 16 September, a day when he attracted forty thousand visitors to the zoo. As the *New York Times* reported at the time, "Nearly every man, woman, and child of this crowd made for the monkey house to see the star attraction in the park—the wild man from Africa. They chased him about the grounds all day, howling, jeering, and yelling. Some of them poked him in the ribs, others tripped him up, all laughed at him."[13]

Although bursting with technological wizardry like giant panda webcams, today's zoos continue to serve as monuments to the urban amusements and entertainments of their pasts. The Central Park Zoo still displays its nineteenth-century bronze statues of a dancing goat, a honey bear, and a tigress and her cubs, along with the zoo's famous 1965 Delacorte clock featuring a set of revolving animals (bear, goat, elephant, hippo, kangaroo), and a pair of monkeys that ring the bell on the hour and half hour. As noted in the last chapter, visitors to the Philadelphia Zoo continue to enjoy amusements that rely on antiquated technologies from the Victorian era: paddleboats, old-fashioned trains, a carousel, camel and pony rides, and a gigantic hot-air balloon.[14]

Of course, today these nostalgic entertainments compete with more contemporary efforts at audience pandering: IMAX films; theme park safaris accompanied by ambient jungle soundtracks; gift shops bulging with stuffed lions, polar bear T-shirts, and chimpanzee-adorned shot glasses; and animal performances that could pass for circus acts. At one of the San Diego Zoo's animal shows, I watched macaws circle overhead to Kenny Loggins's 1980s hit "Danger Zone," and a clapping sea lion named Jake waddle onstage and dive into his pool to George Thorogood and the Destroyers' blues-rock anthem "Bad to the Bone." Sound effects framed Jake's every move. During one performance, a small boy from the audience proceeded to kiss Jake, first to James Brown's "I Got You (I Feel Good)," and then creepily to Marvin Gaye's sex-jam "Let's Get It On." Later the audience met another sea lion, Cabo, who danced, swam, twirled, barked, and bounced from flipper to flipper to Harry Belafonte's "Jump in the Line (Shake, Senora)." Afterward, the audience was thanked for contributing to the zoo's conservation efforts.

In this manner, contemporary zoos may resemble Barnum's circuses more than we would like to admit—although not without pushback from zookeepers, educators, and other conscientious zoo personnel. Over many months of working together at Metro Zoo, Daphne complained to me that zoos ought to prioritize public education around conservation issues above audience demands for lighter fare:

> My mantra is that we really need to be in the education business. I don't think everybody looks at it that way. Outsiders? Some of them get it. But most of them think that it's just a nice time to go and see animals that they wouldn't normally see. The point is making an impact on people, so that they understand conservation and feel connected, and want to *do something* about conservation. To me, that's something that most people don't think about. . . . And if you don't do that, then close your doors, because you are a menagerie, and not a zoo.
>
> Unless we are educating people, we have no right to be in this business. Otherwise, we are Chuck E. Cheese, and sticking animals in cages for no reason.

## If We Could Talk to the Animals

While polar bears, elephants, and other charismatic megafauna excite zoo audiences of all ages due to the larger-than-life sizes of their massive bodies, birds such as macaws, cockatiels, lorikeets, and other kinds of parrots enchant visitors because their ability to mimic human voices makes them extraordinarily easy to anthropomorphize. Part of my training at Metro Zoo included learning how to perform in public with Beaker, the zoo's aforementioned African gray parrot. (The well-trained bird was so competent at mimicry that a note posted in her aviary warned keepers not to tune a nearby clock radio to two local hip-hop stations, out of fear that she might repeat overheard "inappropriate" language when around children in public.)[15] Although she actually bit me not just once, but *twice* during my stint at the zoo, Beaker surprisingly obeyed my commands, to the delight of guests. Her various tricks (they always felt like tricks) included bopping her head when I asked her to dance with me, and responding to "Tastes like?" with an enthusiastic "CHICKEN!" After my successful attempt at performing the latter trick with Beaker, a woman stared at the two of us with astonishment, asking me (us) to do it again. (We did.) By way of positive operant conditioning aided by a pouch of peanuts and other food rewards, Beaker and I had performed the most popular illusion in the zoo world, a fake show of interspecies communication. I spoke to an animal, and it spoke back.

In fact, zoo audiences seem to expect parrots to be able to always speak on command, as if they were ventriloquist dummies or Muppets. One afternoon at City Zoo I was exhibiting Quito, a male blue-and-gold macaw, and a mother with her children passed by. "Does he do anything? Does he talk?" she asked.

"Quito *can* talk, but he doesn't do tricks simply for our amusement," I replied, in keeping with the attitude of the zoo's keepers.

"Oh well, that's too bad," the mother said, walking away with a look of disappointment on her face.

Unlike circus performers, City Zoo's keepers and curatorial staff were not typically in the habit of training their animals to perform like carnival freaks (nor were Metro Zoo's keepers), although the zoo's education staff *had* trained a chicken to peck at a red plastic bowling pin, run through a tunnel, hop over and under a succession of hurdles, and lastly pull a drawstring, revealing a sign reading, "CHICKENS ARE SMART!" (An intern had also trained Barney, a donkey, to dunk a basketball into a netted hoop.) Zoo audiences do not usually expect such stunts, either. Rather, visitors love to watch animals perform what they imagine to be their "natural" behavior, or what Tyler referred to as the animal's "selling point": monkeys swing, elephants spray water out of their trunks, wildcats roar. (Tyler also said that when confronted with an unusually exotic animal visitors often search out its selling point, asking, "Well, what does it *do*?" He joked, "This specific animal is an accountant. If you want me to go back to the mammal shed I'll bring out the attorney.")

Of course, live zoo animals can rarely compete with the entertaining and idealized images that audiences so often associate with them (with the exception of Beaker, and maybe that chicken). City Zoo visitors will sometimes go so far as to *yell* at the zoo's exhibited lions in their attempts to get them to roar as they do in the movies, thus signifying their lion-like quality, or lion*ness*—even though field researchers working in Serengeti National Park report that lions mostly roar at night, and especially just before dawn, rather than during typical zoo visiting hours.[16]

But while zoos and circuses connote different types of urban experiences and industry standards of husbandry and care, both attract audiences expecting to be entertained by animals. On a bright sunny afternoon in November at City Zoo, a father with two older boys walked up to the polar bears' glass enclosure. "Bozos. They're just lying there," one boy said. "Such a rip off," complained the other boy. The father agreed. "Let's go see something else, maybe he'll be swimming when we come back." Boredom is a fairly common reaction to zoo animals at rest:

> A mother and son were standing by the kangaroo exhibit. The kangaroos were lying on a grassy patch. The son was focused on the zoo map, pointing to exhibits he wanted to see. The mother said, "Don't worry about the map. Look at the kangaroos . . . *Hi, Joey!*" The son said to his mother, "I want to see him walk." The mother replied, "They're hanging out, sunbathing!" The son complained, "Let's go." The mother said, "Maybe we can come back later and they'll be more active."
>
> After they left, a young couple walked up to the kangaroo exhibit. As the man read the sign, the woman raised her camera and pointed it toward the red kangaroo. "I want him to stand up and hop, so I can take a picture." The kangaroo did not move. The woman turned away and said, "They're boring. Let's go." The couple walked away.[17]

For some visitors, even watching animals eat enormous amounts of food cannot do the trick. On a bright sunny afternoon in November at City Zoo, a grandmother with two grandchildren stopped to watch the hippopotamuses eating by their enclosure wall. She said, "They knew it was time to eat, so they got out of the water."

"Oh, my god," said one of the boys. "They are so big! Oh, my god." The grandmother read the boys the information on the marker about the hippos.

"They are from Africa, and they weigh about 4,500 pounds each." Without missing a beat, she continued, "Okay, let's go. They are going to stay by the wall eating. Let's go, there's nothing else to see. They're not doing anything."

Given that the entertainment value of zoo animals can be unpredictable from moment to moment, guests will sometimes compensate by attempting to perform their own interpretations of "natural" sounds and behaviors *for the animals*, as when a group of children at City Zoo looked up at its baboon exhibit and squealed, "Monkeys! *Ooh-ooh, ah-ah.* . . ." By literally "talking" to (or for) the animals—"Hi monkey!" "He's itchy, and looking at us!" "I think he heard you!"—the zoo audience plays with the fantasy of forging an emotionally reciprocal human-animal connection with these beasts, just as I did with Beaker. Of course, such encounters can obviously be experienced as deeply meaningful and emotionally satisfying to humans despite the unlikelihood of achieving mutual understanding between people and animals.[18] Then again, sometimes audiences genuinely (if mistakenly) believe that they have actually forged such an interspecies connection with a zoo animal. According to Tyler, Metro Zoo visitors occasionally assume that "they have a way with animals, and all animals innately respond well to them. They'll say of a zoo animal, 'It's looking at *me*, it's paying attention to *me*.'" I asked Tyler for an example.

"People think they have a special relationship with Magic, the male howler monkey. People will think, '*Oh look, he's smiling at me!*' but not understand that when a monkey bears its teeth, it's not *smiling*, but bearing its fangs because you're invading its territory—it's telling you to *fuck off*. You get a lot of that. When an animal is making noises, they go '*It's talking to me!*' It's not talking to you, it's telling you to go away. I think it's interesting, the way that people overstate their relationships with animals."[19]

This desire among visitors to make an interspecies connection with a zoo animal (no matter how fleeting) remained palpable during my days of zoo research, almost to distraction. While volunteering at both City Zoo and Metro Zoo I presented all manner of reptiles, birds, and small mammals to zoo visitors—leopard tortoises, armadillos, hedge-hogs, bunny rabbits, bearded dragons, boa constrictors, screech owls, chuckwallas—and my limited zoological training taught me to answer a variety of visitors' most-asked questions. "*What's its name?*" "*How old is it?*" "*What does it eat?*" "*How big do they get?*" However, most of my time was spent preventing the general public from physically touching the animals, which was against the rules at both zoos. (Typically people would go in for a grope first, and request permission after the fact.) Upon refusal, guests occasionally got huffy—"Do you have any animals we *can* pet?" One spring I exhibited Kaa, the Colombian red-tailed boa constrictor, and a female passerby complained aloud, "I don't want to *look* at it if I can't touch it." More often, visitors would ask if the animal could bite, to which my standard reply was always the same: "*Well, anything with a mouth can bite, sir.*" (Zoos typically enforce prohibitions against touching their animals on public health grounds, or fear that an untraceable illness might spread through their animal collection, or else that these creatures might experience undue stress by being touched by ten thousand strangers a day. However, audiences usually presumed that zoos prescribed such rules out of concern for their *own* welfare and safety as visitors, rather than that of the animals themselves or the public at large.)

Zoos and aquariums have responded to audience demands for animal encounters—opportunities to physically handle the sacred totem, as it were—by establishing touch pools and petting yards where visitors can caress select members of their collections. During my research, Adventure Aquarium in Camden, New Jersey, displayed tanks where families with children could pet horseshoe crabs, rays, and even Indo-Pacific brown-banded and white-spotted bamboo sharks. City Zoo featured a petting yard where children could brush sheep and pygmy goats, and feed them pellets of grain that their parents could purchase at a machine for fifty cents a handful. (The zoo purposely limited the amount of pellets distributed to guests in order to prevent overfeeding.) The most impressive attribute of the petting yard may have been the sheer mayhem it was able to contain within its wooden rails, given the eager enthusiasm of its excitable guests. Actually, sometimes visitors could be a bit *too* excitable. Despite warnings from zookeepers, children would race around the yard, clambering up its fences. They approached goats and sheep resting in the dirt and jostled them with their sneakers, occasionally tossing small rocks and stones at them. One child attempted to milk one of the poor goats; another asked if he could keep the brush. Parents and kids alike howled with equal parts anxiety and amusement when the domesticated yet undisciplined goats aggressively poked at their pocketbooks in search of food, or else urinated on them. It was quite a scene.[20]

Of course, as everyone knows, all sorts of intimate encounters are possible for the right price.[21] For example, SeaWorld San Diego charges guests for special hands-on access to its most charismatic marine animals. In March 2011 I paid to take part in the park's interactive Animal Spotlight Tour, during which I was invited to pet and feed three Atlantic bottlenose dolphins (Cascade, Koa, and Monte) and a Pacific bottlenose dolphin (Gracie). Upon receiving permission to cross over the line alerting more frugal visitors not on the tour to PLEASE STAY BACK, I fed the dolphins mackerel, herring, and sardines, and mimicked the trainers by trying some hand signals for the animals to obey, based on their ongoing protocols of operant conditioning. I pointed up to the sky to elicit bows and jumps and twirled my finger in a circle to get them to spin around. Finally I was invited to pet one near its blowhole, which made it, well, spit at me. (A far more expensive package allows visitors to actually *swim* with the dolphins.) A few days later I paid an additional fee to join SeaWorld's Penguin Encounter tour, where I got to pet an adult macaroni penguin in its frigid twenty-seven-degree exhibit.[22]

The commodification of these kinds of animal encounters has been normalized to a shocking degree. When I politely refused a little girl visiting Metro Zoo who wanted to touch Donatello, an eastern box turtle, she innocently asked, "Do I have to pay to touch it?" Meanwhile, some City Zoo donors simply presumed that their financial contributions entitled them to a physical encounter with its captive animals, even dangerous ones. According to Nicole, the former development assistant at the zoo,

> The donors were mostly interested in getting to interact with the animals themselves, and they all thought they were going to be able to touch the animals. A lot of people asked if they could ride the giraffes, or pet the lions, and they wanted to do all these things they saw on Animal Planet on various shows. . . . A lot of people wanted to use the penguins to propose to their girlfriends—you know, tie a little message to them, or do some sort of surprise like that. . . . We'd get a lot of people who wanted to pet the cheetahs. . . . We got a phone call one day from somebody who was having a pool party, and wanted to rent one of our hippos to be in the pool during the party. They couldn't figure out why we would not rent them the hippo.[23]

Similarly, at a Metro Zoo fund-raiser a tipsy attendee offered Joshua, a guest services employee, five thousand dollars to pet one of the zoo's monkeys. He turned down her proposition, opting to spend the rest of the night stopping other inebriated couples from sneaking into the petting zoo barn.

Other zoo visitors spend their money on close-up encounters with animals by capturing them on film, and not just with their smartphones. On a hot July morning at the Philadelphia Zoo, a middle-aged gentleman armed with a digital Canon and various attachment lenses spent fifty minutes shooting Klondike and Coldilocks, a pair of polar bears relaxing in their poolside Arctic habitat. On another occasion, Scott and I encountered a woman lugging an impressive arsenal of photography equipment around the zoo's Big Cat Falls exhibit. Her camera sat on a tripod and sported a gigantic four-hundred-millimeter telephoto lens. After a few minutes of careful shooting, she showed me her close-ups of a lion's head—you could see the fine details in its face, especially in its eyes, and all without any sign of the glass walls or fencing that held the wildcat within its enclosure. I later saw the woman and her partner, a man with a tripod in his backpack, at an endangered tiger enclosure set off from the main path. The childless couple crowded around the glass with their cameras and accessories, effectively blocking Scott's view of the orange-and-black striped beasts.[24]

The very idea of zoo photography conjures up its own bizarre contradictions. In the wild, the art of "capturing" a photo of an animal or bird is analogous to hunting rare game: a challenge requiring not only enormous technical skill (and luck), but stealth and patience as well, not to mention the agility and expertise necessary to navigate the wilderness itself. Given the difficulties apparent in capturing an elusive creature in the wild on film, the photograph itself then serves as a kind of hard-won trophy, symbolizing a successful hunt. Yet the zoo's animals are enclosed in exhibits specifically designed for public viewing. "Capturing" an animal on film is significantly less of a challenge when the animal itself is already in captivity. (Talk about shooting fish in a barrel.)

Of course, not every American can afford to fly off to Tanzania to arrange a safari among the zebras and blue wildebeests of the Serengeti, much less handle a telephoto lens under such conditions. Still, the goal of the zoo photographer is to frame the animal in its naturalistic enclosure *as if* it were in the wild, without revealing the reality of its captivity or staged surroundings. In this sense the zoo's ersatz habitats serve the same function as the scenic backdrops employed in portrait studios, and in fact many zoos themselves encourage amateur photographers through official camera clubs, including the Smithsonian National Zoo, Kansas City Zoo, and, yes, the Philadelphia Zoo.

Still, it is hard to know how much any of this staging matters to the average zoo visitor, especially those desiring the rapture of making an interspecies connection. Indeed, for many zoo audiences the distinction between the sublimity of the wild and that found at the zoo is not only imaginary, but often ignored altogether. In September 2012, New York state resident David Villalobos jumped from a moving monorail train into a live tiger habitat at the Bronx Zoo. He suffered a broken pelvis, a broken right shoulder, a broken right rib, a collapsed lung, and a broken right ankle from his fall and a subsequent mauling by the exhibit's inhabitant, Bachuta, a four-hundred-pound Siberian tiger. He later explained to police that his leap was motivated by his "passion for cats" and desire to be "one with the tiger."[25] Far less dramatically, one October afternoon at City Zoo a husband and wife watched the giraffes loping around in their exhibit. The woman gushed, "Wouldn't it be great to see them in their natural habitat?"

But her husband just shrugged. "Why? You see them here."

# Deep Blue Something

Few zoos showcase animal performances and the dream of human-animal connectivity more than aquariums and marine mammal parks. In SeaWorld's perennial killer whale spectacular, the lead role of Shamu might be played by any of a number of orcas held in captivity at its three U.S. aquatic theme parks, as has been the case since the character debuted in 1965.[26] One spring I attended SeaWorld San Diego's *Shamu Show: Believe* in a stadium that holds about fifty-five hundred people, just to see it for myself. The most popular seats in the house are its first sixteen rows, dubbed the "Soak Zone" because when these massive, ten-ton creatures perform dives and crash into their pools they drench those awed spectators who dare sit before them. There is no doubt that these powerful and predatory thirty-two-foot killer whales thrill crowds with their sheer size, acrobatic prowess, and formidable presence like few other animals.[27] But SeaWorld wants its audiences to experience not merely stimulation but the sacred itself, or at least a fabricated aura of spirituality.[28] Accompanied by New Age chanting along with orchestral winds, brass, and strings, and a heavily produced video to match, a deep and emotive voice announces, *"Many years ago, one man believed that, indeed, two worlds could come together. This singular idea echoing through these many years has allowed us to build a unique relationship with this magnificent creature. A wondrous connection, reminding us that all things are possible. . . ."*

These two worlds are, of course, human and animal, culture and nature. As celebrity killer whales splash, spin, and flip to the commands of a professional trainer, the performance is intended to inspire audiences by convincing them that this sublime connection between man and sea monster is, in fact, happening before its very eyes—no matter that the only things bringing these two worlds together are years of routinized operant conditioning and the strategic dispersal of food.[29] Yet all at once, the stadium is enveloped in song as the audience swoons to a recording of classical vocalist Nancy Coletti's "Something Far Greater," just as a choir accompanied by pipe organ might stir the collective spirit of a congregation summoned to worship in a glorious cathedral. *"Come!"* the song's climactic chorus commands. *"Touch the face of a mystery!"* As if a prayer, its rapturous lyrics beckon the listener's heart *"to live, to breathe,"* and take that final leap of faith—*"BELIEVE!"*

Cue the diving blackfish, and SPLASH! The audience is ritually cleansed; the baptism complete.

At the AZA-accredited SeaWorld San Diego, formerly owned by beer brewer and theme park operator Anheuser-Busch, windswept images, an ambient soundtrack, and the delicate dance of highly trained aquatic mammals are brought together to evoke the sacred as a mass-cultural corporate entertainment experience. Yet less ambitious zoos and aquariums similarly draw on animal shows along with heavily edited imagery, soundscapes, and technologies of modern mythmaking in order to both generate and exploit audience-driven fantasies of the wild and the "magic" of nature.

For instance, both dolphins and beluga whales perform in live aquatic shows for audiences at the John G. Shedd Aquarium on Chicago's lakefront. Unlike the circus antics surrounding SeaWorld's *Believe*, these performances take a more matter-of-fact approach by emphasizing how trainers marshal positive reinforcement techniques as pragmatic strategies of care.[30] Nevertheless, Shedd's animal shows still manufacture a simulated atmosphere of consecration and awe. Audiences watch marine mammals swim and splash in the facility's three-million-gallon saltwater Oceanarium, while indie-folk band Bon Iver's hauntingly ethereal 2011 song "Holocene," itself a devotional homage to the ineffable vastness of deep time and the natural environment, plays in the background:

And at once, I knew I was not magnificent

High above the highway aisle

(Jagged vacance, thick with ice)

I could see for miles, miles, miles.[31]

Chicago's Shedd and other aquariums engineer multimedia allusions to otherworldly transcendence not only through their animal performances, but artistically rendered exhibits and audiovisual spectacles that sanctify the oceans as swirling pools of mystery and divination. At Adventure Aquarium (owned by Herschend Family Entertainment, a corporation that strives to operate "in a manner consistent with Christian values and ethics"), audiences find themselves awash in New Age cinematic orchestration reminiscent of the *Titanic* film soundtrack, theatrical mood lighting, the constant hum of flat-panel video screens broadcasting documentary marine-life footage, and, of course, the aquarium's highly stylized acrylic tanks themselves, brimming with life. The aquarium's showpiece is its forty-foot-long shark tunnel, in which audiences walk down a corridor surrounded by 180 degrees of water, with sharp-toothed leviathans from great hammerheads to sand tiger sharks swimming overhead. The effect of all this technological hocus-pocus is to create a total media environment where audiences experience the aquarium as a completely immersive underwater fanta-syland. As promotional videos remind guests throughout the day, "You're in Their World Now."[32]

Aquatic zoos, or aquariums, are among the most successful AZA-accredited institutions in the United States, yet many people do not consider aquariums zoos at all. For some, fish seem so lacking in individual personality and phylogenetic similarity to humans that they are hardly even thought of as *animals*, which is why aquariums often fail to elicit the same sense of guilt that turns off some nature-loving audiences from enjoying zoos filled with captive mammals and caged birds. Since fish lack the cognitive and physiological faculties to express emotion in the same way that complex mammals do, it also may be easier for us to naturalize their captivity.[33] We are far more accustomed to seeing fish in captive enclosures—notably as interior decor or household pets—than similarly companionable animals. Golden retrievers get to race around the park off-leash, but goldfish have to stay in the bowl.[34] Even in the wild, we are accustomed to viewing certain bodies of water such as lakes and ponds as self-contained and thus *finite*, and so we naturalize the bounded habitats in which aquariums hold fish captive. Moreover, unlike traditional zoos, aquariums rarely enclose sea creatures behind steel bars, but in glass tanks with modern filtration and water-treatment systems that augment the realism suggested by naturalistic marine environments.[35]

What about highly intelligent marine mammals like dolphins, porpoises, or whales? While even lay audiences might recognize the abnormal repetitive behavior of a bored leopard or grizzly bear pacing back and forth in its zoo enclosure, these same audiences might not recognize the similarly tortured swimming patterns of cetaceans. Why? Aquarium tanks sometimes contain thousands of creatures; exhibition tanks lack visible sight lines from all directions; and although we know that marine mammals swim in oceans vaster than any other terrain on the planet, we may not necessarily recognize their limited living space in an aquarium tank as a kind of deprivation—as long as they just keep swimming. This is a common oversight made by not only aquarium visitors, but professional animal keepers as well. According to animal sciences professor and advocate Temple Grandin,

> I visited an aquarium where a single dolphin was swimming around and around his pool in a repetitive pattern that never varied. He swam the entire length of the tank along the bottom, and then swam up at a 45-degree angle to the corner across the pool. When he got to the top corner, he turned and swam the length of the tank at the surface, and then he followed a second diagonal at a 45-degree angle down to the corner at the bottom of the tank, which put him back where he started. A lot of animals develop circular stereotypies where they move in one plane, but this dolphin had developed a figure eight that used all three dimensions of the tank. His path was unusual enough that the keepers were not aware that this was an abnormal stereotypy.[36]

It should also be mentioned that, unlike traditional zoos, there are far fewer underresourced aquatic zoos that might tarnish the reputation of the entire industry—after all, there are not many roadside aquariums. (New Jersey's unaccredited Atlantic City Aquarium comes close.) Most require so much capital investment that the kinds of institutions that can best survive tend to be sponsored by large corporations, wealthy benefactors, or federal and/or state governments. For example, take the Georgia Aquarium in Atlanta, originally funded by a $250-million donation by billionaire Bernard Marcus, cofounder of The Home Depot. The aquarium's Ocean Voyager exhibit is the largest indoor fish tank in the world. Thirty feet deep at its deepest point, the enclosure holds 6.3 million gallons of water. Almost as big as a football field, it measures 269 feet long and 120 feet wide. Home to thousands of fish including a handful of manta rays and four whale sharks (the females are named Trixie and Alice, after the characters from *The Honeymooners*), Ocean Voyager features 4,574 square feet of viewing area, 185 tons of acrylic windows, and a hundred-foot-long underwater tunnel, more than double the size of Adventure Aquarium's shark tunnel.

(Thanks to Georgia Aquarium's backstage tour held during the 2011 AZA annual meetings in Atlanta, I discovered how shark tunnels work. After admiring the giant exhibit on the first floor I was permitted upstairs, where one can view the tank from overhead—it looks like a gigantic swimming pool in an airplane hangar, with huge rafters and a roof overhead. A guide then pointed out a row of spinning fans and lights affixed directly over the glass tunnel. As the fans ripple the water above the tunnel, the undulating waves distort the audience's view of the rafters and the dank roof above, creating the illusion of actually being underwater and surrounded by sharks, rather than in a tourist attraction just across the way from the World of Coca-Cola in downtown Atlanta.)[37]

Finally, given that the Earth's deep oceans represent some of the last remaining spaces of unexplored wilderness on the planet (although hardly untouched by human activity, thanks to global warming), one might argue that when compared to more traditional zoos, aquariums and marine mammal parks more easily evoke the mystery and sanctification so often associated with Mother Nature. As divinity scholar Mircea Eliade observes in *The Sacred and the Profane,* "The waters symbolize the universal sum of virtualities; they are *fons et origo,* 'spring and origin,' the reservoir of all the possibilities of existence; they precede every form and support every creation. . . . In whatever religious complex we find them, the waters invariably retain their function; they disintegrate, abolish forms, 'wash away sins'; they are at once purifying and regenerating."[38] On the docks of Baltimore's Inner Harbor, the National Aquarium is practically a sacramental temple to the blue and bewitching seas, at least as they are imagined in both traditional and modern myth. Its slogan announces "the National Aquarium inspires conservation of the world's aquatic treasures"—that is, the institution's primary purpose is not to *educate,* nor *inform,* but to *inspire.* In its Glass Pavilion audiences encounter the aquarium's multistory "Maryland Waterfall" that spills its downpour from fabricated cliffs into a tank alive with local fish such as rosyside dace and brook trout. Its signage reads like a beer advertisement for Rolling Rock or Coors Light: "Tumbling over rocky hillsides, mountain streams flow into rivers and cascade down the watershed into the Chesapeake. Fish, salamanders, invertebrates, among others thrive in this cool, oxygen-rich, flowing water."

At the National Aquarium totemic symbols, hallowed images, and liturgical depictions of nature are as prominent as more traditional ocean life exhibits. Venerable proverbs blanket the aquarium's walls: "For all at last return to the sea . . . like the ever-flowing stream of time, the beginning and the end," thus spoke biologist and conservation icon Rachel Carson. Another preaches spiritual communion: "We are all connected by the ocean. Together, Earth's waters form one world ocean linked to freshwater lakes, rivers and wetlands. Countless animals and plants call these waters home. Each of us is affected by the ocean, and each of us leaves our mark on it as well." Inspirational quotations—mostly excerpted poetry—similarly clutter the walls of the Monterey Bay Aquarium: "The sea is as near as we come to another world" (Anne Stevenson); "In one drop of water are found the secrets of all the endless oceans" (Kahlil Gibran); "Life in its jewel boxes is endless as the sand" (Pablo Neruda); "I spin on the circle of wave upon wave of the sea" (again Pablo Neruda). My personal favorite is decidedly less reverent: "I'd like to be under the sea / In an octopus's garden in the shade" (Ringo Starr of the Beatles).

Back in Baltimore, one of the National Aquarium's prominent art exhibits, Aquatics, showcases gorgeous purple images of jellyfish, turtles, and other marine life, all created by photographer Henry Hornstein.[39] In New Age speak, its signage reads,

> More than glass separates us from the strange and wondrous world of water. For while aquatic animals share our planet, their world is profoundly different from ours in ways we are just beginning to understand. Even so, we search for some connection to our aquatic selves.
>
> Henry Hornstein magically transcends the boundaries between our world and theirs. His warm, light-infused images focus on rarely-noticed patterns and textures—organic forms at once strange and yet mysteriously familiar. He invites us to see aquatic life in an entirely new light. And in doing so, reawakens in us our ancient and often enigmatic relationship to the world of water.

In this art exhibit water is not simply habitat and resource, but *wondrous* and *enigmatic*; organic forms are both *strange* and *mysterious* (even if familiar); the marine world *reawakens* us from our existential slumber, forging a *connection* between us and our *ancient* past (one in which humans were presumably more intimately ensconced in their natural environment); and Hornstein's images themselves are *magical* and *transcendent*. Given the lofty aims of the National Aquarium's portrayal of marine wildlife, it is perhaps ironic that its Harbor Market Kitchen cafeteria invites its visitors to feast on fish and chips along with Maryland cream of crab soup, although perhaps not. According to Émile Durkheim, the holy leaders of some ancient clans ritualistically ate their own animal totem during religious ceremonies.[40] In fact, Ryan claims that during his time as an aquarium educator both children and adult visitors would often ask staff if the fish served in the cafeteria had been harvested from the aquarium's own exhibited collections.

## The Savage Mind

Of the nation's 2,764 zoological gardens, petting zoos, sanctuaries, marine-mammal theme parks, and other animal exhibitors subject to USDA inspection under the Animal Welfare Act, only 214 U.S. institutions have earned accreditation from the prestigious Association of Zoos and Aquariums, or AZA. While many of the country's most renowned metropolitan zoos—Zoo Atlanta, San Diego Zoo, Bronx Zoo, Philadelphia Zoo, Smithsonian National Zoo—are longtime members of the organization, not all institutions make the grade. For instance, there is only one AZA-accredited zoo of any kind in the entire state of Nevada, and it is not the Southern Nevada Zoological Park in Las Vegas, nor the Sierra Safari Zoo in Reno. It is the Shark Reef Aquarium, located off the lobby of the shimmering gold Mandalay Bay Resort and Casino on the southern end of the famed Las Vegas Strip. Like the forty-three-story hotel in which it resides, Mandalay Bay's Shark Reef does not skimp on ostentation, but employs bedazzling theatrical techniques to reimagine the aquarium experience as a gaudy theme park adventure. Its marine wildlife tanks are welded to elaborate stage sets featuring hanging gardens, labyrinth corridors, wrought-iron gates, statuary, engraved stone, and a shipwreck (which could not have been easy to find in the surrounding Mojave Desert). The aquarium's creators envision this Disney-inspired experience as a mysterious "journey through an ancient temple slowly being claimed by the sea." It also claims to be "the only predator-based aquarium in the United States," with its Burmese pythons, piranhas, komodo dragons, and one hundred ferocious sharks. (How fitting for Las Vegas.)

Just as nature is socially constructed, so are world cultures, particularly those of the Global South portrayed as exotic by Americans and Europeans. (After all, people don't think of their *own* culture as especially exotic.)[41] Like other casino circuses on the Vegas strip—Siegfried and Roy's Secret Garden and Dolphin Habitat at the Mirage, the Flamingo Hotel's Wildlife Habitat, the 1,700-gallon aquarium tank in the middle of the Seahorse Lounge at Caesar's Palace—Mandalay Bay's Shark Reef aims not for naturalistic *or* cultural authenticity, but for a parody of both, a faux–Southeast Asian spectacle in the heart of Sin City.[42] The Shark Reef's depiction of a sunken Buddhist temple is only one of many instances of how American zoos appropriate religious and ethnic Asian cultures for their narrative and symbolic potential as tired Oriental myths popularized among Western powers over centuries of global dominance.[43] Sometimes zoo exhibits rely on the consecrated materials of Eastern religion to add spice and allure to otherwise ordinary exhibits. At a snow leopard pavilion at the San Francisco Zoo, Tibetan prayer flags hang from the rafters in colors of green, yellow, blue, and red.

Other American zoos represent Asian territories as violent backwaters. At New York's Bronx Zoo I wandered through Tiger Mountain, a woodland trail intended to resemble the Russian Far East. Along the bamboo-lined trail hung a sign in Russian warning of tigers—alas, it was only the zoo's Siberian tiger exhibit. Nearby sat a stage set for a "poachers' truck" decked out with more signs, although these seemed designed to cause outrage rather than faux fear. One was stenciled "Bush Meat," while others promoted animal poaching: "Exploit-a-Park Real Estate: Land! A Real Steal." "Soil B. Gone: Nutrient Suckers: Why leave food for wildlife tomorrow? When you can profit today!" This all led to an overall message about the endangerment of Asian wildcats due to illegal tiger hunting and trapping in the Russian tundra by locals—a lesson that, while honorable, does almost nothing to challenge American zoo audiences about how their *own* consumerist behaviors lead to habitat destruction and biodiversity loss abroad (as will be discussed in chapter 6). Meanwhile, at times the shtick seemed a bit heavy-handed. Tiger Mountain's displayed props included wooden crates full of rifles; a gasoline can; a box labeled "land mines"; rusting tiger traps; a box labeled "Slash & Burn easy pour fire starter"; rolled-up barbed wire; a box of "chickens" labeled "Fowl Play Livestock: Raised on farms that used to be tiger forests"; and a final indictment of indigenous natives: "Killing tigers is illegal, but some traditional medicines still contain tiger parts."

As one might expect, Disney's Animal Kingdom offers a more upbeat if still shameless take on Asia's place in the Western imagination. The centerpiece of the zoo-as-theme park's Asia pavilion is the make-believe township of Anandapur, which seems to evoke India, Nepal, Tibet, Indonesia, and Thailand all at once. Astride a winding, lazy river, its paths are lined with temple ruins (much like Shark Reef Aquarium), and rickshaws loaded down with wooden crates. A souvenir shop sign warns, "No strollers, carts, motorized scooters or livestock allowed in building." High-speed amusement-park rides are similarly themed, including the Kali River Rapids and Expedition Everest—Legend of the Forbidden Mountain. (The latter ride promotes fears of the Yeti, or the Abominable Snowman, which I'm pretty sure isn't a real animal.)

The zoo animals in this part of the park can be found along the Maharajah Jungle Trek, a complicated set of exhibits and enclosures staged as a walking tour of a "famous preserve of royal forest." Like everything at Disney's Animal Kingdom, creative imagineers have invented a narrative for every stop along the trail, as if storyboarding a far-fetched animated film. As one begins the Maharajah Jungle Trek, a sign announces, "*Anandapur Royal Forest*—Since very ancient times the rajahs of Anandapur have hunted tigers in this forest. In A.D. 1544, King Bhima Disampati decreed the forest a royal preserve closed to all save his guests and built a royal hunting lodge whose ruins lie nearby. After 1948, the Royal Forest was given to the people of Anandapur. Today the forest protects not only the remaining tigers and other wildlife but is a valuable watershed of the Chakranadi River and some of the last remaining virgin forest in this region."

Indeed, like the scenes featuring King Louie of the Apes and his insane band of primates in the 1967 Disney film *The Jungle Book*, animals linger among ancient relics in this biogeographically jumbled, pan-Asian mutation of a "forest preserve"—Malayan tapirs, Indian pygmy geese, Argus pheasants, Rodrigues flying foxes, white-cheeked gibbons, and King parrots. As for its tigers—which Disney refers to generically as *Asian* tigers, rather than by their specific subspecies, whether Bengal, Indochinese, or Siberian, as most other accredited American zoos do—audiences are cautioned, "Jungle Trekkers, please apprehend that tigers are frequently encountered in the ruins ahead." (Fortunately, the temple ruins left standing happen to allow for perfect views of the tigers while conveniently protecting visitors from their dangerous teeth and paws.) Elsewhere on the Maharajah Jungle Trek, trails lead audiences past crumbling walls, ornate fountains, and a collection of water urns and ceremonial flags (said to be owned by the "village locals"). A sign marker alerts: "Please—No Climbing: This ancient coral tree is a place of veneration. Scarves and garlands are hung as offerings. Bells are representing prayers that have been answered." Along with facsimiles of religious artifacts, the Trek's pathway is riddled with distressed walls covered with frescoes and stone carvings. They feature elephants, monkeys, snakes, and—if you look closely enough—another animal: hidden silhouettes of Mickey Mouse's famous ears.[44]

Meanwhile, handouts characterize not only the Trek's wildlife but also the indigenous people of Anandapur themselves as one with nature: "Our ancient traditions are centered on compassion for all living things, with a belief in the earth as the common heritage and responsibility of all." This is a familiar trope deployed by American zoos. In their attempts to organize the world into civilization and wilderness (or culture and nature), zoos and aquariums showcase aboriginal people as noble savages and magical stewards of the Earth, a kind of minstrelsy for the anthropogenic age.[45]

To this end, Disney's depiction of the Global South is most outrageous in its major African exhibits both at Animal Kingdom, and the ritzy Animal Kingdom Lodge, a hotel resort with meticulously crafted views of not one but *three* African savannas stocked with giraffes, waterbuck, zebras, impalas, wildebeests, and greater kudu. (Children are given a checklist of grassland species to discover and document from the comfort of their hotel-room balconies—just as if they were on safari in Kenya, except not really at all.) In the park itself, Disney portrays all of Africa as yet another imaginary town, Harambe, which is Swahili for "working together in unity" or "let's pull together." Its animal attractions include the Pangani Forest Exploration Trail, where visitors watch captive gorillas, colobus monkeys, meerkats, and hippopotamuses perform their roles as wild creatures on naturalistic stage sets, and Kilimanjaro Safaris, a tram ride through reproductions of different African ecosystems replete with what might look like free-ranging animals if they weren't carefully secured behind hidden moats and electrified wire.

The predistressed streets of Harambe overflow with stalls and souvenir shops brimming with trinkets for sale—Ghanaian percussion instruments, multicolored bead necklaces, rain sticks, drums with carvings of elephants, thumb pianos. Outside the Mombasa Marketplace, a stenciled sign reads "Please Respect Our Local Customs," apparently without any intended irony. As late-modern successors to Ota Benga and the human circus attractions of the nineteenth and early twentieth centuries, "cultural ambassadors" from a variety of sub-Saharan African nations (including South Africa, Zimbabwe, Namibia, Botswana, Uganda, and the Republic of the Congo) bang drums throughout the park, and in front of the Animal Kingdom Lodge's palatial Jambo House.[46]

Of course, Disney's theme parks offer particularly low-hanging fruit for any cultural analyst, yet even modest zoos feature exhibits that differ only by degree.[47] Designers of the Maryland Zoo's African Journey have created a primitive decor consisting of rock paintings of elephants and rhinos, while individual exhibits feature "tribal" stripes, totemic symbols, and images of masks and arrows in red, orange, black, and white. The Bronx Zoo's African Plains not only exhibits giraffes and lions, but a thatch-roofed Somba Village and African Market. At Metro Zoo, a soundscape of South African music permeates the giraffe enclosure area, just as many zoos pipe so-called world music into their exhibits to create associations between living animals and "primitive" cultures, no matter how contemporary they—and their music—might be.[48]

Zoos also create caricatures of other non-Western native people, including ancient (and by extension, present-day) Australian aborigines. At the National Aquarium's Animal Planet Australia exhibit, placards depict these indigenous societies through a narrative prism of ecological mysticism. "Ancient Aboriginal culture: It is not known how long the Aboriginal people have lived here, but scientists believe they may date back 60,000 years. The Aboriginal people tell stories of the beginning, known as Dreamtime, when their Creation Ancestors—'the first people'—traveled across the country creating the landforms, plants, animals, and Aboriginal people." Another placard reads, "Aboriginal peoples came to this land tens of thousands of years ago. They too have found ways to exist with these dramatic changes in season, calling the outback home." (The sign also displays a quotation by Aborigine Bill Neidjie of the Bunitj clan: "Walking is good. You follow track. . . . You sleep, wake in the morning to birds. . . . You feel country.") Elsewhere at the Animal Planet Australia exhibit, walls of fabricated stone feature rock paintings of animals, and stick figures that (according to its accompanying signage) "depict men and women hunting and gathering food, significant animals, spirit beings, and sacred ceremonies."

But perhaps the most common indigenous group U.S. zoos use to represent ethnic authenticity and the sacramental bond to Mother Earth among exotic primitives is, not surprisingly, Native Americans. Totem poles and teepees adorn zoos throughout the country, sometimes without even explaining why. The Arizona-Sonora Desert Museum displays a rock and botanical garden with circuitous walkways designed to evoke the mysticism of traditional cultures, particularly American Indian tribes. After explaining how "people from both ancient and modern cultures, around the world and throughout time, have looked to labyrinths as archetypical symbols of journey and spiritual renewal," the exhibit signage draws on southwestern Native American tribal groups such as the Hohokam and Tohono O'odham people to make its case:

> Man in the Maze. This common indigenous design of the American Southwest is a unicursal labyrinth that is equivalent to the classical seven circuit labyrinth. Similar designs dating back to the Hohokam appear on the walls of the Casa Grande ruins, although in those depictions the entry is at the bottom. The "Man in the Maze" is commonly used in the basketry of the Tohono O'odham people of Arizona and Sonora (Mexico). O'odham basket weavers often refer to the design as the floor plan to the house of the creator, I'itoi, also known as Elder Brother. Another common O'odham interpretation views the individual moving along the path on a personal journey, gaining knowledge along the way, and ultimately realizing his destiny.

Here the labyrinth is placed in a local indigenous context that evokes the authenticity and spirituality of Native American peoples.[49] The exhibit signage concludes with a quote about the cosmic mysticism experienced among ancient societies, and their sacred relationship to the natural universe: "Walking the labyrinth is another way of tapping into forces beyond our normal conscious mind. It takes us to some ancient part of ourselves, as old as the turning of the planets and stars, as old as the goddess and earth energies, back when night was dark, when people knew the sky, and nature was a part of us and we of it. This is something lost in our modern world, and the imbalance that it causes cries out for resolution. That's why the labyrinth touches so many people so forcefully."[50]

Through these kinds of exhibits, zoos organize the human species into two ideal-types: the ancient primitive of the Global South, and the more civilized citizen of the developed world, the intended audience for such pageantry. Adorned with plant dyes, bearing animalistic totems, displayed alongside stick-figure art, and quoted in broken English, zoo exhibits essentialize aboriginal peoples as childlike—or animal-like—bound to myth and the supernatural. Like savage beasts, they supposedly lack the unsentimental rationality, advanced technology, and enlightened knowingness wielded by modern science and industry. In *Dominance and Affection*, cultural geographer Yi-Fu Tuan suggests the longevity that this tradition of racism has endured in the West:

"The association of dark skin with animality or childishness is a familiar one in Western culture. The dark-skinned person, as someone barely human, is to be harnessed to toil; or, if young and comely, to be treated as an exotic pet. The dark-skinned person, as a perpetual child, is to be fed and clothed, disciplined and trained to perform menial tasks suited to his mental capacity."

Of course, the reality is much more complicated. While these aforementioned indigenous populations most certainly adopted enchanted beliefs and respectful customs regarding their relationship toward the environment and its flora and fauna, the historical record illustrates the extent to which their technologies—notably their farming and resource-extraction practices—were never as simpleminded as zoo exhibits might suggest. In fact, the descendants of Creation Ancestors cleverly engaged in highly rational, technologically innovative, and productive methods of land management and environmental planning to survive, just as we do today, only with far fewer malignant consequences for the Earth.

For instance, Australian aborigines practiced fish farming by employing elaborate dam and canal systems to trap and contain enormous quantities of fish and shellfish. Their efficient strategies of landscape management included deliberate and controlled forest burning, or "firestick farming."[51] Indigenous African tribes may be caricatured as half-naked simpletons in American zoo exhibits, yet these so-called primitives managed to originate copper smelting in the West African Sahara and the Sahel as far back as 2000 B.C. As Jared Diamond argues in *Gun, Germs, and Steel*, "African smiths discovered how to manufacture steel over two thousand years before the Bessemer furnaces of nineteenth-century Europe and America."[52] As for Native Americans, the aforementioned Hohokam not only enjoyed a rich cultural life, but also built "the most extensive irrigation system in the Americas outside Peru, with hundreds of miles of secondary canals branching off a main canal 12 miles long, 16 feet deep, and 80 feet wide."[53] These indigenous societies demonstrated how civilizations could work the land in ways that are both scientifically rational *and* less environmentally costly than more modern methods of industrial agriculture, mineral mining, fossil fuel extraction, and the overall creative destruction of the Earth. If this ability to marshal natural resources in a sustainable manner is truly "something lost in our modern world," it is hardly irrecoverable. Yet by portraying the Global South as a realm of superstition and sacred ceremony, our zoos fail to recognize their ancient societies' ingenuity and innovativeness in navigating their surrounding environments, the vulnerability of biodiverse habitats experienced throughout regions of the developing world today, and our own society's irrational and foolhardy response to the growing environmental crisis.

## Dinosaur Robots and Ghost Hunters

In his 1985 book *Amusing Ourselves to Death*, NYU media professor Neil Postman argued that we live in a show-business age when the entertainment values promoted by television and mass media—the fifteen-second sound bite, rapid-fire editing, celebrity saturation, and an emphasis on electronic images and visual style over the substance of exposition and the printed word—have migrated to the formerly serious worlds of news, politics, religion, and education. Postman observed that even so-called educational TV fare such as *Sesame Street* celebrates empty-headed frivolity to its young audiences through the use of quick cuts, catchy music, and cute puppets:

> We now know that *Sesame Street* encourages children to love school only if school is like *Sesame Street*. Which is to say, we now know that *Sesame Street* undermines what the traditional idea of schooling represents. . . . This does not mean that *Sesame Street* is not educational. It is, in fact, nothing but educational—in the sense that every television show is educational. Just as reading a book—any kind of book—promotes a particular orientation toward learning, watching a television show does the same. *The Little House on the Prairie, Cheers,* and *The Tonight Show* are as effective as *Sesame Street* in promoting what might be called the television style of learning. And this style of learning is, by its nature, hostile to what has been called book-learning or its handmaiden, school-learning. If we are to blame *Sesame Street* for anything, it is for the pretense that it is any ally of the classroom. That, after all, has been its chief claim on foundation and public money. As a television show, and a good one, *Sesame Street* does not encourage children to love school or anything about school. It encourages them to love television.[54]

It is safe to say that the effects of entertainment on everyday life observed by Postman have only grown more pervasive in our digital era of tweets, six-second videos, and pop-up web ads, just as surely as zoological parks have capitalized on the gee-whiz enticements of mass-media entertainment to define the contemporary zoo experience today.[55] At zoos and aquariums around the country, high-definition animated movie experiences, including 4D adaptations of *SpongeBob Square Pants, Ice Age: Dawn of the Dinosaurs,* and *Happy Feet,* divert audiences from actual educational programs (not to mention *live* animal exhibits) by engulfing viewers in mist, artificial winds, and other wet-and-wild special effects. During my visit to the 2011 AZA annual meetings in Atlanta, I was taken aback by the conference's Exhibit Hall, a jam-packed extravaganza featuring the latest in high-tech wizardry and revenue-generating spectacle for amusement-seeking zoo audiences with challenged attention spans. One traveling exhibition firm, Dinosaurs Unearthed, showcased its twenty-five-foot-tall animatronic *Tyrannosaurus Rex* before the crowd. Its life-size dinosaur robots have appeared not only at amusement parks such as Virginia's Kings Dominion and Pennsylvania's Dorney Park, but also at state-of-the art zoos and museums such Chicago's Brookfield Zoo and Philadelphia's Academy of Natural Sciences. (Please bear in mind that Dinosaurs Unearthed was only one of *two* companies exhibiting animatronic dinosaurs at the AZA conference that year, the other being Billings Productions, whose *Megalosaurus* and *Giganotosaurus* have terrorized guests at the Houston Zoo, Cleveland Metroparks Zoo, and Oregon Zoo.)

Also in the AZA Exhibit Hall, an entertainment company called Hurricane Simulator exhibited its coin-operated vending machines, which invite zoo visitors to pay two dollars to stand underneath a superpowered fan that delivers seventy-eight-miles-per-hour category 1 hurricane force winds along with video and sound effects. Elsewhere at the conference, a representative from Sunrise Productions was on hand to seek out branding opportunities with U.S. zoos. Based in Cape Town, South Africa, Sunrise's successful projects include "an entrepreneurial 3D animated character branding and marketing model for high profile sports brands" such as England's Rugby Football Union and Cricket South Africa; commercials for Samsung, BP, Dell, and Hertz; and the CGI-animated children's series *Jungle Beat,* which has been broadcast in over 180 countries.

Of course, it is certainly possible to make entertainment media not only fun but educational and enriching as well.[56] For years the Philadelphia Zoo has been particularly successful designing pop-cultural attractions that provide genuine educational experiences for audiences. In 2012, the Philly Zoo partnered with Dr. Seuss to create the *Trail of the Lorax* (based on the 1971 Dr. Seuss children's book *The*

*Lorax* and its 2012 animated film adaptation), in which children use 3D glasses to solve puzzles as they learn about the endangered circumstances of the orangutan. They eventually arrive at the Lorax Loft, a "fantastical 3,000 square-foot play and learning environment" with elevated views of the zoo grounds. The year before, the zoo partnered with the Jim Henson Company to produce *X-tink-shun*, in which seven puppet characters entertained visitors while teaching children about the plight of endangered wildlife at "Eco-Stages" spread throughout the zoo. The characters were based on six endangered animals—Leo the Golden Lion Tamarin, Alfreda Cheetah, Iggle the Eaglet, Phibi Frog, the "Douc" Langur, and Igor the Tiger—and one extinct creature, Didi the Dodo. The puppet shows rarely failed to enchant young children (including my son Scott at ages four and five), and at the same time managed to teach them about rare monkeys such as douc langurs and golden lion tamarins, hardly celebrity creatures in the menagerie of American childhood culture (or among adults, for that matter). In 2010, the Philadelphia Zoo launched *Creatures of Habitat*, which showcased giant polar bear and penguin sculptures, mosaics, and other installations built from a total of 259,450 Lego bricks, and encouraged children to collect Creature Keeper cards with important facts about deforestation, global warming, and the amount of DNA humans share with apes. (The exhibit has traveled to zoos around the country, including Zoo Miami, New York's Bronx Zoo, and Salt Lake City's Hogle Zoo.) Meanwhile, 4D adaptations of the masterful BBC nature documentary *Planet Earth*, such as *From Pole to Pole* and *Ice Worlds*, have been viewed at New York's Central Park Zoo, the Houston Zoo, Chicago's Shedd Aquarium, and the Mystic Aquarium on Connecticut's southeastern shore. Zoos have also partnered with National Geographic to bring zoo visitors images of exotic wildlife that enlighten audiences without physically removing these creatures from their natural habitats.

Still, it bears remembering that when zoos partner with media companies they often produce cross-promotional attractions that simply make a mockery of their educational missions. For instance, the year the Philadelphia Zoo debuted its *Trail of the Lorax* exhibition, it also replaced its long-running wildcat documentary footage (shown for years in a three-screen theater as part of its Big Cat Falls pavilion) with a movie trailer for the 2012 animated film *The Lorax* starring Zac Efron and Danny DeVito. In 2010 the zoo attempted a far more embarrassing stunt when it invited the producers of the SyFy network's "reality" show

*Ghost Hunters* to film on zoo grounds for an episode that aired later that year. As reported by *Philadelphia Daily News* columnist Dan Gross, "The Philadelphia Zoo is haunted, say staffers who are glad that they may soon have some understanding of what spirit or spirits could inhabit America's oldest zoo." Producers were specifically invited to investigate "several areas where workers have reported flickering lights or apparitions," according to Gross. Kirsten Wilf, a marketing specialist for the zoo, was quoted as asserting, "We feel very strongly that there is some sort of haunting here, and we're looking forward to finding out what," confirming that Jody McNeil Lewis, vice chair of the zoo's board of directors, first had the idea to contact *Ghost Hunters*.[57]

The episode, which eventually aired on 1 September 2010, includes accounts of unexplained and possibly "paranormal" activity at the zoo as reported on-camera by Lewis as well as Andrew Baker, the zoo's chief operating officer, and Desiree Haneman, a primate keeper:

Baker: So this is the Penrose Building. It has a long history: this started out as a veterinary laboratory—in fact it was where we brought animals for necropsies after they died. I was walking up from outside one night, and as I was walking up to the building (and the lights were on), the lights went out. Walked in; nobody else was here.

Haneman: It was around ten o'clock at night, and I was walking up to the Penrose Building. I looked up into what was the library, and there was a woman who had long blonde hair and then all of the sudden she started to have, like, a white light around her, and as soon as we locked eyes she started to slowly back away from the window—at which point I freaked out and ran down toward the hippos to exit.

Baker: So we're in the Solitude; this is the oldest building on the zoo grounds . . . this dates to 1784 when the house was built. In here, one of our board members [Lewis] was down here with her daughter. . . .
    Lewis: I went all the way back into the second storage area, and on my way back my flashlight beam hit a smoky, misty white figure, and I stopped

and kind of went, "What is that?" and I turned around, and there was nothing there. And I couldn't say anything, because I didn't want to scare my daughter.

❧❧

Baker: The Solitude was built by John Penn, the grandson of William Penn, who was the founder of Pennsylvania. He returned to the U.S. briefly from Great Britain, and built this as his country house. We had a report of a woman coming down this staircase from the second floor in eighteenth-century garb. Quite frequently we'd get reports of the attic light going on or going off when there's nobody in the building, and in fact just today we noticed that that door to the attic was locked, and we're not sure how it got locked. It takes a skeleton key to lock, and we don't have a skeleton key.

This being television, the show's ghost-tracking team naturally marshaled "evidence" of the supernatural to confirm that both Penrose and Solitude were indeed haunted. While Baker, a published scientist who holds a doctorate in zoology, did not necessarily agree with the producers' conclusions, he certainly pulled his punches when remarking upon them at the end of the episode:

Baker: I think the fact that it was our two oldest buildings that had the most [paranormal] activity, that was interesting—but in particular, the combination of knocks, some of the voices, the singing, the music, all of which seemed much harder to explain than some of the other events, were most interesting to me. Of course, I'm trained as a scientist, and so I'll continue to look for other explanations, but there's certainly some interesting stuff, and it gives us a little more richness about what the zoo has here on grounds.[58]

The zoo then took things a step further in October of that year, charging adults forty-five dollars each for a three-hour tour of its "most paranormally active locations." As their announcement promised, *"Hear first-hand about the encounters Zoo staffers have had with unexplained sightings and phenomena, and take a chance of having your own 'experience'! Refreshments will be served."*

One might be forgiven for assuming that zoos should be responsible for educating the public about the difference between science and science fiction. But wait, some might protest: perhaps paranormal phe-nomena are so *obviously* imaginary that few Americans audiences would ever take the zoo's posturing on such matters seriously—all in good fun, as they say. Yet here are some sobering if surprising statistics: according to a 2005 Gallup poll, 32 percent of American adults believe in ghosts, while 19 percent say they are not sure. Meanwhile, 37 percent of Americans believe in haunted houses (like Penrose and Solitude?), while 16 percent say they are not sure. Among young adults aged eigh-teen to twenty-nine, the number of those who believe in ghosts and haunted houses increases to *45 and 56 percent*, respectively. In fact, nearly *three-quarters* (73 percent) of Americans admit to believing in at least one of the following paranormal phenomena: extrasensory percep-tion (ESP), telepathy, clairvoyance, witches, astrology, ghosts, haunted houses, channeling, communication with the dead, or reincarnation. (Then again, perhaps this should not surprise us at all. As recently as 1996, 45 percent of respondents told Gallup that they believed that UFOs have visited Earth, with 12 percent claiming to have actually *seen* a UFO.) Of course, it is hard to imagine that the animal behavior-ists, veterinarians, and other scientists employed by the zoo sincerely believe in parapsychology, including those involved with the produc-tion of the *Ghost Hunters* episode. The fact that they willingly assented to participate in such shenanigans only emphasizes the lengths that zoos must go in order to both provide entertainment to consumers and market themselves to the public, and the toll such efforts take on the legitimacy of zoos as promoters of science education.[59]

As for more reality-based media attractions featured at zoos, such as the BBC's *Planet Earth*, it is important to remember that even the best wildlife documentaries do not necessarily capture the realities of the wild but merely our own cultural prejudices—or, as *New York Times Magazine* writer Jon Mooallem observes in his book *Wild Ones*, "an image of nature that's already lodged in our heads." In fact, more often than not wildlife photography and filmmaking draw on the very same camera tricks exploited by the tripod-wielding zoo photographers dis-cussed earlier in the chapter. According to Mooallem,

> It takes extreme amounts of time, money, patience, and luck to catch that sort of iconic material in the wild, and, understandably, some professionals cut corners. Chris Palmer, a veteran wildlife filmmaker who recently au-thored an exposé of the industry, explains how animals from game farms

are routinely used as stand-ins for wild ones, or jelly beans are hidden inside deer carcasses so that trained bears will tear them apart. . . . Wildlife filmmakers, Palmer told me, are good people and often staunch conservationists, but the pressure on them is agonizing, and the ethical lines are blurry. "It's not that you're evil or malignant or malicious," he said. "You're just trying to get the damn shot so you go home and have dinner with your family. So you put the monkey and the boa constrictor in the same enclosure."[60]

Similarly, naturalist and occasional television star Richard Conniff admits in his book *Swimming with Piranhas at Feeding Time* that he has worn a three-thousand-dollar prosthetic model of his left arm when "attacked" on camera by killer bees and an Australian box jelly, and once filmed a show about fire ants in the rough wilds of a spare bedroom in a Tallahassee apartment.[61] More recently, *Mother Jones* magazine reported that Animal Planet's hit reality show *Call of the Wildman* had done the following over its three highly rated seasons: illegally transported a wallaby across state lines; filmed a zebra drugged with sedatives (to the point of falling over) in violation of federal law; prepared phony animal droppings with Nutella, Snickers bars, and rice; solicited animals from farms or trappers and then placed them on camera to be "rescued" by the show's host; portrayed a wild male raccoon as a mother with cubs (three baby raccoons that the crew essentially held hostage from caregivers and then nearly killed); and planted both Mexican free-tailed bats to be flushed out of a Houston beauty salon and venomous cottonmouth snakes to be fished out of a public swimming pool in Danville, Kentucky.[62] Unfortunately, such stunts are not limited to television, as Animal Planet has promoted its video content on interactive ZOOTUBE kiosks located at AZA-accredited zoos around the country, including Zoo Atlanta, the Houston Zoo, the Saint Louis Zoo, Omaha's Henry Doorly Zoo, and, again, the Philadelphia Zoo.

## Chuck E. Cheese with Animals

Notably, the aforementioned zoos in Atlanta, Saint Louis, and Philadelphia are among the most respected in the country, and yet even these industry leaders must meet audience demands for mass-media amusements in a competitive entertainment marketplace. During my stint at

Metro Zoo, I witnessed firsthand how a zoo faced with this challenge can quickly devolve into a circus environment within a shockingly brief period of time.

When I first began working as a docent at Metro Zoo, the zoo's director, Greg Gabon, had previously spent his career working on wildlife conservation issues for a host of zoos and esteemed environmental groups, including Conservation International and the World Wildlife Fund. While I was conducting my research he resigned from the zoo to continue that work, but in truth the zoo had experienced numerous financial difficulties under his administration, and had even been threatened with closure until a raft of emergency donations arrived that kept the zoo afloat. (It was saving money on staffing in the winter by restricting visiting hours to the weekends.) Still, he had accomplished quite a lot at the zoo, including successfully preparing for its eventual reaccreditation from the AZA and supporting a thriving education department and volunteer program.

Given the zoo's near economic collapse, the board decided to replace Gabon with Bob Banner, a local entrepreneur with several successful businesses under his belt, in the hopes that he could bring the zoo back from economic ruin. In his first year as director, he had been enormously successful in reviving Metro Zoo's bottom line, and shortly after his hiring the zoo began making lots of physical improvements to the grounds, increasing its membership rolls, and maximizing customer spending on concessions and amenities. Certainly, this new influx of revenue allowed for capital improvements to some of the zoo's well-worn animal enclosures and exhibits, if such expenditures also competed with renovations to picnic facilities and other recreational areas.

But then a number of changes began happening, gradually at first, and then very quickly. First, the zoo decided to eliminate its volunteer coordinator position, and Daphne was unceremoniously let go after a decade of service. Soon the docents would no longer be allowed to handle animals, and their new duties would include crowd control and serving pizza to guests during fundraisers. Eventually they would no longer even be called docents.

The zoo began hosting more special events, and while such occasions had typically revolved around environmental education in the past, such as overnight campouts and enrichment programs, the zoo organized many of its newer special events around celebrating local professional sports teams and a country music radio station. One Saturday during the Thanksgiving holiday season, the Berenstain Bears visited the zoo to help a local federal credit union celebrate Financial Literacy Day: carnival attractions included Bingo, Plinko, a prize wheel, and a Cash Cube money machine. April brought to Metro Zoo two weekends of Easter egg hunts and brunch buffets featuring that celebrated zoo animal, the Easter Bunny. One early February, Woody, the zoo's resident woodchuck, bit a member of the education staff on the finger. Typically after such incidents, zoos place their animals on quarantine as a health precaution. However, Woody was scheduled to appear as the star attraction for the next morning's Groundhog Day celebration at the zoo, so personnel waited until after the event to properly sequester the furry rodent. (During the event Woody happened to see his shadow, thus predicting six more weeks of winter—a forecast that turned out to be surprisingly accurate that year.)

Docents were asked to work increasingly more private events, including a kids' fair open only to the constituents of a local state senator whose chief of staff was a zoo board member. Vendors included the local sheriff's office and the U.S. Department of State (which passed out copies of a coloring and puzzle book titled *Have Fun with the US State Department*). The state senator was on hand to pass out campaign materials thinly veiled as citizens' newsletters, while the state's game commission promoted the region's "excellent small and big game hunting opportunities"—a strange message to give animal-loving zoo visitors, especially children. (I spent most of the fair fielding questions from visitors about their food tickets, while a woman from the sheriff's office posted by the goat and alpaca barn asked me, "Alright, Dave—what do we do about the smell?") Eventually the zoo built an addition to that barn to house an old-time general store featuring black-and-white photographs and antique knickknacks alongside horse-head bottle openers, plush turtles, and other souvenir schlock, and expanded its entertainment offerings to include Elvis impersonators, Christmastime brunches with Santa, an ice skating rink, and an outdoor adventure park featuring a ropes course and treetop zip lines.

After his own resignation shortly after Daphne's dismissal, I asked Tyler what he thought about the zoo's strange transformation. He explained, "It seems very counterintuitive to the mission. I know the zoo is doing excellent financially, and it's basically because we threw the values completely out the window. . . . It would seem to me that the guest services staff would be just as happy working in an amusement park, like Six Flags Great Adventure. I think they would be just as happy working at a theme park. What they are excited about is working at a big entertainment space."

From the very beginnings of my research, I too had become acutely aware of the tone deafness of some (but certainly not all) of Metro Zoo's guest services personnel. My first eye-opening experience with them involved the zoo's annual Halloween event. They were running a haunted Zookeeper's Adventure train ride around the bison yard, and at the last minute had enlisted me to hide next to an empty wooden enclosure in the pitch-black dark. Armed with a flashlight, I was told to scream in a blood-curdling cry as the train rolled by, "*Oh, my god! The animals escaped! The animals escaped!*" This might have been a harmless spot of fun if not for the inconvenient fact that three days beforehand in Zanesville, Ohio, private wildlife preserve owner Terry Thompson had freed fifty-six exotic animals from their enclosures in his seventy-three-acre yard before committing suicide. The animals escaped into the town, creating widespread panic. Local schools were cancelled the following day as residents were asked to stay in their homes, while law enforcement could do little but track down and kill an astounding forty-nine of these escaped creatures, including eighteen rare Bengal tigers, seventeen lions, three mountain lions, six black bears, two grizzlies, a baboon, and two wolves. A snow monkey also died, having been assumed devoured by one of the wildcats.[63]

Somewhere between the event's Halloween Corn Maze and the *Harry Potter*–themed Diagon Alley pavilion, Metro Zoo's party planners failed to consider the impact of the Zanesville tragedy on zoo visitors and staff, and that myopia continued to grow in later years. Months after she was let go from the zoo, Daphne and I had a heart-to-heart, and she laid out the stakes for me.

Bob Banner, the general director, originally came in as the finance manager, and then he was put up for general director. And we had discussed, "Where is this going?" because it started with claw machines in the café. You know, the machines where you put the quarter in and the claw goes down and picks up [a toy prize]? And we were like, "This is *not* what we are all about." And then there were the ride-'um toys he wanted to put out in the zoo. And all of a sudden there were [an NFL franchise] days, and football teams were coming in, and we were doing this, and we were doing that, and there were bounce-'ums at the zoo. It was something that we talked about. We did not want to turn into Chuck E. Cheese.

Greg Gabon was a more of an old-school zoo guy—you know, conservation, conservation, conservation. And Bob is from another side. And Bob loves the zoo; he loves the animals. But conservation is not his main goal, and rightfully so, because we reached a point in the zoo's trajectory where we had to make some money. Greg kept us alive, barely. But we were at a point where we had this infusion of funds from one large donor, and Bob had a chance to really improve a lot of stuff in the zoo, and it was time to make money.

You can do that without losing sight of the conservation mission. I think with Bob, we were going to have an event every weekend, and we were going to have fun things here, like the moon bounce. We have the moon bounce *all the time*. The moon bounce is *here*! The moon bounce is *there*! What the heck? And I think for those of us who are a little more purist, we were losing sight of what we wanted to do. We cut out all of the conservation donations—that was fine, because we had to survive. Survival is key. But I don't know. It will be interesting to see where it goes from here. . . . But we just didn't want to lose sight of the whole thing, you know, conservation and education. Without it, close the doors. No right to do it. That is what I've always said, and that's what I truly believe. And I think . . . I don't know. It will be interesting to see what he does with the place now.

I just feel like they really don't care about doing anything educational at the zoo at all. I think that what they need right now is to get financially solid, and Bob will do that. But the problem is that he is going to change the whole way the zoo operates. And it's going to be more entertainment programming. And is that a good thing, [even] if it keeps us solvent? I think, you know, you make a deal with the devil, and it's the way *all* zoos are going now.

If they are going to have all this entertainment, then you have to balance that with interpretation and programming for your guests, and they are not doing that. Chuck E. Cheese with animals.

In a digital age of simulated reality, zoos and their audiences together draw on human imagination and cultural myth to construct the natural world as a collective fantasyland inhabited by animated dancing penguins, dinosaur robots, Lego polar bears, and clairvoyant groundhogs. The saturation of the urban entertainment landscape, rife with

IMAX and 3D multiplex cinemas, themed restaurants, brand-name superstores, and downtown shopping plazas, practically requires even accredited American zoos to compete for attention-deficient consumers by enlisting the power of contemporary pop culture from *Ghost Hunters* to *Animal Planet*. Unfortunately the timing of the digital entertainment explosion coincides with our current moment of environmental catastrophe, when zoos should ideally be stirring audiences to action (in keeping with their organizational identity and goals), rather than selling them pizza and circuses.

Therefore, on our next safari stop we will take a critical look at how zoos attempt to fulfill their missions as environmental stewards and educators, with an eye toward recognizing the challenges that zoos face in both confronting the environmental crisis and communicating its urgency to their audiences, especially with regard to the human causes of global warming and climate change. Whether zoos can effectively meet these challenges—and thus serve as cultural and ecological change agents in the Anthropocene—will be determined by their capacity to transform how we think about nature in ways that go beyond conventional strategies of conservation branding and self-congratulation.

Chapter 6

# Simply Nature

## Zoos and the Branding of Conservation

**WHEN THE DEADLY CHYTRID FUNGUS**, or *Batrachochytrium den-drobatidis*, threatened to decimate the Panamanian golden frog population along with other amphibian species in Central America in 2006, keepers and conservationists from the Houston Zoo sprung into action to rescue diseased frogs and treat them in a new state-of-the-art facility at El Nispero Zoo in El Valle, fifty miles southwest of Panama City. As Mathew Fisher of the Department of Infectious Disease Epidemiology at Imperial College London told *National Geographic*, "The amphibian tree of life is being severely pruned. . . . You can't overstate how serious this pathogen is—it is the worst infectious disease ever recorded among vertebrates." However, when it became clear that the El Nispero Zoo facility could not be completed in time to receive the frogs, the Houston Zoo put up three hundred amphibians in style at the Hotel Campestre, where the frogs were "treated to daily cage cleaning and hand-captured insect meals." Moreover, the Houston Zoo was not the only hero in Panama that year. It was reported that researchers from Zoo Atlanta "packed hundreds of the frogs into suitcases and flew them to safety" back in Georgia.[1]

Obviously, most endangered animals will never enjoy zoo-funded hotel luxuries and amenities. Yet over the past few decades, zoos and aquariums have made dramatic moves toward serving as institutional pillars of international conservation, habitat preservation, ecological research, and environmental awareness. Biologists, animal curators, and keepers at the Philadelphia Zoo participate in conservation projects around the globe, working with endangered lions at the Ongava Wildlife Reserve in Namibia, polar bears in Manitoba, Canada, and spiny giant frogs in Haiti. In 2011, more than 180 AZA-accredited institutions and certified related facilities spent $160 million on over 2,670 conservation initiatives in more than one hundred countries.[2]

Still, zoos face a number of challenges transitioning into effective centers of conservation and ecological awareness. Conservation programs can be prohibitively costly to maintain for all except the wealthiest zoos. Even highly capitalized zoos may not be as efficacious as they would like, given both the severity of the environmental crisis at hand and the difficulties inherent in successfully implementing zoo-directed conservation strategies such as global breeding programs. Zoos and aquariums have the potential to provide enormous leadership in educating the public about paramount issues such as habitat destruction, toxic pollution, ocean acidification, biodiversity loss, species extinction, global warming, and climate change—yet even our best zoos require consistent revenue and donor streams, which may limit zoos in their messaging, lest they alienate visitors or corporate benefactors. In certain cases, the environmental lessons communicated by zoos may be so diluted that they may actually do more harm than good. On this leg of our safari we examine both the efforts taken by zoos and aquariums to fulfill their collective mission of species and habitat conservation, and the challenges they face in meeting the goals implied by that mission, despite their best intentions and efforts. Given the alarming warning signs surrounding the perils of human-induced global warming and climate change, let us appraise the capabilities and hurdles zoos face in moving beyond the superficial branding of conservation to becoming dynamic leaders of change in the Anthropocene.

## Going Green

The transition made by zoos from simple animal parks to self-described "conservation centers" can best be explained as an outcome of two watershed moments, one a response to exterior forces outside the zoo industry and the other to a set of changes occurring within the zoo world itself. The American conservation movement dates back to the mid-nineteenth century, as most famously signified by the enthusiastic reception of celebrated works such as the naturalist-artist John James Audubon's beautiful life-size prints of *Birds of America* (1827–38), Henry David Thoreau's *Walden* (1854), and Charles Darwin's *The Origin of Species* (1859); the establishment of the U.S. Department of the Interior (1849) and Yellowstone National Park (1872); and, of course, the emergence of the first zoological societies and their accompanying gardens and parks in the United States—the Philadelphia Zoo (1859), New York's Central Park Zoo (1861), Chicago's Lincoln Park Zoo (1868), Cincinnati Zoo (1873), Buffalo Zoo (1875), and Baltimore Zoo (1876), with the National Zoo in Washington (1889) and New York's Bronx Zoo (1899) soon to follow.[3] The turn of the twentieth century brought about the birth of the Progressive Era conservation movement led in large part by President Theodore Roosevelt, along with Sierra Club founder John Muir. At a 1908 White House conference on conservation, Roosevelt promoted the national adoption of the conservation ethic associated with his legacy as the country's Wilderness Warrior:

> We have become great in a material sense because of the lavish use of our resources, and we have just reason to be proud of our growth. But the time has come to inquire seriously what will happen when our forests are gone, when the coal, the iron, the oil, and the gas are exhausted, when the soils have been still further impoverished and washed into the streams, polluting the rivers, denuding the fields, and obstructing navigation. These questions do not relate only to the next century or to the next generation. One distinguishing characteristic of really civilized men is foresight; we have to, as a nation, exercise foresight for this nation in the future; and if we do not exercise that foresight, dark will be the future! We should exercise foresight now, as the ordinarily prudent man exercises foresight in conserving and wisely using the property which contains the assurance of well-being for himself and his children.[4]

The American conservation movement would continue during the postwar period and beyond. Inspired in part by a series of canonical books warning of the looming environmental crisis, including Aldo Leopold's *A Sand County Almanac* (1949), Rachel Carson's *Silent Spring* (1962), and Paul R. Ehrlich's *The Population Bomb* (1968), American environmentalism blossomed as a number of specific policy and cultural changes emerged during the 1960s and 1970s to reflect the nation's newly revitalized commitment to Roosevelt's conservation ethic. The release of the first photographs of Earth taken from space during the Apollo moon missions helped generate a popular eco-spiritual awakening during this time, as did a greater awareness of the natural (albeit landscaped and entirely man-made) outdoors generated by the suburbanization of the American middle class.[5] Greenpeace was founded in 1969, the Natural Resources Defense Council the following year. Wisconsin senator Gaylord Nelson called for a national teach-in on the environment on 22 April 1970, sparking over twelve thousand events across the country with more than thirty-five thousand speakers, thus inaugurating the first annual Earth Day.[6] That year President Richard M. Nixon signed the amended and strengthened Clean Air Act, and established both the National Oceanic and Atmospheric Administration and the Environmental Protection Agency. During the early 1970s Nixon also approved the Clean Water Act, Federal Environmental Pesticide Control Act, Ocean Dumping Act, Safe Drinking Water Act, and, perhaps most relevant to our discussion here, the Animal Welfare Act and the Endangered Species Act, the latter designed to allow greater federal involvement in habitat preservation and threatened species survival.

Yet during this same time, American zoos faced a real challenge to their own survival as well. A snare of federal and international laws already limited zoos' access to wild animals. For example, in 1972 the Marine Mammal Protection Act restricted the commercial trading of polar bears, walruses, sea lions, seals, whales, and dolphins, and in 1973 the U.S. government signed the multilateral Convention on International Trade in Endangered Species of Fauna and Flora, which protects more than thirty-five thousand species from global foreign trade. Additional proposed regulations (and potential loss of supplemental federal funds) threatened to further constrain the ability of zoos to successfully operate. Meanwhile, public awareness concerning the plight of endangered wildlife grew alongside criticism questioning the welfare and care of captive zoo animals, and the social value of zoos more generally.[7]

Therefore, during the 1970s the American Association of Zoological Parks and Aquariums, the zoo industry's professional association at that time, began to publicly champion zoo-directed captive breeding programs as a conservation strategy. In a world of shrinking biodiversity and habitat loss (as well as dwindling opportunities to legally take animals from the wild), zoos could increase their collections by selectively breeding their own endangered animal stocks, and then claim they were promoting the conservation of threatened species. This clever organizational adaptation promoted not only the preservation of wildlife and the environment, but zoos' *own* self-preservation as well, by heading off public criticism and the threat of heightened federal encroachment.[8]

Today AZA-accredited zoos cooperatively manage the reproduction of their most endangered animals according to centralized Species Survival Plans (SSPs) determined by their corresponding Taxon Advisory Groups (TAGs) within the AZA. These TAGs operate as if they were matchmaking or dating websites for sand tiger sharks or southern ground hornbills—they not only decide which captive animals may reproduce within each participating AZA zoo, but also *with whom* they may selectively breed, particularly since their would-be paramours often reside in other zoos. They accomplish this by devising a "studbook" for each species that, according to the AZA, prevents inbreeding and thus "ensures the sustainability of a healthy, genetically diverse, and demographically varied AZA population." As of 2014 there were forty-six TAGs responsible for managing approximately 550 SSPs for endangered creatures that include the chestnut-mandibled toucan, the speckled mousebird, and the southern hairy-nosed wombat.[9] In recent decades, zoo-directed breeding programs have brought numerous animals back from the brink of extinction, including the Brazilian golden lion tamarin, the Arabian oryx, and the California condor.[10]

Zoo-directed breeding takes place not only in animals' private quarters, but also in the laboratory. Carrie Friese, a sociologist at the London School of Economics, describes how San Diego Zoo Global's research center, Conservation and Research for Endangered Species, keeps a collection of cultured fibroblast cells, blood and tissue specimens, semen and ova, embryos, and DNA from more than 4,300 wild species, all cryogenically preserved in large freezers. (San Diego Zoo Global is the umbrella organization for the San Diego Zoo, San Diego Zoo Safari Park, and San Diego Zoo Institute for Conservation Research.) The facility even has a trademarked name, The Frozen Zoo®. Thus far San Diego Global has artificially inseminated a Chinese pheasant, a red diamond rattlesnake, and a southern white rhinoceros with cryopreserved sperm, all successfully. Friese also explains that a small number of U.S. zoos, including the San Diego Zoo, have experimented with somatic cell transfer technologies to essentially *clone* endangered species. First, scientists add the mitochondrial DNA from a more common domestic animal to the nuclear DNA from an endangered species to create viable embryos. Then they inseminate those embryos into a surrogate mother of that same domestic species. Zoos consider the cloned animals that the mother produces interspecies "chimeras," or "heteroplasmic" animals. For example, in collaboration with two biotechnology firms and some ordinary domesticated cows, the San Diego Zoo successfully cloned two different endangered species of southeast Asian wild cattle: the guar, or Indian bison, and the banteng.[11]

Of course, zoo-directed captive breeding programs may not always be effective in repopulating endangered species in the wild. For instance, in 2013 the International Union for Conservation of Nature (IUCN) listed 11,212 threatened animal species on their Red List as Critically Endangered, Endangered, or Vulnerable. However, as noted above, the AZA manages SSPs for only around 550 animal species, largely because the zoo industry invests far more of its resources protecting endangered *zoo* animals than it does the vast number of other species threatened in the wild. For example, despite their vital importance to marine life and the oceans, the 235 threatened species of coral listed by the IUCN are simply not going to be prioritized by zoos to the same extent as charismatic megafauna such as orangutans, gorillas, elephants, or beluga whales. (Nor are the 2,110 threatened species of fish or 1,898 species of mollusk.)[12]

In fact, the hard truth is that the implementation of selective animal breeding programs (as well as their most immediate consequence—the need to house and care for a growing zoo population) typically requires enormous financial and spatial resources, and so zoos must often make difficult decisions when considering which of their *own* resident species to rehabilitate. For instance, in recent years the Saint Louis Zoo made a conscious decision to give up on trying to save from possible extinction the Mhorr gazelle and lion-tailed macaque, two species long included in the zoo's collection.[13]

Moreover, among those animals traditionally kept in captivity in American zoos, selective breeding can be tricky work, particularly for certain species. Cheetahs are notoriously difficult to breed in captivity due to their complex mating and social behaviors, need for sufficient space to roam, susceptibility to infectious disease, and high selectivity of sexual partners observed among females. (For these reasons, fewer than 20 percent of cheetahs residing in North Americans zoos have successfully bred in captivity.)[14] There is also evidence that juvenile bull Asian elephants learn to mount females only by observing other pachyderms in social groups, and therefore may face difficulties mating while under lock and key.[15]

Other animals are incompetent at mating in captivity *and* in the wild. The female giant panda is sexually receptive only once a year, and often within a brief window of up to seventy-two hours. Meanwhile, according to one zoo biology textbook, pandas "exhibit poor copulatory positioning," as colorfully described by *New Yorker* staff writer David Owen in his 2013 reporting from the Smithsonian National Zoo: "Tian Tian and Mei Xiang are simply 'reproductively incompetent.' A key difficulty is that Mei Xiang places herself in what [the head of the Center for Species Survival] called 'pancake position'—flat on her stomach, legs outstretched—and Tian Tian isn't assertive enough to lift her off the ground. Rather than mounting from behind or pulling her toward his lap, he steps onto her back and stands there like a man who has just opened a large box from Ikea and has no idea what to do next."[16]

As a result, giant panda handlers in China and elsewhere have resorted to giving males Viagra, and showing the animals potentially instructive video footage of other pandas mating—panda porn, if you will. (One zoo similarly tried to teach its male gorilla appropriate courtship and copulation techniques through the power of cinema, but he wound up killing his would-be mate instead.) At other zoos, keepers assist reproductively challenged male mammals with a variety of other biotechnologies, from the use of an artificial vagina to electro-ejaculation, which is exactly what it sounds like.[17]

## Into the Wild

Admittedly, captive breeding programs have a limited impact on the conservation of wild species living in threatened habitats outside the zoo. But during the 1990s, as concerns for vulnerable wild populations grew not only among environmental activists but zoologists, biologists, veterinarians, ecologists, and other scientists working within the zoo industry, zoos and aquariums began leading field conservation efforts to protect endangered animals stranded in disappearing wild habitats around the world. (Signaling a new direction in the developing institutional ambitions of zoos, in 1993 the New York Zoological Society officially changed its name to the Wildlife Conservation Society, or WCS.) This second watershed moment in recent zoo history was nicely articulated in a 1995 manifesto published in the *International Zoo Year-book*, "Beyond Noah's Ark." Penned by then-AZA conservation and science director Michael Hutchins and WCS director William G. Conway, the piece called upon zoos to take up the mantle of in situ conservation in the Earth's most precarious ecological hotspots:

> The zoological community needs to develop an effective track record and to build its reputation within the global conservation community. Too often it has been viewed as an exploiter, rather than a protector, of wild-life. . . . The next few decades will be critical for the survival of many species, especially the larger vertebrates. Historically, zoos and aquariums have seen their primary missions as educating the visiting public and captive breeding for reintroduction. However, in continuing along this traditional path, they run the risk of losing their relevance in a rapidly changing world. Conversely, if zoos and aquariums successfully expand their *raison d'être*, they have an opportunity to become one of the world's largest and most effective conservation networks. Independently zoos and aquariums could not achieve all that needs to be done but should they choose to adopt a broader approach to their conservation activities, they can make a substantial contribution. As [Wildlife Conservation International's David] Western so aptly put it, ". . . saving a species means more than saving its genome; it means maintaining the ecosystem in which a species maintains itself, competes, adapts, or is otherwise exterminated."[18]

Some of this in situ conservation involved sending research teams to global outposts around the world, as when the Philadelphia Zoo partnered with local groups to study vulnerable frog populations in their natural habitats in southeastern Haiti and northwestern Dominican Republic. San Diego Zoo Global manages six field stations in the United States and around the world, including the Cocha Cashu Biological Station in the Peruvian Amazon rainforest, the Hawaii Endangered Bird Conservation Program, the Condor Field Station in central Baja in Mexico, and Cameroon's Ebo Forest Research Stations. Other zoos began assisting with global captive breeding programs and other ex situ interventions abroad, as when the Wildlife Conservation Society sent veterinarians to Grand Cayman to perform health assessments on a captive-reared collection of highly endangered blue iguanas before their Caribbean nonprofit partner, the Blue Iguana Recovery Program, released them into the wild. Still others run Quarters for Conservation programs, for which zoos such as the Knoxville Zoo, Oakland Zoo, Cleveland Metroparks Zoo, and Zoo Boise have collectively raised over one million dollars annually by simply asking visitors for twenty-five-cent donations to fund in situ conservation projects.[19]

In 2012, AZA-accredited zoos collaborated on global conservation projects in 115 countries. However, most zoo-directed field conservation efforts focus on habitat destruction, biodiversity loss, and species endangerment closer to home.[20] The Minnesota Zoo tracks and analyzes movement and activity data from over fifty wild moose in response to a species population drop in the northeast corner of the state.[21] Binder Park Zoo in Battle Creek, Michigan, partners with other local organizations to reintroduce trumpeter swans back into Michigan, monitor Karner blue butterfly populations in the state, and hand-rear endangered piping plovers on Lake Michigan's shores.[22] In 2010 AZA institutions around the country, including the Smithsonian National Zoo and the New England Aquarium, sent animal food, cleaning supplies, vehicles, and teams of experienced personnel to Louisiana, Texas, and other U.S. states along the Gulf of Mexico to rescue and rehabilitate soiled green sea turtles, manatees, whopping cranes, pelicans, and other species impacted by the Deepwater Horizon oil spill, the largest accidental marine petroleum spill in American history.[23]

Has zoo-directed conservation been a success? It depends. Although in 2012 species-specific conservation efforts targeted 650 individual species, 60 percent focused on mammals, particularly charismatic megafauna such as the cheetah, elephant, rhinoceros, chimpanzee, snow leopard, orangutan, giant panda, Grevy's zebra, jaguar, polar bear, and okapi. Meanwhile, only 6 percent focused on saving vulnerable amphibians like the Panamanian golden frog; 4.5 percent emphasized rescuing invertebrates, including corals and butterflies; and 3.5 percent targeted fish.[24]

The bigger issue concerns the scarcity of resources among U.S. zoos. Since budgetary pressures constrain most zoos, there is only so much in situ conservation that even zoos with the best of intentions can afford. For this reason, the bulk of the money spent on zoo-directed conservation efforts comes from the wealthiest institutions in the country, many of which have been mentioned in the preceding pages, including the Houston Zoo, San Diego Zoo Global, Zoo Atlanta, and the Smithsonian National Zoo.[25] In fact, while in 2010 AZA zoos committed a total of $110 million to conservation projects, including those involving direct action, research, or advocacy, $79 million (over 70 percent) came from a *single source*, the Wildlife Conservation Society.[26] This is typical for zoological organizations with colossal resources. Since 1995, the conservation wing of two Florida zoos has spent $27 million funding wildlife conservation projects from Argentina to Uganda. Of course, it helps that the two zoos are Disney's Animal Kingdom and the Seas with Nemo and Friends at EPCOT, both of which help support the Disney Worldwide Conservation Fund. While many nonprofit zoos barely survive on ultra-tight budgets and meager municipal funding, the Disney Corporation enjoyed revenues over $48.8 billion in 2014.

While well-intentioned AZA zoos may contribute what they can to support conservation programs around the world, unfortunately few have the resources to support in situ projects abroad or here in the United States in a way that might be sustainable in the long term. The problem is largely one of institutional constraint. Despite the Bronx Zoo's aforementioned success in saving the American bison from extinction in the early twentieth century, zoos are primarily designed as entertainment attractions for the display of animals, not the organization of monumentally complex global interventions in the wake of the largest environmental and biological catastrophe in human history.

However, even small local zoos can *educate* audiences about conservation-related issues, especially the dangers of habitat depletion, species extinction, and biodiversity loss, and what kinds of structural and behavioral transformations we will have to undertake as a society to have a real and positive impact on the future on the planet. As Daphne from Metro Zoo reminded me, "The big zoos will still be able to fund conservation. Little zoos? Not as much. That's why you have to do education, because that's all you can afford. So at least do that—make yourself worthy of putting animals in cages. If you stick something in a cage, there better be a damn good reason for it."

## Green Roofs and Wild Backyards

In recent years a number of American metropolitan zoos have met this challenge by devising smart exhibitions to educate the public on conservation and environmental issues in both serious and creative ways. We have already discussed one example in chapter 2: the emerging attention some zoos give to insects and other arthropods. Interactive programming revolving around insects helps educators teach zoo visitors about the astounding variety of species living on Earth. While there are about 4,000 known species of mammal in the world, there are over 751,000 species of insect that have been identified by scientists. (Of those insect species, about 290,000, or nearly 40 percent, are beetles.) The display of insects also allows for teaching moments about how ecosystems work, the interconnectivity of life on the planet, and the value of seemingly insignificant bugs and other creepy-crawly critters—and by extension, the importance of protecting a wide range of endangered animals to fight against species extinction and biodiversity loss. According to biologist (and noted entomologist) Edward O. Wilson,

> A large fraction of the plant species depend on insects for pollination and reproduction. Ultimately they owe them their very lives, because insects turn the soil around their roots and decompose dead tissue into the nutrients required for continued growth.
>
> So important are insects and other land-dwelling arthropods that if all were to disappear, humanity probably could not last more than a few months. Most of the amphibians, reptiles, birds, and mammals would crash to extinction about the same time. Next would go the bulk of the flowering plants and with them the physical structure of most forests and other terrestrial habitats of the world. The land surface would literally rot. As dead vegetation piled up and dried out, closing the channels of the nutrient cycles, other complex forms of vegetation would die off, and with them all but a few remnants of the land vertebrates. The free-living fungi, after enjoying a population explosion of stupendous proportions, would decline precipitously, and most species would perish. The land would return to approximately its condition in early Paleozoic times, covered by mats of recumbent wind-pollinated vegetation, sprinkled with clumps of small trees and bushes here and there, largely devoid of animal life.[27]

In addition to the exhibition of insects and other living symbols of the planet's biodiversity and networked ecology, zoos have also recently begun emphasizing the habitats in which species coevolve and adapt. The prime cause of species endangerment and biodiversity loss is habitat degradation, whether the deforestation of the Amazon, the rapid thawing of the Greenland ice sheet, or the acidification of South Pacific coral reef waters. Whether embedded in the Earth's soil or seawater, ecological habitats represent complex sets of symbiotic relationships among vertebrates and invertebrates, plants and fungi, bacteria and protists. In order to get audiences to redefine *species* conservation as the preservation and rehabilitation of natural *habitats*, some zoos now organize fauna and flora collections according to biogeography, climatic zone, or habitat type. Metro Zoo exhibits some of its northern South American creatures—golden lion tamarins, red-footed tortoises, green iguanas, and prehensile-tailed porcupines—in a mixed-species exhibit. Zoo America is divided into five distinct North American regional habitats: Southern Swamps, Northlands, the Great Southwest, Eastern Woodlands, and Big Sky Country. In some zoo habitats, multiple animal and plant species from the same region coexist within the very same enclosure. A notable example is the Arizona Sonoran Desert and its "Cactus Community" in the Great Southwest area of the zoo. The enclosure features regional plants like giant saguaro and cholla cacti as well as animals such as the burrowing owl, Gambel's quail, and roadrunner. To the left of the Cactus Community is the Desert Lizard Community, in which a Gila monster, a collared lizard, and a desert spiny lizard all cohabitate. Further inside the pavilion are nocturnal desert creatures, all displayed in dim lighting.

This marks a shift away from more traditional zoological classificatory strategies, particularly the classical Linnean taxonomy in which animals are categorized by phylum, genus, and species, but not by the wild habitats that sustain them.[28] Nowadays, many zoos present collections catalogued according to both rubrics. The San Diego Zoo's mixed-species habitats organized by biogeography, including the Northern Frontier, Asian Passage, Lost Forest, and Africa Rocks, thrive alongside species-specific displays set aside for the zoo's most charismatic megafauna, notably its elephants and giant panda bears. Still, this is a tremendous step forward for American zoos. As Catherine Brinkley, the zoo consultant, observed during our tour of the Philadelphia Zoo,

> If you want visitors to learn about where animals come from and how they are grouped together, then geographic organization is a better learning experience for them. I understand that it can be better to house all of the reptiles together for sanitation and feeding purposes, but the best exhibits I've seen have been those that managed to put the African snakes near the African lions. There have been some innovative ways of doing that. Otherwise, people walk away confused, thinking that tigers and lions live in the same part of the world, or that all snakes live in the same part of the world.

It depends on what you see as the mission of the zoo. Each zoo has their own mission—but for me, the zoo is a public place for people to learn about an ecosystem. If you can present that ecosystem more cohesively, then you've done a better job of educating people about the environment, and how animals interact with each other, and why they're all important.

Of course, zoos house and exhibit animals together that live in similar bioclimatic zones not only because it contributes to a smarter narrative about wild habitats, but also because it is often logistically efficient to do so. For example, New York's WCS arranges its Central Park Zoo into three climate zones: Polar Circle, Temperate Territory, and Tropic Zone. The north side of the park hosts the Polar Circle, which features indoor penguin and puffin exhibits (and formerly an outdoor polar bear habitat), while the Tropic Zone invites audiences to walk through a fully immersive indoor rainforest brimming with exotic birds, black-and-white ruffed lemurs, emerald tree boas, poison dart frogs, and even piranha. While this configuration organizes the Central Park Zoo's animal collection according to bioclimatic conditions and habitat type, it also productively houses the zoo's creatures according to the amount of heat or cold air they need to survive.

On the other hand, this schema obscures the geographical specificity of wild habitats and ecological niches. Habitats that are equal distances (north or south) from the equator across the globe share similar climate patterns, such as the tropics or polar icecaps. However, while both penguins and polar bears feed on fish in icy waters among (melting) glaciers, they would *never* be seen together in nature, since wild penguins live exclusively in the southern hemisphere while polar bears live in the northern Arctic Circle. Similarly, the Central Park Zoo's Tropic Zone throws together scarlet ibises and keel-billed toucans from Central and South America; long-tailed hornbills and magpie shrikes from Africa; and Victoria crowned pigeons and fairy bluebirds from the Asia-Pacific. To the zoo's credit, signage explains these crucial inconsistencies, but the overall immersive experience of these polar and rainforest exhibits tends to mute such ecological realities nevertheless.

Other zoos emphasize more local or regional wildlife habitats, which show audiences how even their own urban neighborhoods or metropolitan areas might be conceived as ecological environments deserving of preservation and care. Zoo America reminds visitors which of its animals live in its home state of Pennsylvania. Signage explains that four hundred American elk live in north-central counties of the state, and that the snowy owl will migrate as far southward as Pennsylvania in cases of severe weather or a lemming population crash. The zoo's website also highlights its efforts in local conservation efforts such as building nest boxes for American kestrels and distributing them throughout the region, and breeding and releasing peregrine falcons in the state.

Elsewhere in Pennsylvania, the Philadelphia Zoo has an exhibit, My Wild Backyard, that teaches children to appreciate the ecosystems in which they are enmeshed by emphasizing how even man-made outdoor spaces provide habitat for plants and animals that share ecological relationships with one another, and to us. The exhibit features thick gardens enveloped in mulch and forked paths where kids can roam, and signs promise, "If you provide a good mix of food, water, shelter and secluded nooks for raising babies, even your smallest garden could attract wildlife." Signage also urges visitors to plant mixtures of trees, shrubs, and plants in their own gardens to offer a variety of foods to wildlife all year long (especially in winter when food may otherwise be scarce); provide water sources in which wildlife can bathe and keep cool; compost leaves and grass clippings; only use animal-friendly fertilizers, and never pesticides; and cultivate native plants "to attract local critters."[29] The exhibit features local varietals such as wrinkle-leaf goldenrod, smooth witherod, sweet pepperbush, red chokeberry, river birch, sweetbay magnolia, and sparkleberry holly. According to Brinkley,

> My Wild Backyard is a really great exhibit. . . . It gives people an idea of the sorts of plants they can plant in their own backyard to create a biodiverse habitat, so it's an actionable take-home message, which is really important in American zoos, because we tend to show exotic animals that have nothing to with our own backyards. . . . The take-home message in the U.S. is, "Save the rainforest," or "Donate to conservation in Africa," and there is very little about what to do in our own backyard.

As mentioned earlier in the book, the Philadelphia Zoo complements My Wild Backward by displaying magpie, frillback, and fantail domestic pigeon breeds "that fit right into the ledges, nooks, and crannies of city buildings."[30] The zoo's McNeil Avian Center features a multiscreen 4D film about an oriole named Otis who lives in the city's Fairmount Park, and signage that highlights bird species living in the surrounding regional ecosystem of the Delaware Valley, including the American

robin, blue jay, downy woodpecker, northern mockingbird, and northern cardinal:

> Local birds. Meet your avian neighbors. All of the birds pictured here can
> be found year-round in Philadelphia and its suburbs. Many other kinds of
> birds can be seen at nearby sanctuaries and preserves, including hawks,
> shorebirds, and songbirds that migrate thousands of miles each year from
> Pennsylvania to Central and South America and back again. Check out the
> birds on your street or in your backyard. They're beautiful and fascinating
> and they're available for all to watch, just outside our doors!
>
> Many local birds are threatened by habitat loss and other human activi-
> ties. They need our help. . . . Transforming your yard into a naturally bird-
> friendly habitat is easy, fun, and rewarding.[31]

Zoos also teach visitors about conservation by showcasing the environmental sustainability of their own infrastructure, providing models
for how similarly large institutions might achieve a diminished carbon
footprint (and decreased expenditures) by investing in renewable water
and energy systems. At the Philadelphia Zoo, informational placards
explain how its children's zoo exhibits make use of geothermal wells to
control building temperatures, reducing energy use for heating and air
conditioning. Green roofs and gardens absorb rainwater and decrease
runoff, reducing the need for energy-intensive water treatment, while
underground cisterns collect rain that can be recycled as graywater for
flushing toilets in the zoo's restrooms. As for flushing urinals, a sign
strategically placed in one of the men's restrooms reads, "Saving water:
Waterless urinals. The waterless urinals used throughout the Zoo save
over one million gallons of water each year. That's enough to fill your
bathtub 20,000 times!"

The Philadelphia Zoo's newest buildings are also built with energy-
conserving construction supplies such as recycled ceramic tile, coun-
tertops, lumber, and structural steel, as well as green materials fash-
ioned from wheat, sunflower seeds, and cork. Elsewhere on the zoo
campus, many of the zoo's fleet vehicles run on biodiesel fuel. The
Philadelphia Zoo collaborates with local partners to increase its on-site
use of clean-energy alternatives like solar power. Food concession stations distribute biodegradable and compostable tableware made from
bioplastics and containers made from plant-based resins, and recover
waste oil from their kitchens to heat the zoo's animal buildings. These

eco-smart practices not only reduce the zoo's overall environmental impact but also, when publicized, can serve as a model for how to scale up sustainability practices for institutional and even citywide use.[32]

# It's the End of the World
# as We Know It, and I Feel Fine

Sadly, even among the most well-intentioned zoos, messages about conservation directed toward audiences often fall on deaf ears, as was explained to me during a visit to a major zoo in the Midwest, where I had the opportunity to sit down with Barbara, the zoo's vice president for public affairs. "Is there one particular message or a set of particular messages that you wish better resonated with visitors?" I asked her.

"I think it's the importance of conservation and science, and our impact around the world—that's a hard message to convey," said Barbara.

"Why do you think that is?" I asked.

"Because I don't think it's as important as coming to look at an animal for the most part, for the typical visitor. I think that it has been shown, with the research on signs and messages that are in front of people, that they don't read signs. It's not just here—that's something that's happening throughout the country. Reading signs about conservation, and that message, doesn't seem to be appealing."

"Is that a point of frustration?

"Yeah, I think so. I think that's something that we're really trying at every opportunity to make a point of, but it's definitely challenging. And when you look at perceptions from the research, fun and entertainment is number one. '*I'm glad the zoo does conservation, and I'm glad to be educated while I'm here, but that's not why I'm coming to the zoo.*' That's a visitor impression. The members, on the other hand, say that conservation is the reason that they belong—it's just hard when they don't see it. We can talk about the researchers out in Africa, but if they're not seeing it in front of them, they're just not paying attention."

"What do you mean?" I ask.

"I mean, seeing a researcher do work. I think this is something that [other] institutions do well, because they have a lot of visuals or behind-the-scenes things that visitors can watch, like at [the local natural history museum] you can see researchers working on fossils, dead bones. Here, there's not really an opportunity to *see something* that is a conservation message. You can see the animals, and the keepers working with them, but that's conservation of one type, and I just don't think it translates. With enrichment, even though the message is that the zoo provides for the health and well-being of an animal, the visitor experiences it as entertainment—they watch the lion take down a piece of meat because it's entertaining. I think we're doing a lot better at it, because we're consistently getting those messages out where we can. But if they translate or not, I don't know."

Ann, one of the zoo's spokespeople, agreed. "You can't necessarily tell somebody when they're here and looking at the tigers about a conservation message, because they're busy looking at what the tiger is doing, and what's cool at that moment."

Even the most conscientious audiences of metropolitan zoos can be easily distracted by their endless attractions and the cacophony of urban public life. In its Rare Animal Conservation Center, the Philadelphia Zoo used to promote its Cell Phone Recycling Program with a prominent display and a bin where visitors could contribute their old phones.[33] The program supported the Endangered Primate Rescue Center in Vietnam, which protects threatened primates like the douc langur, but when I first peeked inside the bin I found that passersby had dumped their zoo maps, rainforest exhibit handouts, and at least one Hi-C juice box amid the donations. Subsequent visits confirmed that zoo audiences regularly treat these collection bins as garbage cans. (At least they don't just throw their trash on the ground, as revealed in the last chapter.)[34]

Of course, American zoological parks attract a wide range of audiences, and if some visitors merely tune out their conservationist pleas, others actively contest them. For example, Dylan explained to me that on more than one occasion zoo visitors rudely challenged him publicly on the most basic principles of global warming and climate change. I asked Dylan how often this occurred. As he recalled,

It has happened to me twice at City Zoo. I'm surprised that it didn't happen more often, because our interns said that it had happened to them several times in one summer, but only two times can I specifically remember it happening to me. The first time, I was greeting people as they came in, and I was telling them what our conservation exhibits were all about, explaining, "If you go to the penguin exhibit, you can learn about how you can save penguins by choosing the right fish when you go to the grocery store. When you go to the polar bear exhibit, you'll learn about global warming and how you can help save energy." And I didn't get past that when a dad grabbed his little girl (she was three or four years old), picked her up, and said to her, "*You tell this man that there is no such thing as global warming, and it's a load of crap.*" Right in my face, and he was very aggressive when he said it. And then he stormed off. And I was so dumbfounded, and I felt really awkward after that whole encounter.

And then another time, I was at the polar bear exhibit talking about global warming and a guy confronted me. He asked me how many polar bears were left in the wild. I told him it was around 25,000 to 30,000 polar bears. And he said [sarcastically], "*Yeah, that's a lot. Global warming is a load of crap,*" or something along those lines. And then *he* stormed off.[35]

These zoo visitors are not alone in their skepticism of global warming and climate change. Thanks in part to a decades-long campaign (funded in part by the petroleum industry) challenging the scientific consensus surrounding climate change, 42 percent of Americans polled by Gallup in 2014 said they believed that news reports exaggerated the seriousness of global warming, while 40 percent believed that "increases in the Earth's temperature over the last century" were more due to natural causes, rather than human activities like carbon emissions from fossil fuel combustion. Although rising planetary temperatures due to anthropogenic forces have been responsible for recent droughts and food shortages, climbing sea levels, extreme and deadly heat waves, and an increased frequency of high-volatile storms around the world, 45 percent of Americans did not believe that the effects of global warming had in fact already begun (18 percent did not believe they would *ever* happen). Although the consensus among climatologists concerning the realities of global warming is nearly unanimous, 37 percent of Americans remained convinced that most *scientists* were either *unsure* or else *did not believe* in global warming. Meanwhile, only 24 percent of Americans admitted to worrying a "great deal" about climate change (and only 31 percent admitted doing so about the environment more generally), while a majority claimed not to worry about climate change *at all*, or if so then only "a little." In a list of fifteen issues that included drug use, illegal immigration, and the size and power of the federal government, Americans rated climate change second to *last* among their concerns. (The quality of the environment was ranked third to last.)[36]

The international scientific community does not share this skepticism. In its fifth assessment report released in 2013, the authoritative Intergovernmental Panel on Climate Change (IPCC) determined that the "warming of the climate system is unequivocal, and since the 1950s many of the observed changes are unprecedented over decades to millennia. The atmosphere and ocean have warmed, the amounts of snow and ice have diminished, sea level has risen, and the concentrations of greenhouse gases have increased."[37] The IPCC also determined that it is "extremely likely" (with greater than 95 percent confidence) that "human influence has been the dominant cause of the observed warming since the mid-twentieth century."[38] The report reflects current scientific consensus. As reported in the prestigious journal *Science*, Naomi Oreskes, Harvard University professor of the history of science and affiliated professor of earth and planetary sciences, conducted a review of 928 articles published in peer-reviewed scientific journals from 1993 to 2003 with the keyword "global climate change." In a striking demonstration of scientific consensus, *none* of the papers contested the reality of climate change and global warming as promoted by the IPCC and the National Academy of Sciences.[39]

Given their conservation and education goals, informing the public about human-induced global warming and other man-made environmental impacts on biodiversity loss, species extinction, ocean acidification, and habitat devastation could not be a more important mission for American zoos, especially given the skepticism shared among wide swaths of the U.S. population. Yet despite all expressed intentions to the contrary, zoos often present a weak front on global warming, not at all in keeping with the severity of the crisis, or even the wishes of their own keeper staffs and executive board members. Again, as I discussed in chapter 4, zoos must pitch their educational content to a mass audience spanning vast ideological divides and political persuasions. Zoos usually try to avoid alienating these audiences by placing severe limits on the kinds of lessons their educators can promote to the general public, from evolution to global warming.[40] As Paul Boyle, AZA senior vice president for conservation and education, told the *New York Times*, "You don't want [zoo and aquarium visitors] walking away saying, 'I paid to get in, I bought my kid a hot dog, I just want to show my kid a fish—and you are making me feel bad about climate change.'"[41]

# Crisis Without a Cause

Of course, the staff members of zoos and aquariums hardly ignore hot-button issues such as global warming and climate change when on the job. When confronted at that polar bear exhibit by the climate change skeptic who told him that global warming was a "load of crap," Dylan retorted by addressing the surrounding crowd: "I turned to everybody, and said, 'Well, City Zoo recognizes current scientific research that suggests global warming is a serious issue and a concern at this time, and we here take precautions in our daily practices that lessen our impact on carbon emissions and save energy. Furthermore, the Arctic Circle, where polar bears are from, is massive, and 25,000 to 30,000 animals living there is miniscule. It's nothing.'"[42]

As I noted earlier in the book, I got to know Dylan very well by working alongside him at City Zoo, and when it comes to animal protection and environmental conservation more generally, he could not be more earnest. (I would also say the same for the vast majority of zookeepers, educators, and docents I encountered during my research.) But as institutions whose existential survival depends on attracting mass audiences, zoos often downplay the urgency of the environmental crisis for fear of generating unwanted controversy. In fact, this perfectly rational (if shortsighted) business strategy operates at the highest levels of the zoo industry. For example, the AZA's own position statement on climate change actually avoids discussing how human activities contribute to global warming:

> The Association of Zoos and Aquariums (AZA) recognizes that increasing atmospheric carbon dioxide ($CO_2$), as well as other greenhouse gases, is causing changes to the earth's climate. These changes are impacting wildlife in the oceans and on every continent. AZA-accredited zoos and aquariums are trusted institutions through which, yearly, millions of people learn about and value nature and wildlife. By communicating about the impacts of climate change on wildlife and habitats, AZA and its member institutions can play an important role in inspiring people to take personal and civic action that will help decrease atmospheric $CO_2$ concentrations to protect humankind's wildlife heritage.[43]

While the AZA's position identifies the *geophysical* sources of global climate destabilization, nowhere in its statement does the zoo industry identify its *human* causes—not fossil fuel combustion, not intensified agricultural land use or industrial livestock production, not the deforestation of the tropics. This perfectly illustrates how climate change's origins often remain just out of focus—a crisis without a cause—when zoos communicate to the larger public. The reasons are hardly surprising—after all, zoos could not very well attract crowds of visitors while admonishing them for driving their thirteen-mpg SUVs and minivans to their inviting gates.[44]

Yet consumers are not the only revenue source for zoos. Since zoos and aquariums must rely on corporate benefactors to thrive financially, these arrangements often result in zoos bestowing lavish praise on the very companies most associated with the human causes of the environmental crisis. By publicizing their charitable largesse on engraved placards prominently placed near zoo entrances for all visitors to see, zoos play an unwitting role in elevating the public reputations of corporations normally criticized by the larger environmental movement.

While zoos and aquariums accept contributions from thousands of companies, for the purposes of this discussion let us single out just one set of corporate benefactors whose environmental impact includes habitat devastation, species extinction, biodiversity loss, *and* global warming: the fossil fuel industry, as represented by multinational firms such as ExxonMobil, Chevron, and BP. According to the IPCC, combustion from coal, oil, and natural gas has been responsible for the bulk of carbon dioxide emissions since 1750, and is thus among the largest contributors to global warming and climate change.[45] The burning of fossil fuels not only releases greenhouse gases into the atmosphere, but its extraction has historically led to the poisoning of delicate land and marine ecologies. The petroleum giant BP was held responsible by the U.S. federal government for the 2010 Deepwater Horizon incident, when an eighty-seven-day gusher dumped 4.9 million barrels of oil into the Gulf of Mexico, resulting in catastrophic damage to the region, including harm to local marine life—again, the worst maritime oil spill in U.S. history.[46] Two decades earlier, ExxonMobil was held responsible for the 1989 *Exxon Valdez* oil spill off the coast of Prince William Sound, Alaska, the *second* worst maritime oil spill in U.S. history.

It therefore might seem strange that given their track record of well-intentioned, conservation-based efforts, AZA zoos would both accept contributions from the fossil fuel industry and celebrate them accordingly. Yet Chicago's Brookfield Zoo has accepted contributions from BP. The Dallas Zoo has collected contributions from ExxonMobil. The Houston Zoo has benefited from donations from ExxonMobil, Conoco Phillips, Anadarko Petroleum, and Marathon Oil. The Bronx Zoo lists both Chevron and Hess as corporate partners. The San Diego Zoo has taken contributions from ExxonMobil, Chevron Texaco, and ARCO Chemical, a leading manufacturer and marketer of petroleum and chemical products.[47] The Arizona-Sonora Desert Museum has accepted donations from ARCO as well as Pennzoil, Citgo, and Chevron. The Tulsa Zoo and Toledo Zoo have both accepted contributions from the oil giant Sunoco, which the Political Economy Research Institute of the University of Massachusetts named the third biggest air polluter in the United States among the world's largest corporations in 2010, according to its Toxic 100 Index. Of course, it could have done worst, as ExxonMobil was named the *second* biggest corporate air polluter that year.[48]

Admittedly, the fact that zoos enjoy corporate backing from the petroleum industry makes them no different than fine arts museums, symphonies, opera houses, libraries, or other institutions of cultural esteem. Civic nonprofits from the Gilded Age to the present have historically relied on financial patronage from wealthy social elites—including oil barons such as John D. Rockefeller, J. Paul Getty, and more recently Charles and David Koch—and continue to do so today. However, in recent decades global corporations have also played an increasing role in funding local cultural organizations as well as governing them through board participation. Particularly as taxpayer-funded sources of municipal and state funding for nonprofits have withered away, rituals of corporate philanthropy as performances of civic leadership and responsibility have rapidly diffused through the organizational cultures of large companies. Even in a globalized economy, corporations are expected to direct their generosity toward local and regional urban institutional anchors, just as ExxonMobil, headquartered just outside of Dallas in Irving, Texas, contributes sizable donations to the Dallas Museum of Art, the Dallas Opera, and the Dallas Symphony Orchestra as well as the Dallas Zoo—home to the $4.5 million Exxon-Mobil Endangered Tiger Habitat, as befitting its long-standing corporate mascot and tagline, "Put a tiger in your tank."[49]

Of course, the difference between the Dallas Symphony Orchestra and the Dallas Zoo accepting ExxonMobil's contributions is that the petroleum producer's operations do not obviously conflict with the Dallas Symphony's mission "to entertain, inspire and change lives through musical excellence," whereas they are arguably disruptive to the zoo's goal of "inspiring passion for nature and conserving wildlife." The presence of oil company representatives on the governing boards of zoos, as in 2011 when the Board of Directors of the Houston Zoo included executives from petroleum giants Shell and Chevron, may present a similar problem. Intentionally or not, zoos help fossil fuel industry firms polish their corporate identities and conservation credentials by accepting their contributions and subsequently celebrating them on their websites, in their annual reports, and on their grounds—the very definition of greenwashing.[50]

Zoos are hardly ignorant of these issues, and in their defense, decisions over how to manage tensions between institutional funding and integrity of purpose are fraught with complexity. According to Karen Allen, a former public affairs director for the American Association of Zoological Parks and Aquariums and director of public relations at the Audubon Zoo in New Orleans,

> One of our greatest ethical imperatives is to deliver on our claim, our promise, to be conservationists, to back up rhetoric with actions; in short, there must be substance before image. The importance of maintaining conservation and research programs in our institutions speaks for itself. Yet how we are judged as conservationists—and animal welfare advocates—also depends on where we seek financial support. Is it ethical to accept money—money that is critical to both operational and conservation efforts—from individuals or organizations whose actions are labeled unethical or questionable by the public? Regardless of its source, the money spends just as well, but does the end justify the means? Will the public perceive us as conservationists with integrity or as foolishly principled purists if we turn down major sources of support? The time has come for us to address these questions formally and come to a consensus.[51]

Yet despite an avowed awareness of this ethical dilemma, many zoos justify their relationships with donors that engage in environmentally risky practices as part of a rational strategy of not ecological but *organizational* sustainability. Again according to Allen,

> I feel we should judge potential sponsors by where they are going, not by where they have come from. Such a protocol may elicit criticism, but we are in the business of helping to maintain natural diversity, *not sitting in judgment of big business*. If our standards are exceedingly restrictive, we will undermine our own efforts. . . . I am a realist and acknowledge that there is a fine line between maintaining only the purest of principles and ensuring financial solvency.[52]

In fact, zoos rarely judge big business, as illustrated by the AZA's depiction of climate destabilization as a crisis without a truly human cause. Yet it must be mentioned that while many of those businesses may contribute to zoos as well as aquariums and local wildlife preserves, some also help fund conservative think tanks that have intentionally obscured the dangers of human-induced global warming to the American public for decades. During the 1990s, the Global Climate Coalition spent millions lobbying politicians and the public against mitigation policies, falsely arguing that climate change had not yet been scientifically proven as fact. The group's major financial backers included BP, Shell, and ExxonMobil.[53] Actually, ExxonMobil has contributed to at least *forty-two* conservative groups responsible for publicizing climate change skepticism, including the Competitive Enterprise Institute, which once created television ads that intoned, *"Carbon dioxide. They call it pollution. We call it life."*[54]

With the help of these think tanks, the petroleum industry has mobilized their lobbying power and media savvy to fight tooth and claw against the regulation or taxation of fossil fuels for years. The Center for Public Integrity reported in 2004 that the oil and gas industry had spent over $420 million in just six years funding lobbyists, politicians, and political parties in Washington. The biggest contributors were the very same corporations honored on engraved plaques in zoos across the country: BP ($24 million), Shell ($25 million), Marathon Oil ($27 million), ChevronTexaco ($27 million), and ExxonMobil, which spent a whopping $51 million in lobbying expenses in only six years.[55]

Even if we do accept Allen's point of view, that zoos "should judge potential sponsors by where they are going, not by where they have come from," it still bears considering whether a reliance on corporate contributions from fossil fuel interests prevents zoos from educating the public about recent ecological disasters more generally. For example, take the 2010 Deepwater Horizon incident. While AZA-accredited zoos contributed to the heroic cleanup efforts on the ground after the horrific spill (as discussed above), few zoos designed exhibits or programs documenting the devastation along the Gulf Coast or singling out BP as the responsible party, just as zoos rarely alert their audiences about the dangerous environmental practices of ExxonMobil or any other petroleum conglomerate.[56]

How can one expect family-friendly entertainment outlets like zoos to swim in such controversial waters? In fact, there *are* precedents among AZA zoos who have criticized powerful name-brand corporations as teachable opportunities. To take but one example, for years the Hershey Company had been castigated for using non–sustainably produced palm oil in their chocolates, as its mass production has led to the depletion of the Borneo and Sumatran forests of Southeast Asia, where endangered orangutans once thrived. In February 2010 the Cheyenne Mountain Zoo in Colorado Springs, Colorado, warned its visitors of this crisis by shaming Hershey, singling out its many products containing palm oil in its Palm Oil Valentine's Day Candy Guide: Reese's Peanut Butter Hearts, Kit Kat, Cookies 'n Crème, Hershey's Miniatures, Hershey Kisses and Hershey Hugs, and Hershey's Pot of Gold Chocolates. The Philadelphia Zoo led a similar campaign against palm-oil-rich products that same year, and included Hershey's Kit Kat among its targets.[57] Of course, not all zoos embarked on similar campaigns against Hershey—particularly Zoo America, as it is owned by the chocolate giant itself.

## The Branding of Conservation

In 2002, Republican pollster Frank Luntz prepared a briefing memo for the George W. Bush White House. Among his recommendations for the party's political candidates, he suggested, "We should be 'conservationists,' not 'preservationists' or 'environmentalists.' The term 'conservationist' has far more positive connotations than either of the other two terms. It conveys a moderate, reasoned, commonsense position between replenishing the earth's natural resources and the human need to make use of those resources."[58] In fact, zoos across the country draw on what has become the *branding of conservation* to emphasize their commitment to captive breeding, species preservation, and environmental education without alienating their mass audiences. At one time the branding of conservation among zoos *was* truly transformative, as when the New York Zoological Society renamed itself the Wildlife Conservation Society in 1993. But some critics argue that the overuse of conservation as a badge of prestige among zoos and aquariums has today grown so out of hand that the word itself has lost all meaning, much like *green, eco-friendly,* or *sustainability.* At many zoos, long-standing buildings are sometimes relabeled "conservation centers" for no apparent reason. The San Diego Zoo Safari Park features a Conservation Carousel, just as the Elmwood Park Zoo in Pennsylvania attracts children to its playground, Conservation Kingdom. Disney's Animal Kingdom invites tourists to its Conservation Station, located at Rafiki's Planet Watch. (Rafiki is the wise and mysterious mandrill featured in

*The Lion King.*) As former zoo director David Hancocks observes, "zoos can immediately stop degrading the word 'conservation' by employing it so irresponsibly."[59]

In contemporary zoos and aquariums, the branding of conservation serves as part of an overall strategy to advocate for proenvironmental measures as a soft sell, one that deftly avoids challenging audiences to rethink the American way of life, notably our dependence on economic growth, gasoline-powered cars, and cultural consumption. What does this look like on the grounds of a zoological park?

In truth, visitors need look no farther than the zoo entrance, inevitably fortified with an obligatory gift shop—obligatory not only because every zoo has at least one, but also because pedestrian traffic patterns often force visitors to enter and exit through such tourist traps.[60] Zoos typically cram these shops with plastic souvenirs and other non-biodegradable detritus destined to wind up in local landfills or the world's oceans.[61] My nonscientific survey of one major urban zoo's retail shop revealed a selection of baby tiger purses; tiger-print golf balls; tiger coffee mugs; tiger-shaped shot glasses; tiger snow globes; penguin glitter globes; animal masks; pewter thimbles; T-shirts celebrating the Beetles; T-Rex dinosaur puzzles; world music CDs; all manner of stuffed animals (colobus monkeys, ringtail lemurs, river otters, baby elephants); a remote control tarantula with light-up eyes; and toy gas-guzzling all-terrain vehicles painted with silhouettes of endangered animals, including polar bears, lions, and gorillas.[62] No conservation exhibit could convey the wastefulness of our throwaway society as much as these temples to postmodern consumer culture.

Beyond these wild bazaars, zoos feature other attractions that strategically fail to challenge their American audiences where they live. For example, some exhibits emphasize abuses against nature occurring in far-off regions of the world by indigenous locals, rather than environmental threats caused here at home by American consumers, or around the world by multinational corporations like ExxonMobil or Shell. In the last chapter we discussed one example of this kind of messaging—recall that the Tiger Mountain exhibit at the Bronx Zoo bemoaned the killing of big cats in the Russian wilderness as a consequence of natives hunting and trapping tigers for bush meat and traditional medicines, and clearing forests for chicken farms. In Harambe, the fictitious African village featured at Disney's Animal Kingdom, imagineers have covered its artificially distressed train station and marketplace with wheat-pasted signs warning in English and Swahili, "Wild Animal Poaching is a Social Evil!" "*Kuwa macho wajangili!* Keep a lookout for poachers!" "Please report any evidence of poaching to the nearest ranger post." "*Wezi wa pembe jihadharini!* Ivory poachers beware! Think ten times before you dabble in poaching. From now on offenders will not get away lightly! Weigh this against the dream of your profit."[63]

During a weeklong stay at Disney's Animal Kingdom in the wilds of Orlando, Florida, my son Scott and I decided to leave the safety of the Harambe district for the adventure promised by Kilimanjaro Safaris, a dangerous twenty-two-minute tram ride through an endless landscape of African flora and fauna. First we passed a sign notifying us that we were entering Harambe Wildlife Reserve, and then we blazed through the forest where we caught a glimpse of some black rhinoceroses at a watering hole. Our trusty safari guide reminded us, "Black rhinos have a hide that is one-inch thick and protects them from most predators. Unfortunately, their thick hide does not protect them from humans. Humans are their greatest threat. There are less than 4,200 black rhinos left in all of Africa, so not very many of them. Unfortunately even on wildlife reserves like this one, black rhinos are still not one-hundred percent safe from poachers and other illegal hunters, so that is part of our job here—to keep an eye out for those poachers." Poachers! How exciting! Soon we were in the Serengeti grasslands heading for elephant country, when Wilson Mutua, our airborne patrolman, radioed down to us, warning, "Keep your eyes open for me. We picked up a baby elephant that was wandering off the reserve, and we can't find the mother. I'm afraid poachers may be in the reserve." Scott and I kept our eyes peeled for illegal hunters as we passed African elephants, flamingoes, lions, and addaxes, when suddenly Wilson was back on the radio, "There are poachers in the area. Can you help us pick them up?"

"I don't know about that, Wilson," our guide replied. "What do you guys think? Do you want to help Wilson chase down some poachers?" Did we ever!! "Alright, we are going to go for it Wilson—we are on the way." We came upon a set of damaged posts, and our guide seemed flustered. "Oh, this is not good, these gates are not supposed to be smashed open like this, you guys. Wilson, I think you were right about those poachers busting in. We're on our way to help you out, but exactly which way should we be going?"

"The poachers are heading east," said Wilson. "Keep going and you'll drive them to my patrol."

"Alright, we'll do our best. Looks like there are some geysers up here, folks, so hold on. It could get a little bit bumpy." We heard some loud noises followed by screams. "Okay, those were gunshots—we should probably get out of here. Wilson, we have just spotted the poachers' camp up ahead. No poachers there, though; I think we just missed them. Sorry about that."

"Not to worry—you drove them right into my patrol!"

"Well, that is great news. We are glad to hear it."

"The poachers wanted the mother elephant. Thanks to you, they are now both safe. *Asante Sana*, everyone!" Naturally, Scott and I were happy to help, and once we safely made it back to Harambe we paid tribute to our courage with some well-deserved ice cream, and possibly a corn dog.

In all seriousness, animal poaching for the illegal African ivory and rhinoceros horn trade presents very real challenges to species protection efforts on the continent.[64] Yet it is curious that Disney would select the illegal poaching of African wildlife by indigenous hunters as one of its causes célèbres, out of *all* the possible threats to endangered animals and their habitats, including global warming.[65] As zoo consultant Catherine Brinkley wryly remarked when I described how vociferously Disney's Animal Kingdom emphasized illegal poaching in Africa, "What North American person is really in danger of doing *that*?" During our tour of the Philadelphia Zoo, she explained that this type of messaging is fairly common in American zoos:

> Take this tiger exhibit, for example. The signage is all about how the Chinese are selling tiger parts, and how *that's* the part you should be alarmed about with tigers—which is so far removed from anything any American could do about or knows anything about, as opposed to the zoo saying, "Look, we're logging rare woods and sending them over to the U.S., and we're destroying the habitat. Every time you buy a piece of furniture, you've got to think about where that wood comes from, and which animals are affected by it." Zoos could use the tigers to teach a message that would be very pertinent to North Americans, but zoos don't want to be controversial in their *own* culture, in America.

This flight from controversy even extends to the lessons zoos present to their audiences about what they *can* do to promote conservation in their everyday lives. Chicago's Brookfield Zoo urges visitors to step through its Doors of Action to learn how they "can do things every day to save water."

- Start small. Start in your own home!
- Buy a low-flow showerhead. It will save water.
- Waiting for water to turn cold from a faucet wastes water. Refrigerate your drinking water instead.
- Water your garden in the evening so the water doesn't evaporate in the midday heat.
- Run your washing machine only when it is full. You'll save water by running fewer loads.
- A small push in the right place can be a huge step for conservation.

Is this last lesson true? Are small pushes really "huge steps" to alleviating the strain on worldwide water supplies, especially since global warming will lead to a continued increase in extreme weather events from droughts to heat waves? Admittedly, these simple, everyday solutions do have the potential to make Americans more aware of their own consumption, and thus contribute to changes in popular consciousness necessary for the adoption of more substantive water-saving measures. Nonetheless, here is a sobering fact: in California, the nation's most populous state, farmers use 80 percent of all consumed water for agricultural use.[66] In such a context, some critics might question the overall impact of average consumers shifting to low-flow shower heads during a drought but not, say, lobbying the state to more firmly regulate its overdrawn groundwater system.[67] As Aldo Leopold argues in his essay "The Land Ethic" in his 1949 environmental classic A Sand County Almanac, "In our attempt to make conservation easy, we have made it trivial."[68]

Along with their Doors of Action, the Brookfield Zoo invites visitors to cross a Bog of Habits, where they might learn how to make "lots of choices that help waterways and the environment overall."

- Line dry laundry
- Bring your own bag for groceries
- Bike or walk to store
- Turn off lights after use
- Buy recycled goods
- Compost food scraps
- Pack lunch in lunch box
- Buy a cotton shirt that can be easily washed
- Buy some of your clothes at the thrift shop
- Buy refills for cleaners or soaps

Again, as useful as recycling, composting, and biking might be as conscious-raising activities, some scientists might argue that these obviously well-intentioned recommendations are notably small-bore, and not at all responsive to the level of urgency demanded by the environmental catastrophe we currently face. In the journal *Science*, the codirectors of Princeton's Carbon Mitigation Initiative, Stephen Pacala and Robert Socolow, argue that in order to avoid doubling the amount of carbon dioxide built up in the Earth's atmosphere prior to the Industrial Revolution in the next fifty years—largely assumed to be a point-of-no-return Rubicon of about 560 parts per million—we as a global society would have to complete *eight* of the following fifteen goals on a grand scale:

1. Double fuel economy for two billion cars from 30 to 60 miles per gallon.[69]
2. Decrease car travel for two billion 30-mpg cars from 10,000 miles per year to 5,000 miles.
3. Cut carbon emissions by 25 percent in all residential and commercial buildings and appliances.
4. Produce twice today's coal power output at 60 percent efficiency instead of 40 percent (compared with 32 percent today).
5. Replace 1,400 large, 50-percent-efficient coal plants with natural gas plants, which is four times the current production of gas-based power.
6. Introduce carbon capture and storage (CCS) technology at 800 large coal plants (or 1,600 gas plants), so that carbon dioxide can be sequestered and stored underground.
7. Install CCS at new coal plants that would produce hydrogen for 1.5 billion hydrogen-powered vehicles.
8. Introduce CCS at synthetic fuel plants producing 30 million barrels a day from coal.
9. Double today's current global nuclear capacity to replace coal-based electricity.
10. Increase wind power fiftyfold to replace coal-fired electricity. (This would occupy about 30 million hectares, or 3 percent of the area of the United States.)
11. Increase solar power seven-hundred-fold to displace coal-fired electricity.
12. Increase wind power one-hundred-fold to make hydrogen to replace gasoline in high-efficiency fuel cell cars.

13. Increase ethanol production one-hundred-fold to displace gasoline use for two billion cars, using one-sixth of the world's available cropland for growing corn.

14. Reduce all clear-cutting and burning of tropical forests to zero, and reforest 250 million hectares in the tropics or 400 million hectares in the temperate zone.

15. Extend conservation tillage (which emits less carbon from the land) to all cropland around the world.[70]

Meanwhile, according to the California Institute of Technology chemist Nathan Lewis, in order to prevent the most dangerous science experiment in human history—doubling the Earth's atmospheric carbon dioxide and hastening current swings in polar ice melting; ocean acidification and sea life destruction; weather-related disasters such as deathly droughts, cold winter extremes, and tropical cyclones; mass species extinctions; major challenges to growing food; and rising sea levels that could eventually overwhelm New York, London, Shanghai, Venice, and Miami—we must decrease global carbon emissions by close to 80 percent by 2050. (This is in a context of *rising* levels of industrial expansion and fossil fuel use in China, India, and other outposts of development throughout the Global South.)[71]

As Pacala and Socolow argue, tempering the effects of the Anthropocene will require real structural changes in how our societies operate. Their concerns suggest that although the most well-meaning platitudes about recycling and composting might flatter zoo visitors about making "small pushes in the right place," these simple steps might falsely reassure audiences that the environmental crisis can be solved by "starting small," instead of thinking big. Citing book titles such as *It's Easy Being Green, The Lazy Environmentalist,* and *The Green Book: The Everyday Guide to Saving the Planet One Simple Step at a Time,* in 2007 Yale environmental sciences professor Michael Maniates expressed his take on the branding of conservation in the *Washington Post*:

Never has so little been asked of so many at such a critical moment.

The hard facts are these: If we sum up the easy, cost-effective, eco-efficiency measures we should all embrace, the best we get is a slowing of the growth of environmental damage. That's hardly enough: Avoiding the worst risks of climate change, for instance, may require reducing U.S. carbon emissions by eighty percent in the next thirty years while invoking the moral authority such reductions would confer to persuade China, India and other booming nations to embrace similar restraint. Obsessing over recycling and installing a few special light bulbs won't cut it.[72]

In fact, when we look at the kinds of simple steps zoos recommend to visitors, like recycling aluminum cans or taking bike rides, they tend to be oriented toward *individualistic* behaviors, as if the cumulative effect of these voluntary and uncoordinated actions will stop global warming like an invisible hand that will magically thicken the ozone layer all by itself. Yet environmental scientists and other advocates argue that the current crisis calls for not only individual deeds but *collective* strategies as well. Environmental activist and writer Bill McKibben argues for local collective action, noting that although Washington never ratified the Kyoto Protocol or pledged to meet its internationally binding reductions in greenhouse gas emissions, more than nine hundred American cities have independently done so on their own. He has also praised the trustees of Stanford University for divesting its $18.7 billion endowment of stock in coal-mining companies in 2014, noting his expectations that "other forward-looking and internationally minded institutions will follow."[73]

Similarly, the conservationist messages of many zoos tend to overestimate the impact that personal and relatively noncontroversial *consumption*-based fixes have on environmental protection and climate change mitigation. In contrast, a small number of zoos and aquariums emphasize *political* solutions to the environmental crisis instead of merely consumerist choices. For instance, the Monterey Bay Aquarium promotes a variety of public policies aimed at both protecting the environment and developing more sustainable seafood production practices. These policies include using international trade negotiations and agreements to end government subsidies to global commercial fisheries; establishing Marine Protected Areas in key ocean habitats; and implementing tougher aquaculture regulations through intensified government oversight.[74] Other zoos also present political narratives to the public, as when the Philadelphia Zoo began exhibiting Rachel Carson's 1962 anti-pesticide environmental treatise *Silent Spring* alongside its bald eagle exhibit, noting how the book "blew the whistle on widespread environmental pollution and its effect on wildlife" and how subsequent "public outcry led to the banning of DDT in the U.S. and stricter control over chemical dumping."[75]

Admittedly, the zoo personnel I spoke with at City Zoo and Metro Zoo were not overly optimistic that most zoos would want to lean in quite this far—or even could if they wanted to. Clearly, Tyler seemed fairly resigned after working at Metro Zoo for seven years. As he observed of the current state of environmental education at his own former zoo,

> I think a lot of the education that we do at the zoo is kind of a farce. I don't always think that we really ask people the hard questions about what they can do to better serve the world in terms of environmental stewardship. We don't talk about things. . . . I think the general perception is this entertainment focus. People don't want to come to the zoo to feel judged about the way they live their lives. If you come to the zoo and you're telling people about carbon emissions and about the terrible fuel economy of a Hummer, there's going to be that person who is going to feel like the zoo is preaching to them about what kind of car they are supposed to drive.

Perhaps for this reason Metro Zoo holds an annual Party for the Planet celebration each Earth Day, an event where "green" vendors give away plastic toys and other freebees destined to wind up in local garbage dumps and the Great Pacific Garbage Patch: foam stress balls molded to look like globes, pest-control-branded yo-yos, skeletal backbone key chains and crooked novelty pens from a local chiropractor, rubber ducks, and trading cards of famous conservationists, including Rachel Carson.

Through their conservation efforts, zoos attempt to literally mold the Earth's biosphere through environmental interventions, including captive breeding programs and in situ habitat research and preservation. As sites of cultural and symbolic production, zoos also distill and disseminate (or in some cases overlook, or even suppress) specific narratives about the human causes of climate change, industrial pollution, biodiversity loss, and species endangerment. Only the future will determine whether zoos can fulfill their mission as environmental stewards, conservation educators, and public opinion leaders in the Anthropocene. Clearly, zoos face formidable challenges, especially in an urban media landscape where entertainment and cross-promotional branding campaigns take center stage, but the possibilities for positive change are hardly foregone.

Then again, some argue that zoos have problems that loom even larger than their response to the environmental crisis. Whenever I talk to ordinary people about zoos, their remarks invariably lead to what I call the *captivity question*—should animals be held captive in zoos at all, and to what end? The question sometimes cloaks more than it reveals—for example, far fewer animals live in captivity in AZA-accredited zoos than the more than 2.2 million *humans* imprisoned by the criminal justice system in the United States in 2012.[76] No matter: the captivity question permeates all discussion surrounding the ethics of zoos, largely because we collectively view nature (including nonhuman animals) as essentially autonomous and free from human imagination and control, rather than see it for what it is—a cultural construction, a habit of mind. Given the centrality of the captivity question, this next leg of our safari takes us to the beating heart of the zoo, where keepers, educators, volunteers, media relations personnel, and anti-zoo activists fervently debate the morality of zoos and the treatment of animals, great and small.

# Chapter 7

# Wrestling with Armadillos

## Animal Welfare and the Captivity Question

**DURING MY STINT AT CITY ZOO** I raked yards full of dirty hay, shoveled horse and donkey manure, and prepared diets for a menagerie of pigs, rabbits, silky chickens, and macaws. But one of the most memorable of my adventures at the zoo involved wrestling no fewer than three armadillos named Snap, Crackle, and Pop in one afternoon. (To be more specific, I had been assigned to simply bathe the armadillos in a few inches of warm water in a plastic tub, but apparently armadillos don't like to be forced to bathe very much, least of all by me.) First I cleared them all from their aluminum washtub stuffed with newspaper—as nocturnal creatures, they spend their daylight hours burrowing themselves under crumbled-up piles of *USA Today*—and then the wrestling began.

I first bathed Snap, a nine-banded armadillo the size of a large football, who gripped the side of the gray tub for dear life as I poured water over her from a plastic measuring cup. I grasped her tough exterior to keep her from escaping out of the tub, but she continued to splash and squirm. When the bath was finally over, Snap clenched the side of the tub with her claws, and as I tried to pick her up she took the tub with her, water spilling everywhere. After finally yanking her off the tub and placing her safely back in her carrier, I changed the water for Crackle's bath and hoped for the best. Unfortunately Crackle actually *did* manage to escape, crawling out of the tub and across the floor, where I eventually caught her. When her bath was over, she also clutched onto the tub, kicking around wildly as I lifted her up to place her back in her carrier. By the time I picked up Pop, a three-banded armadillo and the smallest of the trio, I realized that the tub needed to be dumped and refilled for a *third* time, so I held her close to my soaking clothes while one of the keepers came to my rescue with fresh bathwater. When it was all over, the armadillos were all clean and safe in their enclosures, and the only one of us still in need of a bath was, of course, me.

Keepers and zoo visitors alike deal with the circumstances of animal captivity in multiple and contradictory ways. The vision of otherwise heroic beasts sequestered in steel cages can disturb even the most die-hard zoo fans. A friend poignantly expressed a kind of ambivalence many of us share about zoos: "I love seeing the animals in real life, but I wonder if it is at the cost of those animals *having* a real life." Of course, in many cases these captive creatures have been badly injured and rescued from the wild, or else confiscated from negligible owners, and kept under a zoo's protective care for the sake of their own survival. Yet the very idea of their captivity under lock and key saddens both zoo visitors and those who avoid the zoo altogether for this reason. Even for those who will never work closely with animals, participants in the social worlds of zoos are always wrestling with armadillos—constantly rethinking the relationship between people and animals, and the human impulse to dominate and conquer even those creatures we simultaneously shower with affection (German shepherds) or else our own humility (Bengal tigers). After all, we keep animals as beloved pets but also wear their skins as designer shoes, race them for high-stakes wagers, and barbeque them for lunch. (Again, at City Zoo concession stands sell fried chicken fingers within a short walk from where flocks of bantam chickens entertain preschoolers, just as the Monterey Bay Aquarium serves its patrons Point Reyes oysters on the half-shell, Oregon pink shrimp and Dungeness crab salad, ale-battered Pacific cod, Manila clam chowder, and coconut curry mussels.) We understandably feel disturbed seeing elephants, gorillas, and grizzly bears tucked away in iron cages and glass boxes, yet our ability to enjoy and celebrate their magnificence in the cities where many of us live persists only as long as they are securely locked behind those thick bars and high-security laminated panes.[1]

The human participants who make up the social worlds of zoos—zookeepers and educators, exhibit designers, media relations staff, ordinary zoo visitors, and anti-zoo animal advocacy activists—publicly grapple with the complex issues surrounding the captivity of their animals. On this leg of our safari we go straight to the heart of the zoo, a hotbed of dialogue and contestation surrounding the captivity question and the morality of zoos. In doing so, the zoo once again reveals itself as not only a man-made habitat for a selection of the world's creatures, but a repository of human culture and meaning making as well—a reflecting pool for our earthly prejudices and desires, the culture of nature.

## The Army of the Twelve Monkeys

In Terry Gilliam's 1995 film *12 Monkeys*, mental patient Jeffrey Goines (maniacally played by Brad Pitt, for which he was nominated for an Academy Award) leads a terrorist group, the Army of the Twelve Monkeys, with whom he hatches a crazy plot to release all the animals from the Philadelphia Zoo into the city. While I encountered no such character during my travels to AZA zoos across America, I did observe a constant chatter among guests critical of the conditions of captive zoo animals—notably charismatic megafauna such as big cats, bears, and large primates. Their apprehension primarily concerned the sizes of the animals' enclosures; questions about their treatment by keepers; and whether they appeared sad, bored, tired, or hot.

Perhaps because of their genetic similarity and physical resemblance to humans City Zoo's captive apes received a great deal of sympathy from visitors, even though keepers gave them free access to outdoor green spaces where they could frolic and groom themselves. One blazing summer day, a woman in her late thirties stood before a gorilla in the ape house and remarked to him, "You look so sad." (This is a common anthropomorphic error, as many mammals, including orangutans, lions, cheetahs, bears, giraffes, and, perhaps most famously, dogs, naturally appear sad to humans, especially when lying with their head on the ground.) She turned to a nearby zoo guest, a young father leaning against a stroller. Conspiratorially, she assured him that the gorillas "know what we are saying." With a sincere nod, the gentleman agreed,

"Oh, definitely, they do." The woman then proceeded to speak in the first-person, as if channeling the thoughts of the otherwise misunderstood gorilla. *"Why am I trapped in here—for your entertainment? Why can't I be where I need to be?"* No one around her said anything, and so she continued, "Look at the look on his face. *'Get me out of here.'* He's stuck doing this every day. I pray that they aren't being mistreated." Moments later, a younger man shouted, "Look, a mouse!" pointing out a rodent inside the exhibit. The woman spotted the mouse, asking aloud, "Are they being mistreated? Why is there a mouse in there? They don't even eat mice. That's just nasty!"

On a sunny Monday a week later, a man in his early thirties navigating the crowds of visitors at City Zoo observed a gorilla lying down on his back with his feet in the air, and asked his companion, "How would you feel if you were in a cage with people staring at you all day?" On a cooler November afternoon, two women passed by a black bear, and one said to the other, "I feel bad for the bear; he's just pacing back and forth. I guess he doesn't like his jungle gym."

Visitors regularly try to evaluate the conditions in which zoo animals live and advocate on their behalf, if to no one in particular. In these moments, people's remarks often reveal more about their *own* discomfort than those of the animals themselves. Visitors particularly express misplaced concern for captive animals during spells of warm weather. On an eighty-nine-degree afternoon at City Zoo, a family approached the orangutan enclosure and the mother said, "They don't want to be outside. It's too hot. They should let them go inside." The father agreed. "I can relate. All that hair in this weather—it's rough." In point of fact, orangutans are native to the equatorial islands of Sumatra and Borneo in Southeast Asia—one of the warmest and most humid places in the world.

On another muggy day with temperatures in the high eighties, a mother and her eight-year-old daughter approached a zookeeper after watching her presentation on City Zoo's western lowland gorillas. The mother asked the keeper, "Why don't you make them a swimming pool out there, so they can cool off and have something to play in?" Confidently, the keeper explained, *"Because gorillas don't like water."* Indeed, most apes cannot swim.[2] Moreover, like orangutans, gorillas are native to hot and humid regions located near the equator, in their case African tropical forests in the Congo Basin.[3] The mother appeared stunned with disbelief, and after a moment she said, "Well, okay. Have you ever thought about giving them drums to play with? Just little drums they could beat on—they might be really creative, you know." The keeper disagreed. *"Umm, no, I don't think we've considered putting drums in their enclosure. That would be dangerous because they would probably just tear them up and try to eat them."* Not giving up, the mother retorted, "Oh well, what about little stuffed animals or something?" The keeper said, again, *"No, they would eat that."*[4]

Zoo visitors around the country also complain when they observe exhibited polar bears collapsed on concrete ice floes, baking in the summertime sun. On an unseasonably hot spring day when temperatures reached into the mid-eighties at the Philadelphia Zoo, a woman in her mid-forties pushed a baby in a Maclaren stroller past the polar bears' poolside exhibit and exclaimed, "He must be so hot! Look at all of that fur. . . . He must be so hot!" I later passed along her concerns to Catherine Brinkley, the zoo consultant mentioned in earlier chapters. She rolled her eyes. "If the polar bear was hot and it were lying in the sun, it would move *out* of the sun and go into the shade. If you give the polar bear the option of swimming and shade, and it chooses the sun, then clearly it knows best." Similarly, Krista, a Metro Zoo keeper, pointed out to me that a stressed-out zoo animal anxiously paces back and forth, chews at its fencing, and devolves into other stereotypic behavior—but the one thing it rarely does is peacefully *sleep*, a likely sign that it feels comfortable and safe in its exhibition space. She complains of the typically incorrect knee-jerk response to "animals that sleep all the time at the zoo" among lay audiences: "They'll have misinterpretations of zoo animals. They think, *'Oh, they must be so bored, they must hate their life. All they do is just sleep.'* When, in reality, that animal is super-comfortable. Animals do not sleep if they are stressed. It's a very vulnerable position for them. So if an animal is sleeping all day, they're comfortable. They're like, 'All right, this is a safe place. I could go to sleep here. It's fine.' It's that kind of stuff that bothers me."

Listening to complaints among zoo audiences reveals an interesting paradox. During my research many people shared with me how much they disliked zoos, how immoral they found the conditions of captive animals on exhibit. Yet accredited zoos in the United States together account for only a miniscule percentage of all captive animals across the country. According to the AZA, in 2014 just over 751,000 animals lived in North America's accredited zoos, collectively. That is certainly a high number, but compare that figure to the number of animals living in captivity on American factory farms. In the U.S. alone, livestock and poultry farms slaughter more than ten *billion* captive land animals for food every year, including nine billion chickens, one hundred million hogs, and thirty-five million cows. On the industrial farms that produce most of the country's food, young pigs live in small thick-wire cages stacked upon one another, allowing urine and feces to rain upon those unfortunate enough to live below the top layer. (Meanwhile, factory farms confine pregnant sows in concrete and steel crates so constraining that the pigs are unable to turn around.) Poultry breeders genetically engineer broiler chickens to grow breasts so unusually large that they inhibit walking, and so they spend their short lives lying about in excrement-laden litter among thirty thousand other birds, many riddled with breast blisters, hock burns, ruptured tendons, and foot sores, or crippled with arthritis and severe leg pain. The typical cage for the nation's three hundred million egg-laying hens provides just sixty-seven square inches of floor space per bird. Since their male offspring serve no value as egg layers, they are unsentimentally destroyed—more than 250 million chicks per year. These are all animals that are subject to conditions of captivity far more abusive than almost any animal living in any AZA-accredited zoo in the United States.[5]

Given how most Americans seem to implicitly approve this system of animal welfare—only 5 percent of Americans consider themselves vegetarians—what explains the relatively louder and more explicit audience disapproval of zoo captivity? I would argue that these attitudes hinge on not only the question of animal captivity but its *visibility* as well. Since zoos allow access to the viewing public, wary visitors and other disenchanted audiences may judge them more severely than private-sector businesses that operate behind closed doors, such as factory farms and slaughterhouses. Like animal-populated circuses, greyhound racetracks, rodeos, and other potentially discredited public entertainments, zoos enjoy a high level of visibility and therefore may face more stigma relative to other institutions that cage and kill captive animals in far greater numbers than zoos, and not just for food production. In 2001, the retail fur industry killed 4.5 million animals in the United States, mostly mink. According to the U.S. Humane Society, animal shelters euthanize 2.7 million healthy and adoptable dogs and cats every year.[6]

The relationship between visibility and stigma is twofold. First, the public nature of zoos allows audiences to inspect and evaluate their operations and conditions of captivity in a way that factory farming does not—indeed, neither industrial farms nor slaughterhouses sell admission tickets to paying customers. Second, even though many educated people *do* know about the poor animal welfare standards associated with agribusiness thanks to a raft of highly publicized exposés—Eric Schlosser's *Fast Food Nation,* Michael Pollan's *The Omnivore's Dilemma,* Jonathan Safran Foer's *Eating Animals*—the otherwise clandestine nature of these conditions allows even those knowledgeable about them to simply *pretend* they do not when in the presence of others, or else selectively *forget* or willfully *ignore* such troubling facts in the face of evidence to the contrary. This is not to say that zoos should not also be held to the highest possible standards of animal safety and care, only that much of the cultural contestation surrounding the captivity of zoo animals stems from the distinctly public nature of zoos, rather than the actual welfare or treatment of their animals relative to other contexts.[7]

If paying guests and annual subscription members routinely express their displeasure for the conditions in which zoological parks and gardens exhibit their animals, animal advocacy activists espouse a far more extensive critique of zoos. I first became aware that anti-zoo activists even existed one afternoon outside the entrance to the Philadelphia Zoo, where advocates protested the transfer of the zoo's elephants—Petal, Kallie, Bette, and Dulary—to a zoo-run breeding facility, rather than a sanctuary where they might live free from the captivity of a small enclosure. While animal sanctuaries are technically institutions of captivity just like zoos, their vast acreages allow animals to wander and forage *as if* they were in the wild, only without fear of predators or starvation. For example, pachyderms living at the Elephant Sanctuary in Howenwald, Tennessee, roam around a 2,200-acre habitat, receive noninvasive veterinary care, and are fed plenty of hay daily. (In fact, one of the Philadelphia Zoo's elephants, Dulary, was eventually sent to the Elephant Sanctuary in May 2007 to live out her days until her passing in December 2013.)[8]

As a steadfast organizer and participant in local animal-welfare campaigns, Sophia has spent over a decade mobilizing against the treatment of captive animals at zoos and circuses around the region. At frequent protests held at the Philadelphia Zoo, she displays signs to counter the placards and exhibits shown inside the zoo's century-old gates. A "Philly Zoo Hall of Shame" recounts the zoo's ten elephant deaths (claiming "most died from arthritis or other captivity-induced conditions") and a 1995 Christmas Eve fire that killed twenty-three primates.[9] A memorial commemorates the 2008 death of Petal, one of the zoo's four elephants, by announcing that hers was "a lifetime of bullhooks, cement, and chains."[10] Posters bearing photographs of captive mammals behind bars read, "The only creature on Earth whose natural habitat is a zoo is the zookeeper," and "The wild, cruel animal is not behind the bars of a cage—*He's in front of it.*"[11]

After attending a number of Sophia's demonstrations, I found many zoo visitors receptive to her advances, and not just because she passed out animal stickers to all well-behaved children. As illustrated above, large swaths of the zoo-going public not only adore animals but harbor concerns for their humane treatment while in captivity. Also, given what we know about how audiences prefer charismatic megafauna over less celebrated animal species, Sophia chose her crusade's mascot—elephants—quite wisely. In a similarly strategic move, Bill, an anti-aquarium activist who participates in demonstrations with the Empty the Tanks Worldwide movement, strictly limits his protests to fighting against the captivity of charismatic marine mammals such as orcas and dolphins. He admits that while many animal advocates believe that *all* fish and aquatic life ought to be free from captivity, a more focused battle to save marine mammals may ultimately prove far more winnable, given the popularity of such intelligent and enchanting creatures among the general public.

Given Sophia's devotion to animal advocacy, I asked her to further articulate her position on zoos for me. As she explained over a lunch of (vegetarian) sushi rolls,

The problem with modern-day zoos is that what they say they do and what they *actually* do are totally different things, because they say they promote education and conservation, and from what I've seen they promote neither. With education, they're actually *counter*-educational. I've been to the Philadelphia Zoo hundreds of times. I've been to other zoos. I've read studies about zoos, about how they supposedly educate. And what I've seen from the peer-reviewed studies—actual, statistically valid studies—is that they are not educational. People generally spend a matter of minutes per exhibit, at the most. They might read the signs that say, "*This is an African elephant,*" or whatever, but [in actuality] zoos are wallpaper for a family outing. The family could be going anywhere. It's a nice place to bring the kids—kids love to look at animals, people like to look at animals. It's fun to see them up close, but it's not educational. It's entertainment.

Secondly: conservation. Zoos are breeding animals to have a captive breeding population. They are not breeding animals to populate the wild. Animals do great in the wild as long as they are protected from humans. They don't need help breeding. They don't need studies about breeding, or things like that. They need protection of their native habitat, and zoos, if anything, spend just a pittance on that, just a drop in the bucket. Most of their money goes for captive breeding, and they call that conservation. It goes to sending their zoo people to foreign countries to learn about animals in that foreign country, but just pennies on the dollar, if even that—[more like] a *fraction* of pennies on the dollar—goes to *true* conservation, which is preservation of the native habitat. So for me, zoos are misleading and actually fraudulent when they say, "We're [for] conservation and education only." They're entertainment, and they're conserving the zoo population, that's basically what they're doing. So that's my problem with zoos.

"Are you morally opposed to keeping animals in captivity?" I asked.

I'm morally opposed to keeping them in captivity for human entertainment, for human use. I think when they're in captivity, like my dogs, there's no choice—they're domesticated, they *have* to be in captivity. That's their life. The elephants at the two sanctuaries in the United States, the one in California and the one in Tennessee, they're in captivity. That's wrong, but they don't have anywhere else to go, so it's the best we can do right now. They should never have come into this system: they should still be in Asia and Africa. We should be working on helping their wild populations instead of having them here, in captivity. I think animals have interests, and I think that we deserve to respect those interests, and the way to do that is to let them live out their lives in nature, and not to bring them into an artificial setting and gawk at them.[12]

In many ways Sophia's understanding of zoos and animal captivity echoes that of the utilitarian moral philosopher Peter Singer, arguably the single most influential intellectual figure of the contemporary animal advocacy movement. His 1975 book *Animal Liberation* is widely considered the foundational text for animal activists who mobilize around the world against factory farms, the leather and fur industries, the exploitation of animals in laboratory testing, and the use of animals for entertainment in circuses, marine mammal parks, horse and greyhound racing, rodeos, and, of course, zoos. In the first chapter of *Animal Liberation*, "All Animals Are Equal," Singer introduces the concept of *speciesism*—"a prejudice or attitude of bias in favor of the interests of members of one's own species and against those of members of other species."[13] Such a position does not mean that *all* cases of animal captivity and even death need be considered immoral—few would protest the suffering of a hedgehog or toad over that of a perfectly healthy person. However, it does insist that the interests of animals should be given equal consideration to the interests of humans, and that the potential human benefits to be gained by inflicting harm on a captive animal must outweigh its costs to that animal. As Singer explains,

> This does not mean that to avoid speciesism we must hold that it is as wrong to kill a dog as it is to kill a human being in full possession of his or her faculties. . . . To avoid speciesism we must allow that beings who are similar in all relevant respects have a similar right to life—and mere membership in our own biological species cannot be a morally relevant criterion for this right. Within these limits we could still hold, for instance, that it is worse to kill a normal adult human, with a capacity for self-awareness and the ability to plan for the future and have meaningful relations with others, than it is to kill a mouse, which presumably does not share all of these characteristics.[14]

Regarding the question of animal captivity in zoos, adherents to this philosophy (and I would suggest Sophia as well) presuppose that the benefits of zoos *could* potentially compensate for the suffering experienced by animals living in cages and tanks—*if* their educational programs *truly* motivated visitors to dramatically reduce society's inputs to human-induced climate change, and *if* their conservation strategies *truly* led to the protection of rainforests, coral reef systems, and other fragile ecological habitats necessary to sustain endangered animals in the wild.[15] As NYU environmental studies professor Dale Jamieson argues, "Whatever benefits are obtained from any kind of zoo must confront the moral presumption against keeping wild animals in captivity. Which way do the scales tip?" (Like Sophia, Jamieson concludes that contemporary zoos have yet to pass this crucial test, given what he contends is their negligible and sometimes counterproductive record on educating the public, generating nontrivial contributions to scientific research, and preserving endangered species.)[16]

While Sophia protests the poor treatment of zoo elephants as particularly inhumane, citing her opposition to animal captivity for frivolous purposes such as human entertainment, I met other animal advocates with even stronger views, who compare captivity to human *slavery* and thus argue for the *total and unconditional abolition* of zoos, regardless of their potential (if not yet delivered) benefits to society. According to Stephen, a thirty-year-old activist,

> I take an abolitionist approach to animal rights. I appreciate some of these single-issue campaigns [such as the welfare of zoo elephants or marine mammals] because they bring a presence to animal rights. [However] as an abolitionist I reject the property status of *all* animals, and the commodity status of all animals. . . . The zoos I've encountered are prisons. They involve physical confinement and deprivation of a wide range of freedoms and behaviors that are so common to these animals when they're in the wild. I think of it as a prison. The Philadelphia Zoo is a prison. Those animals have an interest in their own lives, the interest to make their own decisions, and they can't make those decisions when they're confined in a zoo cage or in a small enclosure.

"Are there degrees of better or worse zoos?" I asked Stephen. "Or is that like suggesting that there could be a more or less humane treatment of slaves?"

"No. All slavery is bad. . . . The Philadelphia Zoo is forty-two acres—it's not very big, it's small, and there are zoos that have more acreage—but that doesn't make them better, and I don't think that enslavement and humane treatment are compatible. I don't think we can say, '*Let's make the zoo bigger and treat [its animals] humanely,*' and everything is all right. That's not the case. . . . Animals have rights, and we should respect those rights."

In contrast to Peter Singer's *Animal Liberation*, Stephen draws on a discourse not of *equal consideration* for humans and nonhumans, but specifically of animal *rights*. This perspective mirrors that of the moral philosophy of Tom Regan, the author of the seminal 1983 treatise *The Case for Animal Rights*, who argues that as *subjects-of-a-life*, many nonhuman animal species—especially those thought to have cognitive capacities and an awareness of the world, such as mammals and birds—share the same moral rights as humans, including the right to be treated not only humanely, but with respect.[17] With regard to zoos in particular, Regan is as adamant as Stephen. In an essay that asks "Are Zoos Morally Defensible?" Regan emphatically answers in the negative:

> Are animals in zoos treated with appropriate respect? To answer this question, we begin with an obvious fact—namely, the freedom of these animals

is compromised, to varying degrees, by the conditions of their captivity. The rights view recognizes the justification of limiting another's freedom but only in a narrow range of cases. The most obvious relevant case would be one in which it is in the best interests of a particular animal to keep that animal in confinement. In principle, therefore, confining wild animals in zoos can be justified, according to the rights view, but only if it can be shown that it is in their best interests to do so. That being so, it is morally irrelevant to insist that zoos provide important educational and recreational opportunities for humans, or that captive animals serve as useful models in important scientific research, or that regions in which zoos are located benefit economically, or that zoo programs offer the opportunity for protecting rare and endangered species, or that variations on these programs insure genetic stock, or that any other consequence arises from keeping wild animals in captivity that forwards the interests of other individuals, whether humans or nonhumans. . . .

In answer to our central question—Are zoos morally defensible?—the rights view's answer, not surprisingly, is No, they are not.[18]

# In Defense of Zoos

Zoos aggressively defend themselves against these sorts of attacks by mobilizing their well-resourced marketing and public relations offices against critics and activists.[19] In a June 2014 article published in the AZA magazine *Connect*, "Caring Together We Can Make a Difference," Scott Higley, a vice president of communications and external affairs at the Georgia Aquarium, writes, "The vocal minority of animal rights extremists often capture public attention. It's time for the majority to be heard."[20] According to Julia McHugh, director of public relations for the Santa Barbara Zoo, animal advocates "don't play fair" when protesting zoos, relying on strategies she characterizes as "propaganda," "personal attacks," "intentional distorting of facts," "lawlessness," "vandalism," "hoaxes," "frivolous lawsuits," "intimidation," and "violence." For this reason, McHugh suggests labeling all animal rights movement activists "extremists," claiming that their ranks are made up of "frustrated," "dissatisfied," and "alienated" individuals "filled with righteous rage."[21]

Zoos (and their keepers, as we will soon see) also attempt to differentiate animal *welfare* from animal *rights* to marginalize all zoo critics as radical and extremist, thereby making zoos themselves seem the most reasonable arbiter of captive animal welfare. Also writing in *Connect*, Jill Mellen, education and science director of Disney's Animal Kingdom, writes, "People involved in animal welfare ensure high quality standards of care for animals under human care. Animal rights advocates are concerned with legal and ethical rights of animals supporting the notion that animals should have some rights or even the same rights as humans. Animal welfare advocates look to science to provide measures for assessing care standards and potential needs for improvement." (In fact, zoo critics and animal activists also argue on behalf of improved animal welfare in zoos, while only a subset of such advocates draw specifically on the language of rights.)[22]

But the staunchest defenders of contemporary AZA-accredited zoos may be the keepers, educators, and docents who devote so much of their time to working alongside the actual creatures deemed in need of rescue by animal activists. In fact, zoo workers sometimes find themselves the target of such activism, both educators, since they serve as the public faces of their institutions, and keepers, whom guests may most closely identify with the captivity of zoo animals (as expressed by Sophia's sign—"The only creature on Earth whose natural habitat is a zoo is the zookeeper"). As zoo proponents with an interest in upholding the honor of their workplaces, occupational and/or lifestyle choices, and relationships with the animals in their care, these workers and volunteers collectively respond to critics by drawing on a set of patterned arguments in defense of zoos.[23]

As one might imagine, staff mobilize specific arguments according to their personal work role and identity in the zoo's larger social world. Educators and docents concentrate on the *messaging* and social or ecological *impact* of zoos by stressing their strengths as centers of conservation and education in a context of rapid habitat destruction and mass extinction, just as their institutions themselves do in their mission statements. Kelsey responds to skeptics she encounters during her rounds at Metro Zoo by pointing out the proactive conservation and reintroduction efforts of major accredited zoos:

> I'm like, "How can you be here and not get the conservation message? We zoos have helped create sustainable populations, and the only way you can create sustainable populations is by (a) having animals in the zoo so you can coordinate breeding, so you reduce inbreeding and they have a greater

chance of survival; and then by (b) having protected habitats in the wild, because it's kind of useless to release an animal into the wild if a hunter just shoots it, or a farmer poisons it. But if you have protected land and then you release it into the wild, then that animal might actually survive and re-populate on its own, and now [zoos have] created conservation success sto-ries for black-footed ferrets, alligators, bald eagles, and California con-dors. . . . They are still critically endangered but they're not extinct yet. That's the point. We are working so hard to keep animals and plants in the environment from becoming extinct. But if they don't agree with zoos? . . . They usually have the impression that zoos take animals *from* the wild, and leave them in cages.

Perhaps unsurprisingly, zoo educators and docents also emphasize the *pedagogical* and even *inspirational* value of zoos, particularly in the wake of the growing environmental crisis. As Leslie, the Metro Zoo docent introduced in chapter 4, remarked, "It just cracks me up. I re-member this woman on a train—I said something about how I volun-teered at a zoo, and she's like, '*I don't believe in zoos.*' And I was like, 'Well, that's fine, and we're all entitled to our own opinions, but you realize the animals' habitats are disappearing. So, if we don't have zoos, you won't learn about them. Their habitats are going to disappear, and they will be extinct. I'd like to see them in the wild, too, but it's not going to happen." Kelsey agrees: "Zoos release animals into the wild, and do this amazing conservation work, and hopefully try to educate people and inspire them to maybe participate, either by donating money, or just recycling, or going to a shelter and adopting an animal." Louise champions a similarly spirited defense of zoos on these grounds. After pointing out her disregard for anti-zoo and animal rights activ-ists—"They also think that carnivores should be fed vegetables!"—she continued,

A zoo is meant to protect and educate. . . . It gives people a chance to see [animals]—how they live, how they react—and get to know them. How many people are going to go on a safari in Africa and get to see elephants or lions? Most people won't. But if you go to a zoo, you can see them. And a lot of it is education; [the zoo] teaches appreciation, really. I really don't know what drives people who are against zoos. I really don't.

If people don't see these animals, they will never see them. Cougars don't really want to be seen in the wild. They don't: people will never see them. They will never get to [where they can say], "Hey, I really like that animal;

I'm going to find out more about it. I never realized they were endangered, and I would like to do something to help them." If you see these animals and you get to know them, and like them, and you want to find out more about them? That is education, and I think that children, adults, anyone—if they don't see these animals, it's like, *"Out of sight, out of mind."*[24]

If zoo educators and docents emphasize the messaging and social impact of zoos, keepers themselves defend zoos against critics by asserting that captive animals lead the best possible lives under their professional care, claiming a mantle as actual deliverers of animal welfare rather than activists fighting for the cause of animal equality or rights (in keeping with the messaging of the zoo industry more generally). According to Lauren, the Metro Zoo keeper, "I think that zoos are necessary, so we do what we can to make them the best place possible for these animals. I know that people think that animals would be happier in the wild or whatever, but that just goes against every sort of research that I've ever seen. How many animals are going extinct or are endangered? [In many cases] there are none left in the wild, or there are a few left in the wild, but yet we still have managed populations in captivity . . . And frankly, in a lot of ways their lives are a lot better. They are living longer, they have a doctor that takes care of them, they get food regularly—all these sorts of things."[25]

"Do you encounter people at the zoo, or even outside the zoo, that are skeptical of zoos in general, or don't think zoos treat animals appropriately?" I asked.

"Yep. And that's one reason why I think that [animal] training and enrichment are so important, because then you can give them an example: *'Look, I just trained my wolf to target and accept a pressure on its shoulders. That means that we can give it vet procedures, we can give it its vaccines without having to knock it out and risk its health, and we are doing these things for the betterment of the animals.'* So you have things that you can spell out—*'Look at this enrichment, look at that.'*"

Krista also finds herself explaining the complicated realities surrounding the care of captive animals to critical audiences, and often with positive results:

I've gotten a lot of, "*Well, why are these animals in captivity? They don't have enough space.*" And I can explain it. For example, our alligator, Scout? Her indoor exhibit is not that big. But for Metro Zoo, it's a good enough space right now *for the size that she is.* I can explain to [visitors] that alligators are ambush predators, and they don't move that much. They're big animals, they're reptiles, so their metabolism is really slow, so they try to conserve as much energy as possible. That's why when you watch documentaries they're all sitting there in the water, not moving. They're ambush predators.

And from that, [many visitors] go, "*Oh, okay. That makes a lot more sense.*" They can appreciate that, yes, the exhibit is small, but that doesn't mean it's bad. I like doing that kind of stuff. Or like, if I agree with them and I think their exhibit isn't satisfactory, then I'll explain to them, "Well, we want to make it better, but this is why we can't." Or, "This is what we are doing to make it better."

Zookeepers save much harsher criticisms for anti-zoo advocates, like those described earlier in the chapter, or activist organizations such as People for the Ethical Treatment of Animals (PETA). Again according to Krista,

[There are] some people who protest zoos to just protest zoos in general, who are just looking for a fight . . . Like I had a woman come in, and you could tell that she was just against zoos. And she was just looking for a way to be like, "*Oh, all your bats are dead.*" And I'm like, "Well, they're not. They're just all males. So once the males get old and die we don't have any babies to come up, so our colony is smaller." And she's like, "*Oh, well—I like the bats.*" And I'm like, "Okay . . . I'm sorry? Hopefully we'll get new bats in the future."

And then we went to the howler monkeys, and I went in, and the monkeys climbed up [onto their high perches, away from me]. It's their habit. And she goes, "*Why don't they like you?*" And I'm like, "Um, they do. . . . I don't know what you want them to do. Jump on my head?" There are those kinds of people—when we come back with professional sounding answers for them, they're just like, "*meh.*"

But the animal protesters like PETA? I feel like they go over the top too much. That kind of stuff gets people angry and actually *turned off* to animal welfare. Like, it is a big deal. But when you go way beyond what everybody thinks is rational, like bombing vet offices?

Animal activists and zoo workers displayed their mutual disregard for one another over the entire course of my research—which is perhaps ironic given how committed both groups are to the welfare of animals.[26] For instance, regarding those advocates who protested the treatment and transfer of their elephants, a former Philadelphia Zoo keeper and educator explained to me, "I think some of them can have valid points. The elephant yard for the Philly Zoo was too small for them . . . so when they were protesting the enclosure and the yard, I understood that, because it was small. But when the anger comes in, like, '*The zoo is terrible and I'm going to, like, release all the animals*,' and then still protesting where they *sent* the animals, because it's not where they would have sent the elephants—that kind of stuff I don't really like."

Zoo personnel also try to change the public perception of animal captivity by replacing particularly charged zoo terms with less damaging euphemisms. For example, Kelsey explained that when working with animals in public, "I avoid the word *cage* because I want people to understand that they aren't being imprisoned, so I'll use *enclosure, exhibit, habitat, carrier*." In defense of zoos, keepers, educators, and docents also emphasize reputable qualities of their own institutions, albeit very selectively. For instance, more than any other detail, keepers and educators at Metro Zoo highlight the care and attention given to those specific creatures in their collection rescued and rehabilitated from the wild, or else confiscated by negligent owners and in need of sanctuary: they include an alligator, a pair of cougars, and all of the zoo's birds of prey—its eagles, owls, falcons, and vultures. According to Kelsey, "A lot of people who come up to me, they might say, '*I really don't believe in zoos*.' And, to me, I find it so ironic that they're at a zoo, telling me they don't believe in zoos. 'If you read the signs, or if you listened when I talked, you would have heard that zoos don't take animals from the wild. The only animals we have from the wild are non-releasable. We have bald eagles with no wings. They wouldn't be able to survive in the wild; that's why we have them.'"

As an example of how zoo workers highlight these and other selective features of their own institutions to justify the existence of zoos more generally, Louise explained how Metro Zoo rehabilitated its beloved cougars, Fred and Ginger, and continues to protect them from the wild:

> A lot of these animals have never been exposed to the wild. They can't be. And I'm going to use our two cougars as an example. When Ginger came to the zoo she had cataracts in both eyes. If she had been born in the wild, her mother would have sensed that she was defective, and she would not have taken care of her. Ginger is a beautiful animal. She is my favorite animal in the zoo. I love her. I could never bear to see her being starved to death, or let her die. She deserves to live. And in a zoo, she can.

Fred was confiscated; he probably belonged to a drug dealer.[27] He is declawed on all four paws. He could *never* go out in the wild. Neither one of these animals could fend for themselves out in the wild. But in a zoo, they have a home. And they are happy. [People] think they are not happy, but they are happy. If you put them in another environment, they are going to be frightened, because it is not familiar to them. But the zoo is familiar, and they are taken care of. All of their needs are met. They're sick? The vet comes. They're hungry? The food goes in. They have each other. A lot of animals need the companionship of other animals. Some don't. But . . . if they were out in the wild, neither one of them would have survived. And a lot of the animals that we have are in that same situation. . . . Seriously, zoos give refuge and sanctuary to animals that really wouldn't have another place to go, or couldn't live in the wild.

Hannah, the former Metro Zoo education program manager, recalls past encounters with skeptical zoo audiences: "You have a lot of people that unfortunately already have the idea that we took this animal from the wild and stuffed it in a cage, as if we didn't care about the animals. This is from a lack of education, and a lack of awareness. At this point, it's a battle." In such instances, Hannah similarly defends the zoo for its commitment to rescuing injured creatures, emphasizing its role as a sanctuary:

[Visitors] would complain and be like, "*Oh, why is that animal just laying there? It doesn't look happy.*" Or they'll be talking to somebody else, acting as if no one else is listening, but they're actually speaking loud enough so that you can hear them say, "*Well, how happy would you be if YOU were stuck in a cage for your whole life?*" [One woman] said that about our peregrine falcon, now that I think about it, because the peregrine was just sitting there by its nest box—it didn't move, it didn't do anything.

I approached her, because I was literally just walking by. "Honestly, ma'am, the peregrine falcon is actually here for a reason. This is its sanctuary. He was a wild falcon at one time but had an injury, and now is missing its entire right wing." [This would explain why the bird didn't move around its cage very much.] And then I go on to explain about the peregrine falcon. "You know, it's the fastest bird of prey, it has been clocked at speeds of up to two hundred miles per hour, and it gets around by flying from point A to point B. This animal would not be able to survive out in the wild. He's out of rehab—that wing, it was so badly damaged, it had to be removed, and the only other humane thing to do was to euthanize it. We actually put this animal on exhibit so you guys could see it up close, learn about it, and appreciate it as an animal in nature." And then I walked away.

One of Hannah's pet peeves is when visitors complain about the size of an animal's enclosure. She explained to me how "it just depends on the species of animal" and its physical condition:

> A perfect example is [another] bird of prey that was missing a wing. We didn't want to just put this animal in an enclosure that was too big for it, or didn't have the perching that [we'd need] for an animal that only had the one wing. So we had to keep it small and desensitize it to the area—but also give it the opportunity to climb, and learn how to do that. And the only way to really do that is to focus it on a very small space, so that if we wanted it to go up the stairs, it would go up the stairs. Eventually, in time, once it got comfortable, then you could open the space a little bit more. It kind of reminds you of someone bringing a feline into one's home for the first time, like a cat. You don't just open the door and say, "Go ahead, have fun, Fluffy—good luck!" You actually want to start them off slow. Same with dogs, too: you want to start them off in a small area, and get them used to the environment that they are going to be in. So you start off small, and gradually get bigger.
>
> Now, as far as an animal that would be stuck in a smaller enclosure, like a larger animal stuck in a smaller enclosure *on purpose*? It would be for its safety and health. At Metro Zoo, the [injured and rescued] American kestrel and the red screech owl had a [voluminous] flight cage. But eventually the owl was put in a smaller mesh cage because of its brain damage. It wasn't able to hop around efficiently enough, and it wasn't getting to its food source enough. Same with the kestrel—it was banging into everything possible, so we put up a different kind of meshing, restructured the enclosure, and then actually made it a little bit smaller, so that it didn't have the ability to hurt itself even more. Like I said, it was more for the safety and health of the animal.

# Bunny Huggers

Zookeepers and other staff have a term for animal activists and zoo critics who, in their estimation, lack a mature understanding of the true needs of animals, both in captivity and in the wild: *bunny huggers*. Just as some moderates denigrate radical environmentalists as "tree hug-gers," Metro Zoo keeper Christina explained, "A bunny hugger is some-body who is concerned about the animal's feelings a little bit too much":

> It's somebody who doesn't understand that sometimes you need to do things for the betterment of an animal's health and well-being, but might not seem like the greatest [favor] to the animal if you're watching it [as an outsider]. It's the same when you take your little kid to get his first haircut—he screams and cries, but you still have to cut his hair. It's the same thing with a lot of animals. You still have to do the things like beak or nail trims that that animal might not want done, but it's for their health and safety and well-being, because they're not wearing them down as properly as they would in the wild.[28]

Lauren contests another popular notion among critics—the idea that captive zoo animals might be better off in the wild. "One of my favorite things to direct people to is, if you've read [Yann Martel's 2001 novel] *Life of Pi*, two sections of that book that are really close to each other: one is that zoo animals are like guests in a hotel, and the other is about animals being 'happy' in the wild, and how false that really is. Those sections are probably the best argument for zoos that I have ever en-countered—a really good explanation of why what we do is good for these animals, and why they wouldn't necessarily be better off in the wild." In *Life of Pi* (a favorite among zookeepers), Martel writes,

> I have heard nearly as much nonsense about zoos as I have about God and religion. Well-meaning but misinformed people think animals in the wild are "happy" because they are "free." These people usually have a large, handsome predator in mind, a lion or a cheetah (the life of a gnu or of an aardvark is rarely exalted). . . . That is not the way it is.
>
> Animals in the wild lead lives of compulsion and necessity within an unforgiving social hierarchy in an environment where the supply of fear is high and the supply of food low and where territory must constantly be defended and parasites forever endured. What is the meaning of free-dom in such a context? Animals in the wild are, in practice, free neither in space nor in time, nor in their personal relations. . . . The smallest changes can upset them. They want things to be just so, day after day, month after month. . . .
>
> Animals are territorial. That is the key to their minds. Only a familiar territory will allow them to fulfill the two relentless imperatives of the wild:

the avoidance of enemies and the getting of food and water. A biologically sound zoo enclosure—whether cage, pit, moated island, corral, terrarium, aviary or aquarium—is just another territory, peculiar only in its size and in its proximity to human territory. That it is so much smaller than what it would be in nature stands to reason. Territories in the wild are large not as a matter of taste but of necessity. . . .

One might even argue that if an animal could choose with intelligence, it would opt for living in a zoo, since the major difference between a zoo and the wild is the absence of parasites and enemies and the abundance of food in the first, and their respective abundance and scarcity in the second. Think about it yourself. Would you rather be put up at the Ritz with free room service and unlimited access to a doctor or be homeless without a soul to care for you?[29]

Among keepers and other experienced zoo personnel, a bunny hugger need not even be a zoo critic, but anyone who romanticizes the sober realities of animal life. According to Heather,

A bunny hugger would not make a good zookeeper. I feel like they don't see the reality of certain situations, especially with dangerous animals. Like, they want to cuddle and snuggle and think everything is butterflies and hearts and all that kind of stuff, but don't see the danger, or they think certain things might be abuse. . . . You know, the people that don't believe in euthanasia and that kind of stuff. We cull a lot of things, like the mice that we breed as feeder animals, and if something is really sick and you can't fix it, and they are just sitting there suffering? I feel like sometimes people try to keep something alive for themselves, not for the animal. Some of the animals that we let stay alive? If I were that animal, I would want to be put down.

I asked Megan, twenty-seven, a former City Zoo keeper who went on to work at a zoo in the Midwest, about the popularity of the term "bunny hugger" as a derogatory slur:

That's a favorite term at City Zoo, because I know a lot of their keepers are not big fans of bunny huggers. The expectation is that they just want to cuddle with the animals. "*Oh, but this animal is so cute and cuddly, and I want to snuggle with it . . . .*" You can't be a bunny hugger and be a zoo-keeper. You are going to have to give them shots. You are going to have to hold them down to trim their beak. Yes, we can train them to do a lot of those procedures, but sometimes you are going to do something that the animal doesn't like, and *"Oh, that poor baby!"* isn't going to cut it as a response. You are going to have to just do it.

I've seen people that work [at the zoo] trying to do a procedure for the animal's health, and this or that education person is like, *"Oh, do we have to do this? The poor baby, blah, blah, blah."* And it's like, "Yes—it's for the animal's own good. I might not like getting shots, but I know it's for my own good." Sadly, I can't tell my animal that. So sometimes I just have to do things it doesn't like, and I can't snuggle with it and make it all better. And so the bunny huggers are just, *"Oh, that poor thing; oh, that cute thing. I want to snuggle with this,"* or *"Oh, FUZZY."* And bunny huggers tend to be more a fan of things that are warm and fluffy [as opposed to] the birds and the reptiles. And that's another thing that I'm not a huge fan of in the zoo world, when all someone wants to work with are the warm, fuzzy things. There are other animals that aren't warm and fuzzy but are just as cool, and just as important.

As a volunteer coordinator, Daphne received frequent requests from would-be interns and docents with these tendencies, and one of her many jobs was to disabuse them of such pie-in-the-sky delusions:

A bunny hugger is someone who just wants to come and pet the animals. They want to come in and pet and play. They think that is what we do. And it's a derogatory term for that. Bunny huggers just want to come in and play and pet. They think we hold these animals and we go in with them and frolic. . . . They think they are going to walk in the door [of the zoo] and climb in the cage.

Oh my god, the emails, the bunny-hugger letters. We had one—oh, Tyler called it the "Crawfish of Destiny" letter. A girl—I think she was in college—wrote a letter about her hopes and dreams. She sent it in with an application to do animal care. And she said in this letter, *"Ever since I was small I have known that I have wanted to work with animals. If a bird fell out of a tree, I was there. If a cat cried out in pain, I was there. Every summer, a crawfish would crawl from a local pond and I would save it."* (Like, every summer this would happen—so we called it the "Crawfish of Destiny" letter.) And she wrote, *"Every man that buys his wife a bird and we hear the little animals screaming in pain . . . ."* And it went on and on, in that vein— that she was destined to save the world's animals. All she really wanted to do was become an intern in animal care. And if it had just been an application, I might have had her in to interview, but that letter! So I tried to call, because I wanted to say, "Honey, *never* send that letter out again. These are things you can't tell people. If you want these positions, you have to be more sensitive—you are going to scare people. You will never get a job in the industry with this letter."

*That* is what we call a bunny hugger. . . . It's not realistic. I would get a lot of those letters: *"Hey, I was laid off and I'm thinking about changing careers and I want to be a keeper. So can I come in and volunteer at your zoo? I particularly want to be in the big cat area."* It's not realistic.

# No Country for Old Men

Disputes surrounding the treatment of captive animals occur not only between zoo insiders and outsiders, but within the zoo industry as well. Zoo workers see themselves as both defenders of zoos as institutions and animal welfare advocates in their own right. They will therefore readily admonish suspicious roadside zoo operators—or even fellow zoo personnel—who they fear may contribute to the maligned reputation of zoos more generally. Given the changing demographics of the zoo profession, conflicts often arise between young college-trained women working as keepers and educators, and middle-aged men, whether zoo directors or aging keepers still bound to yesteryear's standards of husbandry and care.

For example, take roadside zoo operators. Until it closed a few years ago, the just-off-the-highway Exurban County Zoo housed and exhibited its collection of exotic creatures in a cheerless warehouse located in an industrial office park. I visited this so-called zoo twice before it shut down. The range of animals living in cages and plastic boxes subject to this decidedly unnaturalistic indoor environment—African spurred tortoises, ring-tailed lemurs, striped skunks, eagle owls, coati, fennec foxes, Nile crocodiles, green anacondas, lesser yellow-headed vultures, scarlet- and green-winged macaws, vervet monkeys—included Viktoriya, a four- to five-month-old Siberian tiger cub that belonged anywhere but a concrete warehouse.

Exurban County Zoo was notorious among zookeepers who had studied or trained nearby, including some who wound up working at Metro Zoo. Amber described her first visit to Exurban as one in which even she empathized with the anti-zoo activists that regularly protested across its driveway:

I went to the Exurban County Zoo [about three years ago], and it was a warehouse! It was just a warehouse with a bunch of metal cages. Basic bedding, basic, you know, substrates, and then a whole menagerie of animals who—in my opinion—needed more space or more enrichment. I knew a couple of people that worked there that said the owner was very stingy when it came to paying staff or providing for the animals. They did a bunch of traveling shows, too, like birthday parties and such. . . . But I just remember *hating* the setup—everything was in metal cages.

I can see why activists would get upset about [the tiger cubs]. They would only have tiger cubs for a short period of time, and then they would send them to . . . I don't even know where. Just to draw people in. I remember they had an aardvark, which is a really neat animal to see, but I just hated seeing it *there*. I don't remember the exhibit too well, but I think it was just basic wood shavings. A lot of people took undercover videos of it.

It's funny because a lot of people said that you can always get your [required internship work] hours at Exurban County Zoo, but I'd heard so many bad things about that place that I was like, I'm *not* going there. And I'm glad that I didn't.

Whitney, twenty-four, was an intern and later a paid zookeeper at Exurban County Zoo. Although she was appreciative of the professional experience the zoo gave her, she understandably retains ambivalent memories of working for Cody, the zoo's overbearing director:

I started in December of my sophomore year of college, and worked until the beginning of my junior year. I would have been twenty. I started out as an intern, and from there I just became a normal volunteer until the beginning of the summer, when I became an educator/ keeper/ supervisor/ everything-else under the sun.

Cody was an ex-Marine; he had that power in him, that sense of control—*"I'm a big powerful man, and I control things. I run things."* Over the summer, I was probably working anywhere from forty to eighty hours a week. . . . I wasn't actually paid hourly, none of us were. Most of the girls— we only had maybe four or five paid employees, and fifteen to twenty volunteers, mostly college-aged women. . . . It was a set rate every two weeks, no matter how many hours you worked—$200 every two weeks. And I was working forty- to eighty-hour weeks.

It was very helpful because it's very hard to get that much hands-on experience, especially in such a short amount of time. Part of it was also that you get attached . . . These animals, most of them come in when they're very young. Some we got from other zoos—Cody wouldn't usually tell us where they came from. So a lot of them came in very young, and we spent so much time with them—like I said, they're forty-hour to eighty-hour weeks—you're around these animals constantly. And you get to know them very well. Also, when we did get the babies in, the supervisors would often have to do night feedings, and would have to take them home. One of the baby fennec foxes came back to my apartment multiple times, because she couldn't be left alone at night. The lion cubs usually went home with our curator, Alexis. And when we had baby wallabies and kangaroos, they would usually go home with somebody. I had a real special relationship with our sloths—their names were Rufus and Kate. Rufus was Kate's son.

I really, really liked [the sloths], and I was very sad when I found out that Kate had passed away. It had been after I had left, and a friend who was still working there had told me, and it was heartbreaking for me. It was heartbreaking for me with a lot of the animals when they shut down, because I had no idea where they went.

At this point in our conversation Whitney began tearing up, and I asked her if she wanted to take a minute. She said she was okay, but before continuing I took a few minutes to ask her about her various animal-themed tattoos, including a remarkable treble clef with the head, tail, and paw of a lion. After a while we returned to our discussion of Exurban County Zoo, and I mentioned that it was interesting that Cody, the director, had surrounded himself almost exclusively with young women to work eighty-hour workweeks at, by my calculation, a meager dollar twenty-five an hour. I asked Whitney what the workplace dynamic was like.

Cody was very, very pushy. He wouldn't really take no for an answer when it came to, "Oh, I want you to do this, I want you to do that." And it took a lot to stand up to him. [One female employee] would tell him off like no one else, but for a lot of the girls, especially the younger ones, we're not as aggressive—we're more likely to just say, "Okay," and move on with it.

You know, he would run his animal curator around to get his laundry, pick him up from the airport. He had a wife! I don't understand why she couldn't do it, or why he couldn't do it himself. Even just little things that would bother me like when I had to go vacuum his office when I was trying to get the entire zoo ready to be open. He had a very short temper, and I really do believe he liked hiring younger women because he knew that he could project that kind of big, powerful, *"You will listen to me. You will do as I say."*

Given that Cody sounded like a boss from hell, I naturally asked Whitney why any of his employees ever bothered sticking around for any length of time. It turns out Cody was not only a tyrant, but a master manipulator of his keepers' emotions, particularly with regard to their close relationships with his zoo's animals:

> He could be really charismatic when he wanted to, and any time anyone would talk about leaving, he would turn on that charisma. It was, *"Oh, but you know, you do such good work, and you're such a joy to have here, and we'd miss you so much,"* and *"I don't think the animals would be the same without you . . . ."* Things like, I was glove-training the eagle-owl, and he said, *"I don't think Frankie would be able to continue his glove training, and I don't know what we would do with him,"* because he was a monstrous eagle-owl. *"Oh, why would you leave? This animal needs you, what's going to happen to your training with Clarence* [a scarlet macaw]? *It's going to fall to the wayside."* Or you, know, *"Clarence is going to miss you."*

Clearly, as a zoo director Cody may be an endangered subspecies of one, but in fact zoo staff often vocalize their criticisms of older zoo personnel, usually men of a certain age who seem reliant on antiquated methods of animal captivity and care—or sometimes no method at all. Holly complained to me about a former male zoo director who made her quit one of her early jobs in zoo education:

> The executive director that they had at the time was a terrible person and wasn't very good at running a zoo, and respected moneymaking over education. He came from a water parks background, which is just business oriented. However, if a zoo has a mission to educate, education needs to be important as well.
>
> He was just a really bad person to work for. He was incredibly insulting to all of the staff that worked there. No respect for any of them as people, or what they did as their job. He told us that the educators should be teaching all day—if you are working eight hours, you should be teaching for eight hours. He had no idea what it takes to train animals for use in education—what it takes to train a wild animal to go into a carrier, or trained to be

around screaming children, or in a van. So there was no respect for the training aspect.

Once I did an outreach [with a great horned owl] for an insurance firm that had a Take Your Kid to Work Day. So it was about an hour and a half drive on the interstate with traffic, then two programs, and then the drive back. It was a long day. But there was a board meeting going on at the zoo, and as I arrived back, [the director] was like, *"Don't put that animal away. Put it back out to entertain the board members as they are entering."* And I was like, why are we entertaining board members? (They are board members; they have already invested in the zoo.) So, no chance to unpack any of my other animals: I had to immediately take the great horned owl, who had just been traveling all day, out of its carrier to stand outside, greeting our visitors. The way that we use the animals is that if it goes on a program, it doesn't come back out [due to its need for rest]. It should have gone back into its enclosure, and should have been done for the day.

Heather similarly relayed to me another unpleasant encounter with a zoo director whose erratic treatment of the zoo's animals she considered unsound:

When I first started, I was still like, you know, seeing stars in my eyes, so excited: a plucky little educator who finally got a zookeeping job. I love it; this is great. Honeymoon is still happening. So [another keeper] and I were leaving for the day and there was a night event . . . a bunch of politicians coming into the zoo for a boozer. They had kegs and whatever. I am walking down the main path, and our howler monkeys, George and Gracie, are still outside, and I saw [the director] with [another employee] and it looked like they were blowing the path, getting ready for the event, or whatever. But, no, they were using the leaf blower to make the howler monkeys howl. They were blowing soot and exhaust and whatever else into George and Gracie's faces to piss them off enough to howl, to get people to come over. So I freaked out on [the director] because I was like, "Dude, not only is that cruel, and it doesn't smell good, and I don't even want to sit here and listen to it, but if I were to do that, I would have to get approval from the veterinarian, from the curator, and from my lead keeper to be able to use that for enrichment or training."

And he's like, "I'm just invoking their natural instinct to howl at nighttime, this is just getting them going, because I'm another howler monkey howling in the forest." And I'm like, "No you're not! You're an asshole with a freaking leaf blower in their faces! And you are showing people who are *not* animal people, and don't understand what you are trying to do, that it's okay to get their dog to howl by freaking blowing on them with the leaf blower!" I was like, "What the hell! Dude!" And I freaked out. So, I yell, and he turns it off, and I'm walking away, and he turned it back on. And that's when I knew that I just hated him. I was never, ever going to respect him or do anything that he wanted ever again. Oh my god, I was so angry.

Discord over the treatment of captive animals within the zoo world can finally be seen in the tensions that occasionally seethe between young female keepers a few years out of college and aging male keepers facing retirement. The zoo industry was once the domain of tough, manly men like livestock wranglers and ex-cons on work release programs—rugged types who gained respect from their animals by summoning their own inner alpha dog, the leader of the pack. Just as canine trainer and reality TV star Cesar Millan popularized this bossy variant of "dog whispering" among pet owners worldwide in the 2000s, many older male keepers embody this macho worldview in their very being.[30] Lenny, the cantankerous animal curator at Metro Zoo, once explained why the zoo's spider monkeys behaved so gingerly around him, but would bite other keepers: "They may have the sense that I've worked with elephants and tigers," he said, and therefore knew that "I'm going to EAT you if you give me any shit!" The son of a plumber, Lenny and his generational cohort of working-class zookeepers hail from a different background than today's educated female workforce, as Lauren explained:

> Back when Lenny started, this was back when zoos were more a museum that had live exhibits, instead of focused on conservation. And even over Lenny's time they became more focused. But way back then, keepers started as work-release people [from prison]. It was a labor job, your work-release job. . . . You did something that got you in trouble, and instead of going to jail, you got a job at the zoo because it was a labor job. So at some zoos you got people like that—not to say that they're not good keepers, but just that the mentality was that this was a labor job. It was not a care job. It was not animal caretaking; it was [working as a] farmhand—not to say that the keepers that started back then all fit under this category, but I think a lot of keepers fit into that category. But like I said, it was a labor job, they got into it in some fashion or another, and it was just to clean up this museum of live animals.

I discovered this macho perspective for myself while training to handle Miller and Barney, a horse and donkey pair that cohabitated in the same yard at City Zoo. It was a frigid January morning with temperatures in the mid-twenties, and working outside in the cold was simply miserable. Along with Tony and Phil, both zookeepers in their mid-sixties, I shoveled horseshit that had frozen to the ground and replaced water dishes that had iced over, all with splinters and cuts on my hands, snot running down my nose, my nearly frostbitten skin itchy with hay. After struggling to place halters on Miller and Barney and then raking up the yard, Tony asked me if I had hoof-picked the animals yet, which I had never done. After a quick demonstration, it was my turn, and I wasn't sure what to do—honestly, I was kind of scared of the horse. Trying to figure out where to stand, I grabbed one of Miller's forelegs and started scraping the dirt and gravel from his hoof, but clearly the horse was in charge, not me. Yelling *"Professor!"* Tony showed me how I was mistakenly holding Miller's ankle when I should have been grabbing the hoof itself. I poorly attempted to grab a foreleg on the other side, and eventually it was time for me to give Barney a try (although he appeared even less docile than Miller). Sure enough, after a few futile attempts Tony complained that I had it all wrong. I needed to *lean into* the donkey's body, to keep him immobile, and *speak roughly* with him, so he knew just who was boss—*"Like a cowboy!"* Tony said. Frustrated and freezing, I weakly muttered a command to Barney, but Tony yelled, *"No, Professor—not with a Brooklyn accent! Like a cowboy!"*[31]

Rosemary, the educator from City Zoo, tried to describe the perspective of older male keepers like Tony, Phil, and Lenny. "These older zookeepers, especially with the bigger animals—and you see this in dog training, too—it's more about showing your dominance to the animals, and getting them to do what you want them to do. . . . It's more dominance-based, it's a little rougher, a little more cavalier." As Krista from Metro Zoo explained, "In past years, it was more about, if [a male elk] was getting crazy, Lenny would come over with a [makeshift] shield and whip, and make a loud cracking noise to get him to stay away. And when he got too rowdy and was hurting the [female elk], we would knock him out and take off his antlers manually. And that's just what you've got to do—show him who's the boss, pretty much."[32]

In contrast to men like Lenny and Tony, most new zoo hires are educated women with undergraduate and advanced degrees who studied positive reinforcement training and other operant conditioning techniques in their college courses in zoo management and animal husbandry.[33] As numerous keepers pointed out to me, the recent feminization of zookeeping as an occupation has been marked by both a steady rise in women working as keepers and dramatic changes in captive animal care practices. According to Christina,

> It's definitely the difference between old- and new-style zookeeping. For a lot of the old-style keepers—and we consider them very much grandfathered into the keeping field—they don't have the degrees, they don't have the background, they just kind of came into it. For a lot of zoos, being a keeper was a job you got because you couldn't get a job anywhere else. It was people coming out of jail who needed a job and couldn't find one, and juvenile delinquents who needed to do community service. The idea behind a lot of animal keeping back then was that you have to be an alpha male to work with these animals. You have to show them that you're the boss. You have to be the dominant animal in the exhibit.
>
> It's shifted a lot in the twenty or thirty years since then. It has become a more female-dominated field, and it has become more of a *caretaking* field instead of a *controlling* field. We still do understand that for some animals, you have to be that dominant personality in order to correctly work with them, because if they think that they can dominate you, then they will walk all over you. But we also understand that we can use other behavioral techniques and training techniques in order to bend them toward the way we want them to act.

Lauren emphasized to me the contemporary shift in zoo animal handling and care from negative to positive reinforcement training:

> Now we've learned—a lot of those things that we used to wrestle the animal into submission to get them to do? If we just give them a couple of pieces of carrot, we can train them to do it voluntarily. It's a lot less stressful for the animal, and it's a lot less stressful for us. But that requires a lot of long-term work for us. While it might be harder to grab that animal and restrain it for a half-hour to get what you need from it, it doesn't require any work *outside* of that half-hour, and it doesn't require you to change your mindset. Whereas to train that animal to do that behavior, you have to be in that mindset that *this* is what's best for the animal. It's a change in your routine—it's going to require ten or twenty minutes every single day to work with that animal, but it's a lot less stressful for the animal, and it's a lot less stressful for the people. It usually is less risky for the animal—animals can actually [get] something called capture myopathy, where the animal basi-

cally dies from being restrained, in the simplest of terms. Basically it is so stressed out by the vet procedures that it just doesn't come out of it. And so, things like that can be avoided by training.[34]

As an example, Lauren explained her and Krista's recent success employing positive reinforcement and other operant conditioning techniques to train Metro Zoo's two gray wolves to voluntarily accept pinpricks from hypodermic needles as part of their veterinary care. Research demonstrates the importance of these new kinds of training regimes, especially given how traditional methods of collecting blood samples and administering medication to zoo creatures may require keepers and veterinarians to catch and restrain an animal before forcibly stabbing it with a syringe. This can cause unnecessary stress to the animal and possible flight reactions or panic. Veterinary procedures can also induce stress if they require sedating the animal by shooting it with blow darts or a tranquilizer gun, so much so that it can produce an adrenaline rush in the patient that negates the sedation process itself. Moreover, not only can highly agitated animals prove difficult to handle during blood tests, but the stress itself might actually render their blood less useful for determining their health since increased stress can be associated with irregular or even abnormal levels of vitamin E, glucose, creatinine phosphokinase, and plasma cortisol.[35]

Lauren walked me through the long-term training regimen she and Krista introduced to Lucy and Ethel, the zoo's two gray wolves:

One of the reasons the wolf training was so satisfying to me is that there is such a visible improvement in their life. When I first started, the wolves had a history of working with old-school keepers and they wouldn't come in at night, because that's when vet procedures happened—when it was dark and scary inside. They had to basically get chased in by the old keepers with sticks [who would be] shaking the fence and yelling and doing all this stuff. Now we didn't use sticks or anything like that when I started, but we would still stand outside the fence and yell and shake the fence, and basically terrify the wolves to go inside. And it was awful. Krista and I both started and were like, *"This is what we are doing? Why are we doing this?"*

But Krista and I [began] training them. First we were just training them to accept food from us. We would go in there in the morning when they went out [on exhibit], and we would throw them some treats. And they would hide in the back of their stall and they would poop and pee, and then we would leave. They would maybe eat the food after we left, maybe.

But they wouldn't eat it if they were nervous that we would do something to them. [After] a couple weeks of that, we'd throw them food, and they'd eat it after we threw it. A couple more days of that, and they'd eat the food that we threw to them in the back of the stall. They would slink forward a couple inches, and eat the food that was right there. And by a couple months, they were coming up to the front of their stall and taking food *out of our hands*. We were handing them food—grain, scrambled eggs, little bits of meat—and they were taking it from us. [While at first they would have been] huddling, cowering in the back of their stall, pooping and peeing just because we were in there throwing food to them, within a few months [they were] coming up and taking food from us, comfortably.

So now in my sessions with Lucy, I have a target and a long black pole that I use to represent a syringe. When I place the target, she'll stand up on the fence to touch her nose to it. She'll go back and forth to different places around the fence and she'll touch the target, and I say "Good!" and give her food. And then with the pole syringe, I put the pole in and she leans her shoulder into it and presses it. And that is hopefully to get her to a place where we can give her an injection voluntarily.

And just seeing the transition from the wolf that hid in the back out of fear, to the wolf that will hold her shoulder against some black pokey thing that I'm holding out for her for twenty seconds at a time, just to wait for that piece of food. . . . The transition in the stress level of those animals, and how much healthier they are now, is just so, so satisfying. It used to be when a keeper walked past they would start pacing, and they would be running laps and pacing and were nervous. Now, when I walk past, they get up and come to the fence. They sniff around and look at me like, "Do you have food for me? What's going on?" . . . Ethel is not as far advanced as Lucy, but her day-to-day level of stress is as greatly reduced. Ethel isn't comfortable enough to do the target training sessions with us, but she is comfortable enough that she is no longer pacing and cowering in the back and making a mess of herself.[36]

In contrast to Lauren and Krista's careful and patient training regime, Metro Zoo's long-standing lore celebrated Lenny's feats of fearless macho aggression—he rode elephants bareback, netted bobcats, and even instructed his keepers to deflect the advances of a charging big-horn sheep with the business part of a shovel. According to Ashley, "Your safety is always number one. Animal safety is number two. But if [the bighorn sheep] challenges you, then you can't back down, because then he'll always challenge you, and know that he's the dominant one.

So you have to tell him, 'You're not the dominant one—I'm the boss here.' So if he comes at you, you kind of have to whack him with the shovel." In fact, many keepers acknowledged that Lenny's (as well as Tony and Phil's) old-school animal handling practices came in handy when working with powerful megafauna, and as keepers they respected the experience and hard-earned wisdom they brought to the job. (As I told Tony one day, "At the zoo, in these yards, you are the professor.")[37]

Nevertheless, when walking the zoo grounds of either City Zoo or Metro Zoo one could not help but feel an inevitable changing of the guard, the last gasps of the endangered alpha men among the zoo's human primates. As Corinne, a former City Zoo keeper, observed, "Some of the newer-school keepers—most of them happen to be women—think that [Tony and Phil] may be chauvinistic, or cavemen, the way they do certain things." According to Joanna, a former City Zoo keeper,

> With the older people, I think they just start to feel a little outnumbered. I see them back away into a little protective bubble, because you have these masses of younger kids coming in, trying to take their jobs, waiting for them to go, and depending on the individuals, sometimes they're trying to undermine them and get them out. . . . I guess there was a lack of respect. I grew up in a military environment where you respect your elders—you respect their authority. Even if you don't like them, you showed them respect. I saw a lot of people disrespectful to Phil.

As for Tony, his "protective bubble" was no mere metaphor. The only time I witnessed him lose his cool in front of a young volunteer was the day he screamed at a female high school student who had snuck into his tiny office to make a private phone call. He shared this office with Phil, and they mostly just used it as a refuge where they could eat their lunch in peace. But as experienced alpha male keepers in their sixties who answered to two younger curators, both educated women in their thirties, these men must have felt like that office was the very last remnant of their authority at the zoo, their final territorial claim. A few years later, Tony retired, with Phil sure to follow.

## Death of a Bison

In early September 2011, Tropical Storm Lee made landfall on the Gulf Coast before heading northeast, spreading widespread torrential

rain and panic from Louisiana all the way to New York. By the time
its gale-force winds and downpour finally hit central Pennsylvania
on Wednesday, 7 September, Lee dumped over twelve inches of rain
over Dauphin County, flooding Spring Creek and everything in its
path, including Zoo America in Hershey. The floodwaters saturated
the zoo, eventually trapping two bison, a pair of siblings named Ryan
and Esther. Left beyond all hope of rescue, one tragically drowned in
the flood, and by five o'clock that evening the second had to be
euthanized.

In the following days, zoo administrators publicly responded to the
incident by denying any wrongdoing or responsibility for the tragedy,
insisting that while preparations for the flood had begun well in ad-
vance of Lee's arrival, including evacuations of the zoo's animals to
higher ground, few could have predicted the shocking enormity of the
storm. According to an official announcement Zoo America posted online the day after the bison deaths,

> Unfortunately, no one could anticipate a weather event that went from
> inches of rain to feet of flooding in a matter of a few short minutes. While
> we were able to ensure the safety of the vast majority of the animals in the
> Zoo, flood waters rose too quickly in the area occupied by two of the Zoo's
> bison and we were not able to rescue them.
>
> Faced with the prospect of watching the extended suffering of the bison
> and their eventual death due to drowning, the Zoo staff chose the most
> humane path possible and euthanized the bison.
>
> This is a very sad day for us, as we've built strong bonds with all the ani-
> mals in our care. We can tell you that each one of us feels this loss very
> deeply.

Online response to news of the incident was emotionally intense and
impassioned, but widely varied. Some attacked the zoo for malpractice;
others forgave the administration for handling a challenging and un-
foreseeable crisis with professional decisiveness and also compassion.
The death of a zoo animal generates not only mournful grief and re-
gret, but the potential for public outrage as well. Such circumstances
reveal layers of contestation surrounding the treatment of animals
under conditions of zoo captivity, and also the emotional attachment
humans feel toward zoo creatures themselves. They also illustrate the
defensiveness of zoos as organizations vulnerable to criticisms surrounding their commitment to animal safety, conservation, and care.

As I noted in chapter 1, zoos are marked by the ubiquity of death and dying. While zoos devote an abundance of resources to veterinary care and animal safety, demographic research shows that captive zoo animals in fact exhibit the same mortality rates as animals in the wild.[38] Ordinary zoo deaths occur so frequently that according to John Sedgwick's book *The Peaceable Kingdom*, an account of the author's year-long stint at the Philadelphia Zoo, "The vets have developed a lot of terms for dying, since they see so much of it at the zoo: 'vaporlock,' 'flameout,' 'crash-and-burn,' 'going to the Great Exhibit in the Sky.' 'I'm not a religious person,' says Keith [Hinshaw, a senior veterinarian at the Philadelphia Zoo]. 'I try not to make too much of it. If death breaks you up too much, you ought to think about another job.'"[39]

There are numerous explanations for why zoos might have significant death rates. Many American schools of veterinary medicine do not teach students how to care for exotic wildlife, as they focus primarily on companionable pet animals, riding horses, or domesticated farm animals used for agricultural labor or food production. Sick animals often die regardless of what zoo veterinarians do anyway, because "in most cases the vets don't know that the animal is sick until the disease is well advanced and beyond curing."[40]

In addition, the number of zoo animals routinely (if humanely) euthanized annually in AZA zoos cannot be underestimated. Zoo veterinarians often put elderly animals and those with serious injuries or untreatable illnesses out of their misery, as they do with those animals whose contagious diseases might trigger an epidemic that could risk the health of the entire zoo collection. During dangerous escapes, keepers, police, and security personnel often shoot to kill rather than rely on tranquilizer darts since, according to Lauren, "those drugs take a long time to kick in, and they sometimes *don't* kick in if the [animal's] adrenaline is up. If the animal is really pumped up, the drug might not work at all, or it might work very, very slowly." Zoos cull other animals on a more recurring schedule, such as City Zoo's duck population, which rapidly grows as wild birds crossbreed with the zoo's resident duck collection in their open-air exhibits and ponds. (Also, remember the tweeting baby chicks that the zoo incubates and hatches so they might entertain children for two weeks before they are sent to the commissary to be gassed and fed to the zoo's carnivorous birds of prey.)[41]

Perhaps not surprisingly, zoos are usually reluctant to publicize the death of one of their own animals, certainly when compared to the number of *birth* announcements made by their press offices ("Baby

Okapi Born at the Zoo!" "Gorilla Birth: It's a Girl!" "Meet Some of Our Babies!"). Like Zoo America's bison incident, captive animal deaths give ammunition to zoo critics and anti-zoo activists, and provide opportunities for frenzied outrage in the mainstream press and online social media. According to Ann, the midwestern zoo spokesperson introduced in the previous chapter, "Life and death is a constant thing at the zoo, and people don't necessarily understand that. So you have some species whose life expectancies are a few days, and that's just something you deal with."

But some animals' deaths cannot easily be hidden from zoo guests, such as a pair of seventeen-hundred-pound bison. On such occasions, public outrage can be fast, furious, and global, as it was on Sunday, 9 February 2014, when media outlets reported that Copenhagen Zoo officials in Denmark decided as a population control measure to kill a perfectly healthy animal—a reticulated giraffe named Marius—with a shotgun blast, after which its remains were fed to the zoo's lions and other wildcats. According to the *New York Times*, nearly thirty thousand people signed online petitions in support of saving Marius, while others responded to Copenhagen Zoo officials with death threats.[42]

Therefore, in cases where animal deaths cannot easily be overlooked, zoos issue carefully written press releases to the public. According to Ann, "When it's a high profile animal where there's a public interest in that animal, you tell that story."[43] Public relations staffers compose these press releases as carefully worded obituaries that always take great pains to portray the zoo itself as blameless and beyond reproach, as if to immediately inoculate itself from inevitable public outrage. For example, in June 2008, the Philadelphia Zoo issued the following press release announcing the death of Petal, the aforementioned elephant:

### THE PHILADELPHIA ZOO SAYS GOOD-BYE TO PETAL THE AFRICAN ELEPHANT

Petal, 52, the oldest African elephant in an American zoo, and a resident of the Philadelphia Zoo for more than a half-century, died this morning. Tests are being conducted to determine the cause of death and results are expected over the next few weeks.

Dr. Andrew Baker, Vice President for Animal Programs, said Zoo elephant care staff arriving shortly before 7AM were alarmed when they found Petal, who typically slept standing up, lying in her stall on her right side.

Veterinarians and other animal care staff were immediately called. The team attending Petal tried for nearly two hours to comfort her and to help her to her feet. She died at approximately 9:15AM.

"It is too early to know exactly what happened, but it would appear after viewing our in-stall video monitoring system that Petal's right rear leg buckled suddenly early this morning, causing her to collapse," explained Baker. "Petal had been in excellent health through her years at the Zoo, showing no recent signs of illness or decline. We will know much more in the days ahead."

Petal lived with two other African elephants, Kallie, 25, and Bette, 24. "Our elephant care staff will continue to monitor Kallie and Bette as they both had close bonds with Petal."

"Petal will be warmly remembered by staff and generations of Zoo visitors as one of our most beloved animals. This is a great loss to the Philadelphia Zoo family and the greater Philadelphia community," said Zoo CEO Vikram H. Dewan. "She was a great ambassador for the endangered and threatened wild elephants of Africa and Asia. She will be missed."[44]

There are a number of interesting things about this press release, but let us for now focus on just two of them. First, notice that the release immediately portrays Petal's death as a *natural* death due to her old age, foreclosing any possibility that the elephant might have died before its preordained time while under the zoo's care. The very first sentence announces Petal's fifty-two years of age, assuring readers (in this case, both reporters and zoo members) that she was "the oldest African elephant in an American zoo." At the same time, despite being of such a noteworthy and superlative old age, Petal was reported to be "in excellent health through her years at the Zoo, showing no recent signs of illness or decline." This illustrates how zoos frame animal deaths as ordinary and inevitable, much like medical professionals manage human deaths in hospitals.[45]

Second, notice that the press release reports that zoo personnel took heroic measures to try to save the animal. Veterinarians were *immediately* called, and animal care staff worked tirelessly for nearly *two hours* to save the pachyderm, but to no avail, lest an angered public believe otherwise. According to UCLA sociologist Stefan Timmermans, the mobilization of aggressive symptom management is an important component of a *good* death, in which surrounding caregivers and loved ones comfort the patient until their passing.[46]

Zoos handle the public relations surrounding the *euthanasia* of a captive animal slightly differently. In June 2008, the Philadelphia Zoo released a similar announcement for another animal death. Its opening lines read,

### TWIGGA, 29-YEAR-OLD PHILADELPHIA ZOO GIRAFFE, WAS A FAVORITE OF KEEPERS AND VISITORS

Twigga, who at 29 was among the oldest giraffes in the U.S., has been euthanized. She had been on pain management medications related to her age-related arthritis for some time. Based on Twigga's comfort and behavior, the Zoo's veterinary team decided this week that humane euthanasia was now the best course for her. Giraffes typically live into their late teens to early twenties.

Twigga, born at the Knoxville Zoo and a member of the Philadelphia Zoo family since 1979, was an extraordinarily gentle giraffe, a favorite of Zoo staff and guests alike, and raised six calves over her years here. At 15 feet, she was tall for a female giraffe.[47]

Readers learn that Twigga "was among the oldest giraffes in the U.S.," and thus died a natural death rather than having faced a tragic or otherwise untimely demise. Lest Twigga's fans not realize that twenty-nine is a sufficiently ripe old age for a giraffe, the press release informs them that giraffes "typically live into their late teens to early twenties." Of course, just as zoo staff took heroic measures in attempting to resuscitate Petal, the zoo here proudly announces that veterinarians similarly employed heroic efforts to keep Twigga alive, managing her pain for "some time" (readers do not learn how long) prior to eventually pulling the plug on her as a gesture of sympathy. This is in keeping with some of the ideals surrounding a *dignified* passing, in which death is seen as preferable to a future of endless suffering. (Obviously in the case of zoo animals such deaths cannot be truly considered "dignified," since unlike in cases of physician-assisted suicide the patient is incapable of making an autonomous decision to die.)[48]

Pleas of justification commonly appear in zoo obituaries delivered to the media, suggesting a kind of institutional defensiveness against adverse public opinion. When Pattycake, the first gorilla born in New York City, died at forty years of age at the Bronx Zoo in March 2013, the zoo's press release noted that "upon her death, Pattycake was the 31st oldest gorilla of the 338 presently residing in North American zoos. She surpassed the median life span for gorillas in zoos, which is 37 years of age."[49] Later that year, when Gus the polar bear died at the Central Park Zoo at the age of twenty-seven, the zoo's press release reminded mourners that "the median life expectancy for a male polar bear in zoos is 20.7 years." The death itself was portrayed as both dignified and unavoidable: "Gus was euthanized yesterday while under anesthesia for a medical procedure conducted by WCS veterinarians. Gus had been exhibiting abnormal feeding behavior with low appetite and difficulty chewing and swallowing his food. During the procedure, veterinarians determined Gus had a large, inoperable tumor in his thyroid region."[50] By euthanizing Gus while he was under anesthesia, the press release emphasizes that the zoo's veterinary staff took pains to keep the furry patient in a state of "closed awareness" in which death quietly slipped away, a death without dying.[51]

The language expressed in these press releases is hardly accidental, although public relations officials explain it away as a nod toward *transparency*, rather than strategic communication. When I asked Ann what might be included in a press release announcing the death of a prominent animal beloved by the community, she explained, "We just had one last month—our lion, Sleepy. He was eighteen, which is four years past the median life expectancy for an African lion, and transparency is very important. So we sent out a release mourning him, but saying that the keepers made the difficult but appropriate decision to euthanize Sleepy: his quality of life was compromised, this was his age, these are the facts."[52]

Of course, it may be argued that the logic of human communication itself predisposes zoos toward emphasizing the longevity of their animals' lives as well as the dignity of their deaths. University of Wisconsin sociologist Douglas W. Maynard argues that in everyday talk we emphasize good news while simultaneously diminishing the psychic punch delivered by bad news. By cushioning the impact of bad news in ordinary conversation while celebrating the good, we collectively reproduce what Maynard calls "the benign order of everyday life," perhaps to protect ourselves from common sources of grief, including illness and death.[53] Given that media relations offices operate in the same interactional universe as the rest of us, perhaps these press releases reveal not public relations strategies but simply the conversational discourse of unplanned ordinary life.

There is likely some truth to this, although I suspect it does not explain the whole story. Zoo animal obituaries are notable for their rarity, and consequently few people ever read great numbers of these press releases in one sitting. Yet read in bulk, they demonstrate how zoos rely on a crude system of boilerplates, a soulless fill-in-the-blank response to the deaths of these otherwise beloved creatures.[54] In May 2008, a polar bear named Olaf died at the Denver Zoo, after which the zoo posted this press release:

### DENVER ZOO MOURNING LOSS OF POLAR BEAR OLAF
#### Famous Polar Bear Dad of Klondike & Snow

Denver Zoo is deeply saddened to announce the death of Olaf, a 22-year-old male polar bear. Olaf, father of famous polar bears Klondike and Snow and Ulaq and Berit, died Saturday, May 3, at Denver Zoo.

"This is a very sad loss for Denver Zoo and our community. Olaf will be missed by all of us including the many families and children who have visited him over the years," says Denver Zoo President/CEO Craig Piper.

Over the past several weeks, keepers noticed that Olaf's eating habits were fluctuating and he was acting lethargic. He had good days and bad days. On Friday, May 2, veterinary staff anesthetized the bear for a medical examination and to biopsy a swollen area of his abdomen. Unfortunately, biopsy results showed that Olaf was suffering from an aggressive and terminal form of liver cancer. Necropsy results determined Olaf had a massive tumor in his abdomen. Having lost his quality of life to an irreversible medical condition, Olaf was humanely euthanized on May 3. The longevity of polar bears is 20–25 years.

"Often with wild animals, they do not show symptoms of illness until they are quite ill. This was the case with our valiant Olaf, who showed no signs of illness until quite recently, despite having terminal liver cancer. He was such a good bear," says Senior Veterinarian Dr. David Kenny. Dr. Kenny was Olaf's veterinarian for 21 years.[55]

So far Olaf's obituary reads like many of our other examples of animal euthanasia portrayed as dignified, emphasizing that the "longevity of polar bears is 20–25 years" (Olaf was 22); he was "humanely euthanized," having lost his "quality of life to an irreversible medical condition"; he "will be missed by all of us." Then in 2010 the Denver Zoo euthanized a second polar bear, Voda. Although released nearly two years after Olaf's obituary, the press release announcing Voda's death somehow reads as familiar:

## DENVER ZOO MOURNING LOSS OF POLAR BEAR "VODA"
### Excellent Bear and Caring Mother Brought Joy to Millions of Zoo Visitors

Voda, a beloved 23-year-old female polar bear at Denver Zoo passed away April 27. The bear had been under close veterinary observation for chronic renal disease.

*"This is a very sad loss for Denver Zoo and the entire community.* Voda was an amazing ambassador for her species and helped educate millions of zoo visitors about how people can take action to help ensure these animals do not become extinct. *She will certainly be missed by us all,"* says Denver Zoo President/CEO Craig Piper.

Voda had been under veterinary care for renal disease. Over the last few days zookeepers and vets saw Voda take a turn for the worse as she began to refuse food. When she would eat, she was not able to keep it down. On the evening of April 27 veterinary staff anesthetized the bear for a medical examination and found that Voda's kidneys were continuing to decline at a rapid rate. *Having lost her quality of life to an irreversible medical condition, Voda was humanely euthanized. The longevity of polar bears is 20–25 years.*

"We were watching Voda very closely. Unfortunately, she was getting progressively worse and during an emergency veterinary evaluation, we found that her health was declining rapidly. We knew she was not going to get better. It's always difficult to make these decisions. *She will be so missed,"* says Senior Veterinarian Dr. Felicia Knightly.[56]

Two months later the Denver Zoo euthanized a *third* polar bear, Frosty. By now readers should be able to reverse-engineer his death announcement without much assistance.

## DENVER ZOO MOURNING LOSS OF POLAR BEAR "FROSTY"
### Bear Was Oldest Male Polar Bear in North American Zoos

Frosty, *a beloved 25-year-old male polar bear at Denver Zoo passed away* June 24. *The bear had been under close veterinary observation* for liver cancer. At the time of his death, Frosty was the oldest male polar bear in North American zoos (born Nov. 11, 1984).

"We hadn't had him very long, but he made his mark on our hearts," says Curator of Primates & Carnivores BJ Schoeberl. "He was such a dynamic bear. *He will be greatly missed."*

Frosty *had been under veterinary care for* liver cancer. *On the morning of* June 24 *veterinary staff anesthetized the bear. Having lost his quality of life to an irreversible medical condition,* Frosty *was humanely euthanized. The longevity of polar bears is 20–25 years.*

*"This is never an easy decision,* but it was the right one. *We watched* Frosty *closely and saw his condition deteriorate rapidly. His quality of life had deteriorated* to the point that he was not able to sustain a healthy weight and he *was not going to get better. We'll miss him* terribly," *says Senior Veterinarian Dr. Felicia Knightly.*[57]

# Funeral for a Friend

Whether for an elephant, giraffe, or polar bear, in the event of an animal's passing zoos rely on their public relations engines to minimize criticism and reputational damage. But in the social world of the zoo, keepers, educators, volunteers—and yes, even public relations personnel themselves—sincerely mourn for the animals under their daily gaze. To these animals they attach deep feelings of identification and sympathy, both as individuals and as a community, almost as if these creatures were human kin.[58]

When working at Metro Zoo I witnessed outpourings of grief for expired animals on a number of occasions, and I was always struck by how powerfully such events impacted keepers and other staff. According to Amber,

> It's hard to not feel upset. It's almost like a member of your family, like your zoo family. It could be relatable to the family dog that you've had for years. I mean, especially the older keepers that have been there forever, and that have worked with these animals for years. . . . But that's probably the hardest part: seeing an animal that you've spent time with and cared for, and eventually they've reached their end. And it's sad. . . . When we [get] the call, you know, "*Can all the keepers come to the clinic. . . .*" And as soon as you hear that, you know. That's when it starts to hit you.

Although typically suspicious of the oversentimentality expressed toward animals by so-called bunny huggers, these same no-nonsense zookeepers themselves proved capable of experiencing deep emotional heartbreak in the event of a beloved creature's passing. They often channeled their sadness through impromptu rituals of mourning, articulating their bereavement through ceremony and shared expressions of loss. According to Lauren,

> Usually we cry, and everybody gets to say their goodbyes. It's kind of like a wake. You know, if they have been humanely euthanized or if something has happened, usually one by one we pass by, and we give it a scratch on the nose, or a rub on the shoulder, and say goodbye, and say you were a good such-and-such. We'll all say our goodbyes, and a lot of us will be crying. When Seamus [the zoo's spotted jaguar] passed, the keepers that were there that day after work—we all took a shot in his honor, a shot of really, really potent rum. Heather poured some out on the ground for Seamus, and we all took a shot. It was probably the most awful, painful shot that I'd ever had, but one of the things that makes the job worth it is that connection that we have to those animals. But having that connection makes it really painful when they're gone.[59]

As part of these rituals, zookeepers will sometimes physically lay hands on the deceased animal—sometimes for the very first time, in the case of highly dangerous carnivores like jaguars and cougars that even keepers never personally handle. According to Heather, "You get to touch them, to pet them for the first time. It's cathartic. Some of us get their paw prints, because it's the only chance you have."[60] Through these kinds of rituals, keepers enable themselves to continue working in an environment in which death is inevitable and common. Eventually, departed animals like Seamus are incorporated into the collective memory of the zoo, as when Louise penned a heartfelt poem honoring the jaguar in memoriam and emailed it to the zoo's staff and volunteers. As she later explained to me, "I cried all through the time that I was writing it because each and every word is from my heart, which was breaking." It is narrated from Seamus's point of view.

> My ears heard the laughter and joy of children as they stopped by each day to say hello. I may not have paid much attention, but I knew they were there.
>
> In the world I shared with my beautiful companion, in stormy days and quiet nights, there is an empty space along the paths that she now walks alone. I miss her.
>
> I see the friendly faces of my friends who greeted me every day though marred with tears. I knew you loved me, I knew you cared, even if we never touched.

If you thought that I was the regal king of a rainforest that I have never seen, I hope it inspired you to learn more about its fragile and endangered future.

I was so much more than just another jaguar in an exhibit. I am the symbol and the spirit of each and every one of my species that has gone before me who paid the price of indifference and the carelessness of greed.

My life was not lived without meaning, if everyone who knew me has seen the significance that all of Earth's living creatures have in the Circle of Life. I was a proud part of that circle, and the greatest joy that I have known was sharing my place in the same circle that holds each and every one of you, my friends.

*I will always live as long as you remember me.*

While Louise wrote her tribute to Seamus during a period of mourning, eventually the collective memory surrounding a captive animal's passing takes the shape of lighter fare, such as the social banter of zookeepers and other staff—thus easing the laborious work of caring for the surviving zoo collection. As Lauren explained, "Sometimes we joke about the big pasture in the sky. Yeah, I guess between us, we'll joke about what that animal is doing in the afterlife, like, you know, 'Oh, I'll bet Billie [a Tennessee fainting goat] is prancing in that big pasture in the sky.' Or, 'I'll bet Pinky the porcupine has got lots of peanut butter where she is.'"

~⦿~

The human participants that make up the social world of the zoo—zookeepers and educators, media relations staff, ordinary zoo visitors, and anti-zoo animal advocacy activists—wrestle with complex issues surrounding the captivity of animals. In moments of sympathy, anger, mourning, and even self-promotion, people cannot help but attribute sentiment and meaning to the zoo's creatures, lending shape to our collaborative construction of nature. Although they maintain vastly different social roles and political points of view, all seem to voice a tender empathy toward creatures from polar bears to porcupines, at least under the protective canopy of the zoo. Whether expressed as moral outrage or heartfelt poetry, this warmth and compassion is a likely prerequisite for any assertive collective action to be made on behalf of the Earth's biosphere and its uncertain future. By way of concluding our safari adventure, let us consider how zoos can be part of the solution to the environmental crisis in the shadow of the Anthropocene.

# Chapter 8

## The Urban Jungle

### The Future of the American Zoo

**ON A TUESDAY MORNING** in early September, I raced to my new volunteer job at City Zoo just in time to unload and stack hay bales with Tony and some of the other keepers and interns. After that I left for the shed where red-footed tortoises lived among porcupines, owls, and a turkey vulture. As country music blared in the background, I bathed and inspected the tortoises, refilled their water bowls, and spot-checked their enclosures. Next I raked under the nearby rabbit hutches, and cleaned and lined the macaw cages. Finally I headed to the animal kitchen for my next assignment with a new volunteer partner. At eleven o'clock I had been scheduled to work in the parrot yard alongside Chelsea, and since I was a new recruit and eager to please, I waited patiently for her outside the kitchen. And waited. Eventually, I had the sense to ask one of the volunteers to point out Chelsea to me, and he laughed. He took me back into the kitchen, and there was Chelsea—a ring-necked dove, my partner for the rest of the morning.

Zoos reflect how humans construct the natural world, both literally and figuratively. With its 4D multimedia spectacles, educational programming, and staged encounters with anthropomorphized animals bearing pet names, the zoo provides a miniature model for how humans distill the chaos of the ambient environment into legible representations of collective meaning and sentiment that project our creativity and wit as well as our chauvinism. At the same time, zoos are monuments to man's mastery over his environment, and his willingness to shape the world according to his every whim. (We are nothing if not a species of nature makers.) The zoo's built environment is simulation and reality all at once, an ersatz naturalistic environment for wildlife and a sustaining habitat in its own right. Like the Earth itself, the zoo is rife with slithering snakes and sneaky squirrels, black jaguars and blue jays, soil-saturated substrates and steel trees, all held together with electrified fencing, irrigated water systems, and joint compound. Their carefully designed exhibits, cultivated gardens, and captive collections all help to illustrate the recurring impact and reconstitution of the Earth's geophysical and biological reality by humankind in the age of the Anthropocene. Perhaps more than any other urban attraction in the American metropolis, zoos cannot help but represent nature as a man-made creation, a product of collective imagination, ecological stewardship, and not a little chutzpah. By understanding zoos, we become better able to reconsider our presuppositions and priorities on a warming planet beset by mass species extinction, broiling heat waves, sprawling forest fires, arid droughts, and the polluting of the Earth's waters from the Gulf of Mexico to the Great Pacific Garbage Patch.

Like the Anthropocene itself, zoos reveal the illusiveness of the symbolic boundaries we use to differentiate culture and nature. At the zoo, adults talk to live monkeys and apes while children play with inanimate statues of elephants and reptiles. Zookeepers feed perfectly healthy flamingos grain pellets doused with beta-keratin, just so they can retain the pinkish color they sport in the wild. Other keepers and volunteers mourn deceased jaguars as if they were people, with memorial odes and ceremonies of bereavement. They give their orangutans Kevlar-lined fire hoses as enrichment to bring out behaviors that naturally occur in the Sumatran rainforest. They plant gardens on the rooftops of buildings and put up Panamanian golden frogs in hotels. Endangered animals thrive in zoos where artificial woodlands prove healthier habitats than the denuded landscapes of yesterday's depleting wilderness. Zoos attract people to their gates to observe eels and elands, but also each other. They invite guests into stadiums in which killer whales and sea lions dance for food, and public gardens where Earth's most evolved primates both work and play.

It is in this manner that zoos and aquariums serve as repositories of human culture and collective belief. They feature stylized environments woven from synthetic materials, scientific fact, and symbolic art. They provide landscapes of meaning making in which audiences construct nature through the zoo's performances and productions—whether entertaining, educational, or both—as well as through cultural lenses shaped by traditional prejudices and modern myth. Zoos are also sites of cultural contestation, stages for confrontation and debate surrounding the morality of animal captivity, the scientific reality of evolution and climate change, and the authenticity of children's encounters with feral squirrels and Lego penguins. Finally, zoos are defined by their cultures of community, in which tireless female animal keepers campaign for respect while shoveling bison yards for little pay; educators enjoy the rewards of bringing hope and delight to audiences among the young and aged; and junior docents and adult volunteers find havens of sociability and mutual support among peers who kiss Russian blue Dumbo rats, greet each other with awkward jellyfish hand gestures, and road-trip to zoos across America. Like nature itself, the zoo is a social and cultural accomplishment, a product of human domination and expressiveness, sense and sensibility.

But zoos reflect not only the culture of nature, but the contemporary city itself. Both zoos and cities are alchemies of man and beast, art and science, concrete and dirt, order and chaos. With its dense human populations and companion animals, networked ecosystems of termites and mass transit, sewer rats and migratory birds, disease and contagion, man-made parks and community gardens, night fishermen and rooftop beekeepers, the city is a jungle. In the urban metropolis, the boundaries between office towers and polluted sky, busy harbors and stoic sea, and civilized society and wilderness itself are forever blurred, just as in the zoo.

The future of both our cities and their zoos are also deeply intertwined as each tries to reinvent itself in the wake of the worsening environmental crisis. Just as human history and natural history have become one and the same in the Anthropocene, the fate of urbanism itself mirrors than of the Earth's fragile biosphere, especially given the projected severity of the climatic consequences of human-induced global warming for cities around the world: rising sea levels, increased flooding, encroaching wildfires, dangerous heat waves and droughts, ferocious superstorms, and other freakishly extreme weather-related disasters.[1] American and European cities as well as those rapidly developing megalopolises in the Global South—Beijing, Shanghai, Mumbai, Manila, Delhi, Mexico City, São Paulo, Jakarta—will therefore have to evolve at the pace of climate change to reduce their overall carbon footprint while strengthening their ability to survive the next century by implementing stricter efficiency standards, cleaner alternative fuel technologies, and smarter investments in green infrastructure, costal shore protection, the availability of safe drinking water, and the reforestation and preservation of nearby ecological habitats.

Yet along with cities, zoos, aquariums, and other self-proclaimed centers of conservation will similarly have to adapt to meet the challenges of the anthropogenic age, both to slow the accelerated pace of species extinction and biodiversity loss and to educate their mass audiences about the severity of the crisis at hand, along with providing models for innovation and collective change for society at large. It begins by extending the promise of zoos toward three institutional ends—to serve as *sanctuaries* for protecting local as well as nonnative wildlife, as *schools* to teach the public about all aspects of the Earth's biosphere, including the growing climate change crisis, and as *showcases* for our cities that can provide models for urban living on a warming planet.

## Zoos as Sanctuaries

In the past three decades zoos have done a great deal to rehabilitate themselves in the public imagination, by both providing animals with naturalistic and enriching environments and actively contributing to the preservation of endangered species and their shrinking habitats both regionally and globally. Zookeepers spend less time bullying their animals, preferring to draw on positive operant conditioning and other training techniques to transport them and administer veterinary care. Some zoos also care for and even exhibit injured or abused animals incapable of living in the wild, while others send their retired animals to enjoy their twilight years on thousands of acres of protected habitat in sanctuaries. The zoo industry should take the initiative of scaling up these enlightened practices to all of our zoos.

Overall, we need fewer, and better resourced, zoos that adhere to the highest of standards. We ought to start by gradually ridding ourselves of all roadside animal attractions, circuses, and non-AZA-accredited zoos that exhibit animals more exotic than domesticated farm animals. The U.S. Department of Agriculture's Animal and Plant Health Inspection Service already regulates these cavalier zoos. It has the authority to dramatically raise its quality standards to meet or even exceed those of the AZA, and over time shutter those zoos that fail to achieve those benchmarks. These governing and regulatory bodies should also require zoos to maintain a high level of capitalization to prevent potential declines in animal care and public safety (or outright bankruptcy) during economic downturns in these times of financial and ecological instability. Such zoos will also be better prepared to invest in land and coastal waters to provide more living space for their animals and protect more of the continent's remaining healthy landscapes from overdevelopment.

Zoos can also become more efficient with the space they already have by reconsidering their species mix. All zoos should consider dramatically lowering the number of species represented behind their gates, if only to better direct resources toward smaller but smarter and better curated animal collections. Zoos should also continue to experiment with displaying multiple species together in shared exhibits, not only for reasons of efficiency but also to educate the public about integrated ecosystems and habitats. A final strategy might be to invest in smaller and less cuddly animals that take up less space overall but are no less fascinating to behold, notably zookeeper favorites such as fruit bats; boas and python snakes; tiny but gorgeous amphibians such as blue-and-black poison dart frogs; and invertebrates of all kinds, including insects, millipedes, shellfish, and spiders. Notably, the Central Park Zoo does all of these things very well, and others should learn from this storied city zoo. While it sits on only six and a half acres in Midtown Manhattan, the Wildlife Conservation Society manages this postage stamp of a zoo by keeping its collection down to a limited number of species (including penguins, snow monkeys, and harbor seals), and all from a few select climatic zones. In many cases single exhibits house entire menageries unto themselves: the Central Park Zoo's indoor tropical rainforest features black-and-white ruffed lemurs, fairy bluebirds, emerald tree boas, golden lion tamarins, dart frogs, two-toed sloths, long-tailed hornbills, red-footed tortoises, piranhas, bats, leaf-cutter ants, and a crested wood partridge in a fake tree.

If zoos really want to be known for their contributions to animal welfare, they could begin tomorrow by offering their largely female zookeeping staffs a well-deserved raise. Zookeepers are no longer displaced farmhands and ex-cons: they are college-educated and highly trained professionals who perform the necessary backbreaking labor that allows zoos to exist at all. In terms of veterinary care, they are the first-responders during traumatic events, and the careful eyes that observe inconspicuous symptoms among their animals before the vets are even called. Keepers fight tooth and claw for their jobs and don their dirty uniforms with dignity and pride. If zoos can afford to pay their CEOs and other administrative executives so extravagantly, they can certainly pay their own zookeepers a professional middle-class salary. They are the heart and soul of the zoo's social world, and it is because of their service that their institutions not only survive but thrive.

Finally, as a society we should strongly reconsider whether we should continue to keep the Earth's most intelligent nonhuman mammals under constrained conditions of zoo captivity at all. Any such conversation would have to begin by rethinking the status of chimpanzees, gorillas, orangutans, bonobos, gibbons, elephants, dolphins, and whales.[2] Obviously, such animals could not be released from captivity into the wild; indeed, for some of these creatures there is no longer any remaining wilderness in which to release them. Ideally they would be sent to sanctuaries where they would enjoy lives of relative freedom of mobility, veterinary care, scheduled feedings, and opportunities for reproductive breeding. These sanctuaries could even be funded and sustained by a consortium of AZA zoos that would share the facilities collectively.

It is undoubtedly true that great apes, pachyderms, and cetaceans are among the most beloved of zoo creatures. But in recent years the Philadelphia Zoo, Detroit Zoo, San Francisco Zoo, and Chicago's Brookfield Zoo and Lincoln Park Zoo have all voluntarily closed their elephant exhibits out of concern for their ability to develop adequately sized living enclosures for their considerable physical and psychological needs. Indeed, these animals require institutional commitments that reach beyond what even wealthy zoos can afford, and to include intelligent apes and marine mammals in such a calculus would be more than compassionate: it would speak volumes about the future of American zoos and their renewed commitment to an ethic of environmental stewardship and respect for all life on Earth.

## Zoos as Schools

As for reaching out to the public, zoos have invested heavily in their quickly obsolescing entertainment infrastructures built in the name of education, including giant 4D theaters, audiovisual kiosks, and panda webcams. But the best investments zoos can make in their educational programs are *human* resources—well-paid, whip-smart educators; enthusiastic and outgoing docents; and guest services personnel who know and care as much about environmental learning as they do Halloween festivals and birthday parties. In the American zoo of the future, *every* staff person will have to contribute to educating the public, if only by modeling proactive behaviors conducive to building healthy relationships with the city's urban/natural ecology and its biodiversity. (Perhaps zoo executives could start by leaving their wingtips and high heels at home, and stop holding champagne-swigging fund-raising soirees on zoo grounds.)

Zoos should also let their keepers and educational staffers off-leash when it comes to the informal gag rules typically placed on their communications with the public. In the Anthropocenic age, no zoo should allow "evolution" to become the forbidden "e-word," nor should educators be discouraged from teaching the public about topics that should never be considered controversial, even for small children, whether natural selection, global warming, mating and sexuality, animal death, or the food chain. (Trust me, kids can handle knowing what carnivorous owls eat, perhaps more than their parents.) Most of all, given that a majority of Americans live in a state of skepticism regarding the human causes of climate change, zoos should be taking the lead on instructing the public on the environmental crisis, quite possibly the single most important issue of our lifetimes.

In fact, flagship zoos could do something about this immediately, if they were so motivated. They could dedicate their IMAX theaters to showing serious documentaries on the climate change crisis, from Davis Guggenheim's *An Inconvenient Truth* to Showtime's Emmy-winning series *Years of Living Dangerously*, instead of exclusively juvenile content like *Happy Feet*, *SpongeBob SquarePants*, and other animated shorts aimed at toddlers. Zoos could also recommend relevant books (perhaps even this one) on these vital environmental issues, and make them available for purchase in their jam-packed gift shops—classics like Charles Darwin's *The Origin of Species* and Rachel Carson's *Silent Spring*, and also more contemporary books on the environmental crisis written for general audiences, such as Bill McKibben's *The End of Nature* (1989), Al Gore's *An Inconvenient Truth* (2006), Alan Weisman's *The World Without Us* (2007), Thomas L. Friedman's *Hot, Flat and Crowded* (2008), and Elizabeth Kolbert's *The Sixth Extinction* (2014). Unlike the regular fare sold in zoo "trading posts" and "zootiques," these books may not include cute photographs of koalas or ocelots, but might nevertheless enlighten parents and assist them in speaking to their children, family and friends about our simmering planet. Considering that most visitors have to exit the zoo through these souvenir-stocked tourist traps anyway, they would be sure to come across them while perusing the vast selection of plush clownfish toys and gorilla shot glasses such stores usually have on offer. The zoo would also provide an ideal environment for hosting lectures, local field trips, and reading groups on these kinds of topics, just as they sponsor and lead zoo photography clubs and eco-tourism adventures. Just as great urban libraries like the New York Public Library and the Philadelphia Free Library serve as centers for general education and community engagement around reading, zoos could serve a similar function as civic hubs of conservation and environmental education.

Finally, I argued earlier that zoos could be run more efficiently by dedicating their resources toward exhibiting fewer species in greater depth, along with smaller and less popular animals than those large charismatic mammals typically featured in a zoo's Big Twelve. Exotic insects and other arthropods—even the Madagascar hissing cockroach—could be used to teach about how dynamic ecologies work, and provide opportunities for young people to encounter animal life up-close. Meanwhile, more common critters native to American habitats could teach zoo guests more about the wildlife in their own backyards, while mixed-species exhibits would emphasize the symbiotic relationships within such ecosystems. An emphasis on small amphibians such as the Panamanian golden frog would also allow for a wide-ranging set of exhibits on the vulnerability of delicate ecosystems around the world. In fact, during the course of my research I came across an unusual manifesto recommending something close to a number of these ideas. It had a funny title: "How to Exhibit a Bullfrog: A Bedtime Story for Zoo Men." If the reader will patiently indulge me just once more, here is an excerpt from this strange essay:

> He inaugurated my dream with a question. I shall simply call him "M."
>
> "How is it," he asked, "that you are trying to buy a pygmy chimpanzee for $5,000 when you don't have a proper exhibit of bullfrogs?"
>
> "Bullfrogs!" I replied. "What in the world will common bullfrogs do for the Zoo?"
>
> "Oh," said M. "You want to do something for the Zoo! I thought you might want to do something for the zoo visitor. But no, you would prefer to spend thousands on a creature that only another zoo man will appreciate

rather than present an educational exhibit of a fascinating creature who lives in your own backyard. . . ."

"Zoo men don't have any inspiration. Why, an entire zoo could be devoted to the bullfrog; a major building is hardly adequate to present the excitement. The lessons this animal has to teach—if only the zoo man had imagination! . . ."

"That's all very well," I replied, "but it is also a zoo's responsibility to show many kinds of animals to give the zoo-goer some idea of the wonderful variety of animal life. We can't settle for fifteen species."

"True, true," M. conceded; "but do you need fifteen hundred? What possible excuse have you for acquiring more animals than you can meaningfully exhibit? . . . You must give your visitors a new intellectual reference point, meaningful and aesthetically compelling; a view of another sensory and social world."[3]

Two points should be made about this manifesto. First, there is nothing new about its message: in fact, "How to Exhibit a Bullfrog" was written and delivered in the mid-1960s at the beginning of that era's conservation movement. (Still, even today its prescriptions are refreshing and highly fitting for *this* generation's still nascent environmental call to arms.) Second, this essay's author was no fringe member of the zoo world (much less an idealistic sociologist) but William G. Conway, who at the time was the ultimate zoo insider—the director of the New York Zoological Society and the Bronx Zoo. His words were wise decades ago and are just as relevant for contemporary American zoos if they are to evolve into the schools of conservation that future U.S. cities and their citizens will desperately need, particularly if they are to make informed decisions about their own environmental security and the Earth's ecological future.

## Zoos as Showcases

We are used to thinking about zoos as collections of miniature replicas of natural habitats around the globe: the jungles of Brazil, the frozen tundra of Antarctica, the grassland savannas of Tanzania, the islands of Galapagos. Yet these worlds have been forever altered in the Antropocene, and many are disappearing at a rapid clip. For Americans, what will remain are our congested cities and their metropolitan areas, along with what is quickly becoming the new urban normal: brutal heat waves, extreme hurricanes, wild tornados, and intense flooding from encroaching tides. In just a few generations, rising sea levels may devastate highly populated coastal cities around the country: New York, Boston, Los Angeles, New Orleans, San Diego, Seattle, Miami. American cities will need to develop innovative strategies for both dramatically reducing greenhouse gas emissions and adapting to urban life on a hot planet, and zoos and other public attractions can provide models for their hometown cities to follow. To a certain extent some zoos already do this. In chapter 6 we learned that the Philadelphia Zoo teaches visitors about conservation by showcasing the environmental sustainability of its own infrastructure, demonstrating how local residents, businesses, public institutions, and entire municipalities might achieve a diminished carbon footprint by investing in geothermal wells, green roofs, solar farms, and other renewable water and energy systems. Since many zoos already employ this green infrastructure for their current operations, perhaps they could also showcase these technologies for their cities, and provide the inspiration, knowledge, and support necessary to scale up such operations for local or regional metropolitan use.

Zoos can also provide models for their cities by insisting that their oil-producing corporate benefactors—ExxonMobil, Chevron, Citgo, Hess—as well as their municipalities invest in sustainable energy technologies that could be introduced to the public on zoo grounds. Zoos could become showcases for a city's first windmill farm, smart power grid, bike- and car-share program, or neighborhood greening initiative. Solar-paneled food trucks could serve visitors locally sourced produce, while zoo garden paths might showcase more native and wild plants. Hybrid city buses could offer free rides to zoo visitors, while zoo parking lots could have battery chargers for electric cars and fry-o-diesel fueling stations. These showcases could serve cities as models of both sustainable and symbolic infrastructure for urban living.

## The Future of the American Zoo

American zoos are not perfect institutions, by no means; nor are they morally bankrupt. Caught between multiple and competing priorities, zoos must educate the public about human evolution, biodiversity loss, and global warming while providing enough satisfying entertainment to attract audiences willing to hear their message. They must devote resources toward building ever more sophisticated environments for their animal inhabitants while simultaneously paying their professional human workers a livable wage. They must invest in their own economic and organizational sustainability, but without forgetting the importance of the Earth's *ecological* sustainability.

Facing these tensions has never been easy, but it is hardly impossible. In *The Zookeeper's Wife*, Diane Ackerman recounts a heroic tale of how after the German invasion of Poland during World War II, the director of the Warsaw Zoo and his wife secretly sheltered hundreds of Jews in empty animal exhibits, along with some of the zoo's remaining hyenas, otters, and lynxes.[4] Her book is a compelling story of war, resistance, compassion, and survival, yet it also provides a lesson about the American zoo of the future: if an ordinary city zoo and its cages could be repurposed as a safe haven for Jewish Poles marked for death during the Holocaust, then we can make our zoos anything we need them to be, whether sanctuaries for protecting wildlife, schools to teach the public about the Earth's biological riches and vulnerablities, or showcases for urban life on a warming planet. American metropolitan zoos may be built upon edifices of fiberglass and cast-in-place concrete, but they still can be reimagined as special places where we might seek and even discover deeper understandings of the human constitution of the natural world and its uncertain future, right here in the tangled urban jungle of the Anthropocene.

# Acknowledgments

Beginning at its conception, a book is like a complex living organism, requiring a healthy mix of genes, constant nourishment, a sustainable ecological habitat, and eventually the ability to reproduce (or at least *be* reproduced). This book in particular went through a long process of fertilization punctuated by endless strolls through Chicago's Lincoln Park Zoo as a graduate student in my twenties, and eventually family trips to the Philadelphia Zoo as the father of a young boy with an interest in animals. Its moment of birth actually occurred as a joke made during one such visit to the zoo with my Penn colleague and fellow ethnographer John Jackson, along with our respective kids. Like many social science professors of our subspecies, we could not help but talk about work even as we escorted our children past African elephants, Amur tigers, and Colombian black spider monkeys. At one point I stopped before an animal enclosure and started comparing the world of a zoo to a blues nightclub (the topic of my first book)—both feature backdrops of staged authenticity, tourist-packed audiences, a cast of performers and promoters, and souvenir shops full of trinkets for sale—and how great it would be to go behind the scenes and see how it all worked from the inside out—from blues to zoos, as it were. I meant it as a joke, but John didn't even crack a smile (a common response to my humor), and told me with a straight face, "You *have* to do that. You have to see that through." And thus a joke became a book, and I thank John for his encouragement.

Of course, he had done the easy part—now what to do? As readers know by now, I eventually took a job volunteering at two different metropolitan zoos, City Zoo and Metro Zoo, and I must thank the friendly and conscientious staffs at both institutions, particularly the zookeepers, educators, and fellow volunteers who made conducting fieldwork in their enriching environments so rewarding. I must also thank those zoo professionals and other informants who gave generously of their time to meet with me to elaborate on the complexities of zookeeping and exotic animal training, veterinary care, zoo education and conservation, exhibit architecture and design, public relations, volunteerism, and animal advocacy activism. Finally, an extra helping of mealworms goes to all the animals whom I had the pleasure of handling—the snakes, tortoises, sheep, rabbits, alligators, goats, lizards, owls, horses, cockroaches, bantams, and countless others. Obviously, you never gave me informed consent to work with you; I can only offer my thanks, and hope for your forgiveness. I will always cherish our time together. (Except maybe for Beaker, who bit the bejesus out of me—again, not just once, but twice. No, no, Beaker, I thank you as well—we made a great performing team, you know, except for the biting part.)

The book's nourishment was sustained over time by a number of generous benefactors and colleagues. The University of Pennsylvania supported me with sabbatical leave as well as numerous financial awards, including a University Research Foundation Award, an Andrew W. Mellon Faculty Fellowship associated with the Penn Humanities Forum, and course development grants from the Netter Center for Community Partnerships and the Benjamin Franklin Scholars program. The School of Social Science at the Institute for Advanced Study in Princeton, New Jersey, also gave me salary support as well as the most enriching environment a scholar writing a book could hope to inhabit.

These two institutions offered my project not only financial sustainability but also a set of symbiotic relationships among supportive troops of sociable human primates. The Sociology Department at the University of Pennsylvania has been my academic and intellectual home since 2001, and I thank Tukufu Zuberi and Emilio Parrado for their leadership and encouragement during the life of this project, as well as a number of close supportive colleagues, including Randy Collins, Kristen Harknett, Jerry Jacobs, Grace Kao, Annette Lareau, Robin Leidner, Sam Preston, Jason Schnittker, and Melissa Wilde. Katie Breiner, Betsie Garner, Lauren Johnston, Marlena Mattei, Lauren Springer, Stephanie Vogel, and Natalie Volpe all provided essential research assistance, and I remain grateful for their commitment to the project. They are a talented bunch, and all have tremendous futures ahead of them. I also thank the Penn undergraduate students enrolled in my fall 2014 seminar on zoos and the culture of nature for their insights and thoughtful observations throughout the semester.

Penn faculty members and students Lainie Bailey, Catherine Brinkley, Philip Cochetti, Randy Collins, Jim English, Betsie Garner, Annette Lareau, and Junhow Wei all read or discussed drafts of manuscript chapters during its many iterations and gave generous feedback when I needed it, as did a raft of colleagues from beyond Penn's ivy walls: Meredith Broussard, Marjorie DeVault, Brian Donovan, Alice Goffman, Tim Hallett, Philip Howard, Colin Jerolmack, Victoria Johnson, Caroline Lee, Gemma Mangione, Lauren Rivera, Tyson Smith, and Eviatar Zerubavel. (In addition, Lainie Bailey and Gemma Mangione both graciously read the entire manuscript before its final revision.) The book has also benefited greatly from remarks made by participants at talks given at Arizona, Chicago, the Institute for Advanced Study, Michigan, Northeastern, Princeton, Purdue, Rutgers, Stanford, Umeå University in Sweden, Yale, and the University of Pennsylvania's Urban Ethnography Workshop and Penn Humanities Forum.

In addition, several friends and colleagues accompanied me to various zoos across the country from Philadelphia to Chicago to Las Vegas, sharing their insights and wit along the way: Catherine Brinkley, Sean Davis, Scott Hanson, Kristen Harknett, Colin Jerolmack, Vivian Levy, Dawne Moon, Christena Nippert-Eng, Guy Raviv, Lauren Rivera, Michael Rosenfeld, Jason Schnittker, Dinh Tran, Fred Wherry, Matt Wray, Jane Zavisca, and of course John Jackson. Bryant Simon provided his New Jersey Shore house as a base of operations for my time spent conducting research at the Cape May County Zoo. My parents' apartment just outside New York City served a similar function for research conducted at the Bronx Zoo, Central Park Zoo, and New Jersey's Bergen County Zoo. For their friendship I also thank Mike Cimicata, John Doyle, Dave Gerridge, Jerome Hodos, and Jay Kirk.

At the School of Social Science at the Institute for Advanced Study I gained a brand-new set of colleagues during the 2013–14 academic year, and while I cannot thank them all here, I do wish to single out those whose constructive feedback went above and beyond the call of duty: Didier Fassin, Wendy Griswold, Joe Hankins, Dale Jamieson, Joe Masco, Ann McGrath, Sverker Sorlin, and Richard York. Fellow sociologists Beth Popp Berman, Nitsan Chorev, and John Padgett also provided helpful suggestions on the project, as did the participants and organizers of the school's working groups in Ethnography and Theory, and the Environmental Turn in the Human Sciences.

Of course, the survival of any manuscript comes from its reproduction, and for that I must thank Meagan Levinson, Eric Schwartz, Fred Appel, Ryan Mulligan, Ellen Foos, Joseph Dahm, Andrew Harvard, Maria denBoer, and the entire team at Princeton University Press, including two not-very-anonymous but extraordinarily generous reviewers, Gary Alan Fine and Shamus Khan, for an all-around rewarding experience.

Finally, I could not have even begun, much less finished, this book without the unconditional love, encouragement, and thirst for life that my wife, Meredith Broussard, and our son both give me every day. Thank you both for agreeing to accompany me across the country on this zany adventure—especially my son, who never really had much of a say in the matter, but always managed to endure our obsessive zoo visits with grace, along with the occasional bribe of a stuffed animal or junk food. For his sake, I promise my next book will be about doughnut shops and video games.

# Notes

## Introduction: The World in a Zoo

1. On the continued strength of culture/nature distinctions in modern life, see Jennifer Price, *Flight Maps: Adventures with Nature in Modern America* (New York: Basic, 1999), 160–65.

2. Another product of that era was the development of sociology itself, which also drew heavily on iterations of the nature/culture distinction: community and society, folk and urban, primitive and civilized, agriculture and industry, and the country and the city; for example, see Georg Simmel, "The Metropolis and Mental Life" (1903), in *On Individuality and Social Forms*, Donald N. Levine, ed. (Chicago: University of Chicago Press, 1971); Robert E. Park, "The City: Suggestions for the Investigation of Human Behavior in the Urban Environment," in *The City*, Robert E. Park, Ernest W. Burgess, and Roderick D. McKenzie, eds. (Chicago: University of Chicago Press, 1925); Louis Wirth, "Urbanism as a Way of Life," *American Journal of Sociology* 44 (1938): 1–24.

3. Vernon N. Kisling, Jr., "The Origin and Development of American Zoological Parks to 1899," in *New Worlds, New Animals: From Menagerie to Zoological Park in the Nineteenth Century*, R. J. Hoage and William A. Deiss, eds. (Baltimore: Johns Hopkins University Press, 1996).

4. These successive iterations occur just as fractals in nature replicate their basic symmetry and self-similarity without end; on the properties of fractals and their application in the sociology of culture and knowledge, see Andrew Abbott, *Chaos of Disciplines* (Chicago: University of Chicago Press, 2001), 3–33.

5. Mike Davis, *Ecology of Fear: Los Angeles and the Imagination of Disaster* (New York: Vintage, 1999); Michael Sullivan, *Rats: Observations on the History and Habitat of the City's Most Unwanted Inhabitants* (New York: Bloomsbury, 2004); Sandy Bauers, "Philadelphia's Raccoon Problem," *Philadelphia Inquirer*, 2 November 2011; Hillary Angelo and Colin Jerolmack, "Nature's Looking Glass," *Contexts* 11 (2012): 24–29; Colin Jerolmack, *The Global Pigeon* (Chicago: University of Chicago Press, 2013).

6. Jared Diamond, *The Third Chimpanzee: The Evolution and Future of the Human Animal* (New York: HarperCollins, 1992), 23; Eviatar Zerubavel, "Lumping and Splitting: Notes on Social Classification," *Sociological Forum* 11 (1996): 424.

7. Arnold Arluke and Clinton R. Sanders, *Regarding Animals* (Philadelphia: Temple University Press, 1996), 132–66; Hal Herzog, *Some We Love, Some We Hate, Some We Eat: Why It's So Hard to Think Straight about Animals* (New York: Harper, 2010), 58–59; Hugh Raffles, *Insectopedia* (New York: Pantheon, 2010), 141–61.

8. William Cronon, "The Trouble with Wilderness; or, Getting Back to the

Wrong Nature," in *Uncommon Ground: Rethinking the Human Place in Nature,* William Cronon, ed. (New York: Norton, 1996), 70.

9. Ibid., 79.

10. On the cultural and social construction of nature and the collective meanings people attach to nonhuman animals, landscapes, weather events, and other empirical features of the ambient environment, see Clifford Geertz, "Deep Play: Notes on the Balinese Cockfight," in *The Interpretation of Cultures* (New York: Basic, 1973), 412–53; Michael Mayerfeld Bell, *Childerley: Nature and Morality in a Country Village* (Chicago: University of Chicago Press, 1994); Gary Alan Fine, *Morel Tales: The Culture of Mushrooming* (Cambridge, MA: Harvard University Press, 1998) and *Authors of the Storm: Meteorologists and the Culture of Prediction* (Chicago: University of Chicago Press, 2007); Eric Klinenberg, *Heat Wave: A Social Autopsy of Disaster in Chicago* (Chicago: University of Chicago Press, 2002); Jerolmack, *Global Pigeon.*

11. Candace Slater, "Amazonia as Edenic Narrative," in Cronon, *Uncommon Ground,* 114–31; Charles C. Mann, *1491: New Revelations of the Americas before Columbus* (New York: Vintage, 2011), 319–59, quotes appear on 318, 354.

12. Anne Whiston Spirn, "Constructing Nature: The Legacy of Frederick Law Olmstead," in Cronon, *Uncommon Ground,* 95–99, quote appears on 97.

13. N. Katherine Hayles, "Simulated Nature and Natural Simulations: Rethinking the Relation between the Beholder and the World," in Cronon, *Uncommon Ground,* 410.

14. Jon Krakauer, *Into the Wild* (New York: Anchor, 1997), quote appears on 174.

15. "IPCC, 2013: Summary for Policymakers," in *Climate Change 2013: The Physical Science Basis. Contribution of Working Group I to the Fifth Assessment Report of the Intergovernmental Panel on Climate Change* (Cambridge: Cambridge University Press, 2013); also see Al Gore, *An Inconvenient Truth: The Planetary Emergency of Global Warming and What We Can Do about It* (New York: Rodale, 2006); Bill McKibben, *The End of Nature* (New York: Random House, 2006) and *Eaarth: Making a Life on a Tough New Planet* (New York: St. Martin's Griffin, 2011); Jan Zalasiewicz, Mark Williams, Will Steffen, and Paul Crutzen, "The New World of the Antropocene," *Environmental Science and Technology* 44 (2010): 2228–31.

16. As reported in the journal *Science,* "recent extinction rates are 100 to 1,000 times their pre-human levels in well-known, but taxonomically diverse groups from widely different environments"; see Stuart L. Pimm, Gareth J. Russell, John L. Gittleman, and Thomas M. Brooks, "The Future of Biodiversity," *Science* 269 (1995): 347; also see F. Stuart Chapin III et al., "Consequences of Changing Biodiversity," *Nature* 405 (2000): 234–42; Elizabeth Kolbert, *The Sixth Extinction: An Unnatural History* (New York: Henry Holt, 2014).

17. Dipesh Chakrabarty, "The Climate of History: Four Theses," *Critical Inquiry* 35 (2009): 197–222.

18. Zalasiewicz et al., "New World of the Antropocene," 2228–31, quotes appear on 2228; also see Paul J. Crutzen, "Geology of Mankind," *Nature* 415 (2002): 23; Will Steffen, Paul J. Crutzen, and John R. McNeill, "The Anthropocene: Are Humans Now Overwhelming the Great Forces of Nature?," *AMBIO: A Journal of the Human Environment* 36 (2007): 614–21; Will Steffen et al., "The Anthropocene: From Global Change to Planetary Stewardship," *AMBIO: A Journal of the Human Environment* 40

(2011): 739–61. Also see Elizabeth Kolbert, "Enter the Anthropocene—Age of Man," *National Geographic*, March 2011 and *Sixth Extinction*.

19. Steffen et al., "Anthropocene," 41, 56–57.

20. Eviatar Zerubavel, *Time Maps: Collective Memory and the Social Shape of the Past* (Chicago: University of Chicago Press, 2003).

21. This kind of research is known in the social sciences as ethnographic field-work, or *participant observation*. The sociologist Erving Goffman, "On Fieldwork," *Journal of Contemporary Ethnography* 18 ([1974] 1989): 125–26, once defined participant observation as collecting data "by subjecting yourself, your own body and your own personality, and your own social situation, to the set of contingencies that play upon a set of individuals, so that you can physically and ecologically penetrate their circle of response to their social situation . . . so that you are close to them while they are responding to what life does to them . . . as a witness to how they react to what gets done to and around them." In my role as a volunteer worker, I conducted this type of immersive and participatory ethnographic fieldwork at these two zoos for a combined total of six hundred hours, with my time divided between manual zoo work and educational encounters with the public. During this time I jotted down notes, recorded voice memos, and took digital photographs while on site, and when offsite I wrote thousands of pages of detailed field notes documenting my observations and reflections. In this book I refer to these zoos pseudonymously as City Zoo and Metro Zoo, in keeping with both disciplinary conventions in the social sciences and those formally approved by the Institutional Review Board (IRB) of the Office of Regulatory Affairs at the University of Pennsylvania. I also changed certain superficial details of each zoo to further protect their identities.

22. Courtney Bender, *Heaven's Kitchen: Living Religion at God's Love We Deliver* (Chicago: University of Chicago Press, 2003), similarly observes the culture of kitchen work and collaborative food preparation among volunteers preparing meals for those who face difficulties cooking for themselves. (In Bender's case, clients are people debilitated by HIV/AIDS, rather than zoo animals.)

23. As with numerous public institutions, many zoos are transitioning to public/private partnership models, with varying results.

24. For the most current list of AZA-accredited zoos, see https://www.aza.org /current-accreditation-list/. Zoo statistics appear on the AZA website: https://www.aza .org/zoo-aquarium-statistics/.

25. This is likely due to dominant fears among Americans concerning the (very rare) abduction or molestation of children by male strangers; Barry Glassner, *The Culture of Fear: Why Americans Are Afraid of the Wrong Things* (New York: Basic, 2009), 29–40; Hanna Rosin, "Hey! Parents, Leave Those Kids Alone," *Atlantic*, April 2014. In addition, Arlie Russell Hochschild and Anne Machung, *The Second Shift* (New York: Penguin, 2003), observe that mothers are burdened with the majority of household labor, including parenting. They therefore may be more likely than fathers to take their children on zoo outings. For all these reasons, public areas of city zoos are gendered spaces in which the very presence of adult men (but not women) is considered problematic until validated by an accompanying child or family; on the gendering of public spaces more generally, see Erving Goffman, "The Arrangement between the Sexes," *Theory and Society* 4 (1977): 315–16; Doreen Massey, *Space, Place, and Gender* (Minneapolis: University of Minnesota Press, 1994), 186.

26. On the deployment of children in ethnographic research in sociology, see Hilary Levey, "Which One Is Yours? Children and Ethnography," *Qualitative Sociology* 32 (2009): 311–31, who argues that children can be instrumental as "wedges" that facilitate an adult ethnographer's introduction into a field research site.

27. These research assistants were all enrolled as undergraduate or graduate students at the University of Pennsylvania, and I thank them all for their painstaking work throughout the project: Betsie Garner, Marlena Mattei, Lauren Springer, and Natalie Volpe.

28. I conducted face-to-face interviews, which ran from forty-five minutes to four hours, with twenty-six key informants in all. Nearly all interviews were digitally recorded with informed consent (except for one interview for which I took copious notes by hand, as the informant requested that I not record our conversation). In keeping with both ethnographic conventions and those formally approved by the IRB of the Office of Regulatory Affairs at the University of Pennsylvania, I use pseudonyms to protect the identity of each of my informants, with exceptions where noted in cases where individuals agreed to be identified by name. I also attended the 2011 AZA Annual Conference in Atlanta, where I met with zoo industry personnel from around the country; attended numerous sessions on zoo design and exhibition, animal management, conservation, and zoo branding; and toured both public and backstage areas of Zoo Atlanta and the Georgia Aquarium with conference participants.

29. Like both prisons and certain kinds of hospitals, zoos function as total institutions for their captive animal populations; see Erving Goffman, *Asylums: Essays on the Social Situation of Mental Patients and Other Inmates* (New York: Anchor, 1961); Michel Foucault, *Discipline and Punish: The Birth of the Prison*, Alan Sheridan, trans. (New York: Vintage, 1979).

30. According to Amy Sutherland, *Kicked, Bitten, and Scratched: Life and Lessons at the World's Premier School for Exotic Animal Trainers* (New York: Penguin, 2007), 60, in 2000 women represented three-quarters of all zookeepers nationwide. On the recent feminization of other animal-related fields such as veterinary medicine, see Leslie Irvine and Jenny R. Vermilya, "Gender Work in a Feminized Profession: The Case of Veterinary Medicine," *Gender & Society* 24 (2010): 56–82; Anne E. Lincoln, "The Shifting Supply of Men and Women to Occupations: Feminization in Veterinary Education," *Social Forces* 88 (2010): 1969–88.

31. Paula England, Michelle Budig, and Nancy Folbre, "Wages of Virtue: The Relative Pay of Care Work," *Social Problems* 49 (2002): 455–73. The U.S. Bureau of Labor Statistics reports that the mean annual wage of a nonfarm animal caretaker in May 2012 was $22,370, with a mean hourly wage of $10.75 per hour. These low wages persist in a context that debases women and devalues their work. There is growing evidence that female-dominated occupations pay significantly less than jobs that employ men. Women are also far more likely than men to work in occupations (like zookeeping) that involve the direct delivery of care, and such jobs carry a relative wage penalty. Given its cultural associations with being a mother, professional caregiving is often crassly thought of as little more than housework that females are considered predisposed to perform out of love and compassion, rather than for earned pay. Zookeeper wages have thus remained stagnant even as job requirements have demanded increasingly higher levels of education, work experience, and skill. The college-educated figure comes from J. Stuart Bunderson and Jeffery A. Thompson, "The Call of the

Wild: Zookeepers, Callings, and the Double-edged Sword of Deeply Meaningful Work," *Administrative Science Quarterly* 54 (2009): 35.

# Chapter 1: Where the Wild Things Aren't: Exhibiting Nature in American Zoos

1. On the development of the "new naturalism" in zoos, see Jon Charles Coe, "What's the Message? Education through Exhibit Design," in *Wild Mammals in Captivity: Principles and Techniques*, Devra G. Kleiman, Mary E. Allen, Katerina V. Thompson, and Susan Lumpkin, eds. (Chicago: University of Chicago Press, 1996), 167–74; David Hancocks, *A Different Nature: The Paradoxical World of Zoos and Their Uncertain Future* (Berkeley: University of California Press, 2001), 111–48.

2. American and European audiences have long preferred such idealized natural environments to encroaching industrialized landscapes, a product of nineteenth-century Romanticism. For this reason, the twentieth century has been marked by successive attempts at providing American cultural consumers with naturalistic experiences through immersive exhibition and spectacle. Early examples include Carl Hagenbeck's Arctic panorama suggestive of a polar landscape filled with sea lions and penguins at the 1904 St. Louis World's Fair; Carl Akeley's arrestingly realistic dioramas in his Hall of African Mammals at New York's American Museum of Natural History; and the expertly landscaped African Plains exhibit at the Bronx Zoo and the 1,500-acre Jackson Hole Wildlife Park in Wyoming, both launched by zoo president Fairfield Osborn in the 1940s. See Gregg Mitman, "When Nature *Is* the Zoo: Vision and Power in the Art and Science of Natural History," *Osiris* 11 (1996): 117–43; Hancocks, *Different Nature*, 104; Elizabeth Hanson, *Animal Attractions: Nature on Display in American Zoos* (Princeton, NJ: Princeton University Press, 2002), 142; Jay Kirk, *Kingdom under Glass: A Tale of Obsession, Adventure, and One Man's Quest to Preserve the World's Great Animals* (New York: Henry Holt, 2010). On the culture of enchantment surrounding the natural world, see James William Gibson, *A Reenchanted World: The Quest for a New Kinship with Nature* (New York: Metropolitan, 2009).

3. W.J.T. Mitchell, *The Last Dinosaur Book: The Life and Times of a Cultural Icon* (Chicago: University of Chicago Press, 1998), 48–50; Edward O. Wilson, *Biophilia* (Cambridge, MA: Harvard University Press, 1984), 51.

4. According to Terry L. Maple and Lorraine A. Perkins, "Enclosure Furnishings and Structural Environmental Enrichment," in Kleiman et al., *Wild Mammals in Captivity*, 213, this is a familiar problem with a variety of charismatic zoo animals, including lions. For instance, in the wild howler monkeys spend 70 percent of their day at rest; and walruses, 67 percent.

5. Natural history museums and zoos are obviously very different types of culture-producing organizations, yet their similarities are nevertheless striking. Many natural history museums feature live animal collections, and exhibit butterflies, frogs, and other small animals. (And like museums of natural history, some zoos also display dinosaur fossils as well as other animal reproductions.) Natural history museums incorporate education, research, and typically conservation in their institutional missions, and many employ former zoo workers. On the history of U.S. natural history

museums, see Mitchell, *Last Dinosaur Book*; Steven Conn, *Museums and American Intellectual Life, 1876–1926* (Chicago: University of Chicago Press, 1998), 32–73, and *Do Museums Still Need Objects?* (Philadelphia: University of Pennsylvania Press, 2010), 138–71; Kirk, *Kingdom under Glass*.

6. As Erving Goffman reminds us in *Frame Analysis: An Essay on the Organization of Experience* (New York: Harper & Row, 1974), 130, the theatergoer "collaborates in the unreality onstage. He sympathetically and vicariously participates in the unreal world generated by the dramatic interplay of the scripted characters. He gives himself over." On the production of staged or manufactured authenticity, see Dean MacCannell, *The Tourist: A New Theory of the Leisure Class* (Berkeley: University of California Press, 1999); David Grazian, *Blue Chicago: The Search for Authenticity in Urban Blues Clubs* (Chicago: University of Chicago Press, 2003).

7. Nature making should not be confused with the concept of "naturework" in Fine, *Morel Tales*, 2, which refers to the process by which consumers (rather than cultural creators) inscribe the natural environment with meaning.

8. Culture/nature boundaries are socially constructed and negotiated in the context of lived experience and interaction; see Bell, *Childerley*; Fine, *Morel Tales*; Jerolmack, *Global Pigeon*.

9. According to the website of the Association of Zoos and Aquariums (AZA), "In the last 10 years, AZA-accredited zoos and aquariums formally trained more that 400,000 teachers, supporting science curricula with effective teaching materials and hands-on opportunities. School fieldtrips connected more than 12,000,000 students with the natural world"; see https://www.aza.org/public-benefits/.

10. See https://www.aza.org/visitor-demographics/. Hochschild and Machung, *Second Shift*, observe that mothers are burdened with the majority of household labor, including parenting, which may explain why mothers are more likely than fathers to take their children to zoos.

11. Annette Lareau, *Unequal Childhoods: Class, Race, and Family Life* (Berkeley: University of California Press, 2003).

12. James Parker, "The Beast Within," *Atlantic*, June 2011, 42–45.

13. For instance, Hanson, *Animal Attractions*, describes some of the expeditions conducted in Africa and the Dutch East Indies to collect wildlife for the Smithsonian National Zoo during the early to mid-twentieth century.

14. Temple Grandin and Catherine Johnson, *Animals Make Us Human: Creating the Best Life for Animals* (Boston: Mariner, 2010), 265.

15. On the collaborative and participatory process of designing and building zoo exhibits among a variety of experts and institutional stakeholders, see Gwen Harris, "Building a Drain at the Lowest Point: The Role of a Zoo Keeper during Exhibit Design and Construction," *Animal Keepers' Forum* 41 (2014): 138–40. The collaborative and interactional work of nature makers reminds us that zoos are not merely organizational and industry actors but "inhabited institutions"; see Tim Hallett and Marc J. Ventresca, "Inhabited Institutions: Social Interactions and Organizational Forms in Gouldner's *Patterns of Industrial Bureaucracy*," *Theory and Society* 35 (2006): 213–36; Amy Binder, "For Love and Money: Organizations' Creative Responses to Multiple Environmental Logics," *Theory and Society* 36 (2007): 547–71; Tim Hallett, "The Myth Incarnate: Recoupling Processes, Turmoil, and Inhabited Institutions in an Urban Elementary School," *American Sociological Review* 75 (2010): 52–74.

16. On the collaborative and social nature of art and cultural production more generally, see Howard S. Becker, *Art Worlds* (Berkeley: University of California Press, 1982); David Grazian, *Mix It Up: Popular Culture, Mass Media, and Society* (New York: Norton, 2010).

17. Coe, "What's the Message?," 171.

18. Carolyn Marshall, "Tiger Kills 1 after Escaping at San Francisco Zoo," *New York Times*, 26 December 2007; Andrew Grossman, "Snake Escape Makes Zoo Squirm," *Wall Street Journal*, 28 March 2011.

19. David Hancocks, "The Design and Use of Moats and Barriers," in Kleiman et al., *Wild Mammals in Captivity*, 199; "Park Is Sued over Death of Man in Whale Tank," *New York Times*, 21 September 1999; Phillip T. Robinson, *Life at the Zoo: Behind the Scenes with the Animal Doctors* (New York: Columbia University Press, 2004), 77; Sarah Rainey, "Drunk Zoo Visitor Attacked by Monkeys after Climbing into Pen," *Telegraph*, 18 November 2011; Colin Moynihan, "Man Mauled by Tiger at Zoo Is Charged, *New York Times*, 22 September 2012; Adam Vaughan, "Man Pours Beer over Tiger as London Zoo Lates Parties Get Out of Hand," *Guardian*, 18 July 2014.

20. Stephen Bitgood, Donald Paterson, and Arlene Benefield, "Exhibit Design and Visitor Behavior: Empirical Relationships," *Environment and Behavior* 20 (1988): 487.

21. I was shown these behind-the-scenes holding cells during the 2011 AZA Annual Conference in Atlanta, when I was invited to tour various backstage areas of Zoo Atlanta and the Georgia Aquarium along with other conference participants.

22. Hancocks, "Design and Use of Moats and Barriers," 197–98.

23. Ibid., 194.

24. Signage at New Jersey's Cape May County Zoo not only urges guests to refrain from feeding its animals, but also highlights five common foods toxic to several species housed in its collection: chocolate; avocado; raw onion; macadamia nuts; and cherry, plum, apple, and apricot seeds. As for other dangers posed by visitors, Vicki Croke, *The Modern Ark: The Story of Zoos: Past, Present and Future* (New York: Scribner, 1997), 99, reports that once during a trip to the Bronx Zoo she witnessed "a small female gorilla recovering from a head wound she received from a zoogoer who had thrown a rock at her."

25. Bitgood, Paterson, and Benefield, "Exhibit Design and Visitor Behavior," 489.

26. Given their spatial constraints, city zoos face serious challenges when attempting to keep elephants in captivity. In the wild, African elephants residing in arid regions of the northern Namib Desert in Namibia travel an average of about 16 to 17 miles per day, and even under less extreme environmental conditions, elephants living in the wild still generally travel between 3.1 and 6.2 miles daily; see Katherine A. Leighty, Joseph Soltis, Christina M. Wesolek, Anne Savage, Jill Mellen, and John Lehnhardt, "GPS Determination of Walking Rates in Captive African Elephants," *Zoo Biology* 28 (2009): 17. For this reason, in recent years the Detroit Zoo, San Francisco Zoo, Philadelphia Zoo, Toronto Zoo, and Chicago's Brookfield Zoo and Lincoln Park Zoo have all voluntarily closed their elephant exhibits out of concern for their ability to develop adequately sized living enclosures for their pachyderms' physical and psychological needs.

27. Susan G. Brown, William P. Dunlap, and Terry L. Maple, "Notes on Water-Contact by a Captive Male Lowland Gorilla," *Zoo Biology* 1 (1982): 243–49.

28. Croke, *Modern Ark*, 15; Mark A. Rosenthal and William A. Xanten, "Structural and Keeper Considerations in Exhibit Design," in Kleiman et al., *Wild Mammals in Captivity*, 223. On the inconsistencies with which we evaluate the relative treatment of animals in different contexts, see Herzog, *Some We Love*.

29. Dian Fossey, *Gorillas in the Mist* (Boston: Mariner, 1983), 47–48. It is therefore perhaps not surprising that, as Croke, *Modern Ark*, 15, observes, "Gorillas in Atlanta, Georgia, destroyed $20,000 worth of plantings in the first month of living in their new outdoor enclosure."

30. Bitgood, Paterson, and Benefield, "Exhibit Design and Visitor Behavior," 482–84.

31. John Seidensticker and James G. Doherty, "Integrating Animal Behavior and Exhibit Design," in Kleiman et al., *Wild Mammals in Captivity*, 180–90, esp. 187.

32. Kathy Carlstead, "Effects of Captivity on the Behavior of Wild Mammals," in Kleiman et al., *Wild Mammals in Captivity*, 328.

33. More alarming stereotypic behaviors include chain and wood chewing, self-mutilation, regurgitation, and coprophagy; see Maple and Perkins, "Enclosure Furnishings and Structural Environmental Enrichment," 215; Croke, *Modern Ark*, 59; Kathy Carlstead, "Determining the Causes of Stereotypic Behaviors in Zoo Carnivores: Toward Appropriate Enrichment Strategies," in *Second Nature: Environmental Enrichment for Captive Animals*, David J. Sheperdson, Jill D. Mellen, and Michael Hutchins, eds. (Washington, DC: Smithsonian Institution Press, 1998), 172–83; Grandin and Johnson, *Animals Make Us Human*.

34. Sandy Bauers, "Philadelphia Zoo's Big Cat Crossing Gets a Test Run," *Philadelphia Inquirer*, 8 May 2014; Andrew Baker, "Zoo 360," *Connect*, January 2015, 36–43.

35. Bitgood, Paterson, and Benefield, "Exhibit Design and Visitor Behavior," 480–81; Maple and Perkins, "Enclosure Furnishings and Structural Environmental Enrichment," 215; Croke, *Modern Ark*, 27; Kathleen N. Morgan, Scott W. Line, and Hal Markowitz, "Zoos, Enrichment, and the Skeptical Observer: The Practical Value of Assessment," in Sheperdson, Mellen, and Hutchins, *Second Nature*, 153.

36. Maple and Perkins, "Enclosure Furnishings and Structural Environmental Enrichment," 218.

37. Joyce Chen, "Orangutans Latest Users of Apple's iPad; Orangutan Outreach Hopes to Facilitate Zoo-to-Zoo Skype," *New York Daily News*, 3 August 2011; Jan Uebelherr, "Orangutans Go Ape over iPad Apps," *Milwaukee Journal Sentinel*, 30 August 2011.

38. Coe, "What's the Message?," 171.

39. Some of the ASDM's enclosures also employ a patented type of quasi-invisible fencing called Invisinet, a mesh made from exceedingly thin, multistranded, bronze-colored stainless steel cables. According to Peggy Pickering Larson, *Arizona-Sonora Desert Museum: A Scrapbook* (Tucson: Arizona-Sonora Desert Museum Press, 2002), 44, the cables are hand-knotted into mesh by inmates at the Arizona State Prison in Tucson—where humans live in captivity, rather than animals.

40. Larson, *Arizona-Sonora Desert Museum*, 46.

41. Jon C. Coe, "Design and Perception: Making the Zoo Experience Real," *Zoo Biology* 4 (1985): 203–4; also see Coe, "What's the Message?," 171.

42. Catherine Brinkley allowed me to use her real name for this book.

43. Donald Jackson, "Landscaping for Realism: Simulating the Natural Habitats of Zoo Animals," *Arnoldia* 50 (1990): 14–17.

44. Donald W. Jackson, "Landscaping in Hostile Environments," *International Zoo Yearbook* 29 (1990): 14.

45. Jackson, "Landscaping for Realism," 19.

46. Ibid., 19–20.

47. At least beta-keratin supplements seem like a less risky coloration device than the petrochemical-based artificial dyes (such as canthaxanthin) used in salmon farming; see Marian Burros, "Issues of Purity and Pollution Leave Farmed Salmon Looking Less Rosy," *New York Times*, 28 May 2008. Of course, our preferences for pink flamingos in zoos may simply arise from the ubiquity of *plastic* pink flamingo lawn ornaments on American suburban yards during the second half of the twentieth century; see Price, *Flight Maps*, 111–65.

48. This might seem like a rather commonsensical approach to handling discussions of sex in front of children, yet it was not always so. According to Steven Conn, *Do Museums Still Need Objects?*, 161, in the 1940s Chicago's Museum of Science and Industry featured an exhibit aimed at children titled "The Miracle of Growth," which offered "a visually arresting—indeed, frank—display of human reproduction."

49. On the bizarre mating rituals of a variety of animal species, see Olivia Judson, *Dr. Tatiana's Sex Advice to All Creation: A Definitive Guide to the Evolutionary Biology of Sex* (New York: Holt, 2003).

50. For the curious: According to the Zoo Miami website, the Indian rhinoceroses in their collection can devour a hundred pounds of hay, produce, and food pellets in a single day.

51. There is functionality to this schedule as well, given that many of the zoo's raptors (such as the owls) are nocturnal, and therefore feed at night.

52. George B. Schaller, *The Serengeti Lion: A Study of Predator-Prey Relations* (Chicago: University of Chicago Press, 1972), 208–9.

53. Iliana V. Kohler, Samuel H. Preston, and Laurie Bingaman Lackey, "Comparative Mortality Levels among Selected Species of Captive Animals," *Demographic Research* 15 (2006): 413–34.

54. Of course, in backstage areas zoos regularly make use of pest-control devices such as ant traps and flypaper to capture and exterminate insects, sometimes of the same genus and/or species as those animals represented in the zoos' exhibited collections.

55. While the exhibit portrays its mini-laboratory as if it were practically an amusement park for rats, in fact multitudes of rats and other rodents, including rabbits and guinea pigs, have died in laboratory experiments conducted worldwide—and not only in the biomedical sciences and the pharmaceutical industry, but also for decidedly non-lifesaving purposes such as the testing of cosmetics, bleach, shampoo, ink, food coloring, weed-killers, and floor polish; see Peter Singer, *Animal Liberation* (1975; repr., New York: Avon, 1990), 25–94. According to the U.S. Humane Society, "In China alone, 300,000 animals die each year in cosmetic tests." See http://www.hsi.org/issues/becrueltyfree/facts/infographic/en/, accessed 9 June 2014.

56. On pigeon handlers in both New York and Berlin, see Jerolmack, *Global Pigeon*.

57. As repositories of culture, zoos and aquariums often exhibit creative works typically associated with art museums; on the artificial designation of both art objects and artistic exhibition spaces, see Becker, *Art Worlds*.

58. Marcus Eriksen et al., "Plastic Pollution in the World's Oceans: More Than 5 Trillion Plastic Pieces Weighing over 250,000 Tons Afloat at Sea," *PLOS ONE* 9 (2014); also see John Schwartz, "Study Gauges Plastic Levels in Oceans," *New York Times*, 10 December 2014.

59. On the Great Pacific Garbage Patch and the contamination of the Earth's oceans with plastic trash, see Alan Weisman, *The World without Us* (New York: Picador, 2008), 140–61; Charles J. Moore, "Choking the Oceans with Plastic," *New York Times*, 26 August 2014, A23.

60. The signage surrounding *Laysan Albatross* reads, "These seabirds mistake plastics for food. Plastic trash drifts far out to sea—to the North Pacific Gyre, where Laysan albatross feed. When the birds mistake bottle caps, straws, and lighters for food, they can choke or get sick. They also pass this 'food' to their chicks, who can starve with bellies full of plastic. We can help by using less plastic. If we stop using plastic water bottles and foam food containers and coffee cups, they can't end up in the North Pacific Gyre. When you buy fewer plastic items overall, you're making a difference for Laysan albatross and other ocean animals."

# Chapter 2: Animal Farm: Making Meaning at the Zoo

1. Émile Durkheim, *The Elementary Forms of Religious Life* (1912), Karen E. Fields, trans. (New York: Free Press, 1995), 99–140. On the contemporary usage of animals as totems symbolizing ethnic group identity, see Jerolmack, *Global Pigeon*.

2. On the growing culture of enchantment surrounding the natural world, see Gibson, *Reenchanted World*. On the valorization of authenticity in contemporary American popular culture, see Grazian, *Blue Chicago* and "Demystifying Authenticity in the Sociology of Culture," in *Handbook of Cultural Sociology*, John R. Hall, Laura Grindstaff, and Ming-Cheng Lo, eds. (New York: Routledge, 2010), 191–200.

3. Edward O. Wilson, *The Diversity of Life* (Cambridge, MA: Belknap, 2010), x. Admittedly, estimates of this figure dramatically vary depending on the source and calculation methods used.

4. Bitgood, Paterson, and Benefield, "Exhibit Design and Visitor Behavior," 482.

5. Andrew Metrick and Martin L. Weitzman, "Patterns of Behavior in Endangered Species Preservation," *Land Economics* 72 (1996): 1–16 and "Conflicts and Choices in Biodiversity Preservation," *Journal of Economic Perspectives* 12 (1998): 21–34. The quote can be found in George Orwell, *Animal Farm* (New York: Harcourt, Brace, 1946), 112.

6. Of course, firms also draw on the iconic power and charisma of animals in selecting their names: Jaguar luxury cars, Red Bull energy drinks, Puma sneakers. On the use of cultural symbols in corporate branding, see Warren Dotz and Masud Hu-

sain, *Meet Mr. Product: The Art of the Advertising Character* (San Francisco: Chronicle Books, 2003); Naomi Klein, *No Logo* (New York: Picador, 2009).

7. "Letting Go of the Elephant," *Lede/New York Times*, 3 May 2007.

8. On the uses of animal collections as royal trophies in the ancient world, see Croke, *Modern Ark*, 129–36; Hancocks, *Different Nature*, 7–9; Eric Baratay and Elisabeth Hardouin-Fugier, *Zoo: A History of Zoological Gardens in the West* (London: Reaktion, 2004), 17–19. Exotic wild animal collecting as a means of expressing power and machismo among dictators and autocrats continues in the modern world as well; for instance, see Lawrence Anthony with Graham Spence, *Babylon's Ark: The Incredible Wartime Rescue of the Baghdad Zoo* (New York: Thomas Dunne, 2007). Notably, Russian president and strongman Vladimir Putin harbors a well-known obsession with tigers, which he prominently features in publicly reported events; see Leslie Kaufman, "Meeting Aims to Turn Tiger Fascination into Conservation," *New York Times*, 21 November 2010.

9. On this particular naming practice among zoos, see Irus Braverman, *Zooland: The Institution of Captivity* (Stanford, CA: Stanford Law Books, 2013), 9.

10. Juliet B. Schor, *Born to Buy* (New York: Scribner, 2005), 19, 25.

11. Klein, *No Logo*; Schor, *Born to Buy*; Grazian, *Mix It Up*, 64–65.

12. Allison J. Pugh, *Longing and Belonging: Parents, Children, and Consumer Culture* (Berkeley: University of California Press, 2009).

13. J. K. Rowling, *Harry Potter and the Sorcerer's Stone* (New York: Scholastic, 1998), 26–29. Although the snake winks at Harry Potter in both the book and film versions of the tale, snakes do not actually have eyelids, and are thus incapable of winking. (Also, snakes cannot talk, and wizards are not real.)

14. My thanks to Penn sophomore Lainie Bailey for making this observation on our class field trip to the Smithsonian National Zoo.

15. Even zoo professionals compare their resident animals to mass media images, as when a keeper from Elmwood Park Zoo in Norristown, Pennsylvania, publicly remarked aloud that one of the zoo's prairie dogs "looks like Gus," the wisecracking groundhog from an old Pennsylvania Lottery advertising campaign.

16. On our collective cultural fascination with dinosaurs, see Mitchell, *Last Dinosaur Book*. The fossilized remains of extinct animals can found at other American zoos as well. For example, the San Diego Zoo depicts fossils of extinct creatures from the Pleistocene—Columbian mammoths, saber-toothed cats, dire wolves. Although not as ancient as dinosaurs, they all roamed the wilds of Southern California during the last Ice Age.

17. Bitgood, Paterson, and Benefield, "Exhibit Design and Visitor Behavior," 482–84.

18. Jon Mooallem, *Wild Ones: A Sometimes Dismaying, Weirdly Reassuring Story about Looking at People Looking at Animals in America* (New York: Penguin, 2013), 60.

19. Polar bears also share neotenic features that add to their aforementioned status and irresistibility among zoo visitors; see Mooallem, *Wild Ones*, 61.

20. Rachel Donadio, "Masterworks vs. the Masses," *New York Times*, 28 July 2014, observes of the tourists visiting the Louvre, "Inside the museum, a crowd more than a dozen deep faced the *Mona Lisa*, most taking cellphone pictures and selfies."

21. Bitgood, Paterson, and Benefield, "Exhibit Design and Visitor Behavior," 480–82.

22. Metrick and Weitzman, "Patterns of Behavior in Endangered Species Preservation."

23. In the late nineteenth century American women adorned their clothes and hats with not only bird feathers, but also the stuffed heads, wings, tails, and entire bodies of woodpeckers, orioles, bluebirds, jays, owls, sparrows, pheasants, and other fowl; see Price, Flight Maps, 57–109; and Kirk, Kingdom under Glass.

24. Durkheim, Elementary Forms of Religious Life, 133. This should come as little surprise, since even in contemporary times we celebrate many animals as heraldic totems while treating the flesh-and-blood creatures themselves quite poorly. For instance, Americans find horses endlessly charming, even mythic—Black Beauty, Seabiscuit, Mr. Ed—hence the popularity of thoroughbred racing. Yet the dangers to horses in competition are so considerable that over five thousand horses died at U.S. racetracks between 2003 and 2008. (In 2007, the death toll exceeded three horses per day; see Herzog, Some We Love, 172.) Perhaps most famously Eight Belles, a two-year-old filly, placed second at the 2008 Kentucky Derby before collapsing, after which she was euthanized; see Bill Finley, "Triumph, and Then Tears, at the Derby," New York Times, 4 May 2008.

25. While art critic and cultural historian John Berger, About Looking (New York: Pantheon, 1980), 23, proposes that "adults take children to the zoo to show them the originals of their 'reproductions,'" I would counter that regardless of their parents' fair intentions, children themselves experience zoo animals as rather meager representations of far more fabulous pop-cultural touchstones from animated fish to stuffed pandas.

26. On the totemic status of the bald eagle in American culture and public life, see Wilbur Zelinsky, Nation into State: The Shifting Symbolic Foundations of American Nationalism (Chapel Hill: University of North Carolina Press, 1988), 199–201. In this manner the bald eagle serves as a solemn ritualistic symbol in America's civil religion; see Robert N. Bellah, "Civil Religion in America," Daedalus 96 (1967): 1–21.

27. Virtual encounters with animals mediated through photographic images happen elsewhere at the zoo as well. Both City Zoo and Metro Zoo allow visitors to "adopt" a variety of zoo animals, which in the latter case consists of making a fifty-dollar donation in return for a photograph of the animal, along with an official-looking certificate of adoption, fact sheet, and sometimes a plush toy as well. During my stint at Metro Zoo, Scott and I adopted a peregrine falcon named Courage. (Among the fun facts we learned about the peregrine falcon: it is the fastest member of the entire animal kingdom, capable of flying over two hundred miles per hour during high-speed dives, and is the official bird of Chicago.)

28. Mary Douglas, Purity and Danger: An Analysis of the Concepts of Pollution and Taboo (London: Routledge, 1966). On socially produced cognitive logics that emphasize rigidity and boundary enforcement, see Eviatar Zerubavel, The Fine Line: Making Distinctions in Everyday Life (Chicago: University of Chicago Press, 1993), 33–60.

29. Natalie Angier, "A Masterpiece of Nature? Yuck!," New York Times, 9 August 2010. Other types of zoo animals that defy easy classification include chimeras produced in laboratories through interspecies somatic cell nuclear transfer; see Carrie

Friese, "Classification Conundrums: Categorizing Chimeras and Enacting Species Preservation," *Theory and Society* 39 (2010): 145–72.

30. Geoffrey Brewer, "Snakes Top List of Americans' Fears," Gallup, 19 March 2001.

31. Sigmund Freud, *The Interpretation of Dreams* (1899; repr., New York: Oxford University Press, 1999), 260, and *Five Lectures on Psycho-Analysis* (1909; repr., New York: Norton, 1961), 10–11.

32. Edward O. Wilson, *In Search of Nature* (Washington, DC: Island, 1996), 9, 22–23, 119.

33. Brewer, "Snakes Top List."

34. Ibid.

35. Raffles, *Insectopedia*, 44.

36. Edo Knegtering, Laurie Hendrickx, Henry J. Van Der Windt, and Anton J. M. Schoot Uiterkamp, "Effects of Species' Characteristics on Nongovernmental Organizations' Attitudes toward Species Conservation Policy," *Environment and Behavior* 34 (2002): 378–400. According to the authors, only gastropods, or snails and slugs, are valued less than insects.

37. Research demonstrates that the closer the distance between visitors and an exhibited animal, the more likely visitors are to stop by the exhibit; see Bitgood, Paterson, and Benefield, "Exhibit Design and Visitor Behavior," 487–89.

38. In addition to spiders, the female members of over eighty other species engage in similar varieties of sex cannibalism, including certain species of scorpion and praying mantis; see Judson, *Dr. Tatiana's Sex Advice to All Creation*, 96.

39. Michael Pollan, *The Botany of Desire: A Plant's Eye View of the World* (New York: Random House, 2002).

40. Russell Leigh Sharman and Cheryl Harris Sharman, *Nightshift NYC* (Berkeley: University of California Press, 2008); Jerolmack, *Global Pigeon*; also see Davis, *Ecology of Fear*; Sullivan, *Rats*.

41. As discussed in an earlier chapter, many charismatic animals, including lions, howler monkeys, and ocelots, are naturally inactive creatures, and spend a majority of their daylight hours at rest; see Maple and Perkins, "Enclosure Furnishings and Structural Environmental Enrichment," 213; Jill D. Mellen, Marc P. Hayes, and David J. Sheperdson, "Captive Environments for Small Felids," in Sheperdson, Mellen, and Hutchins, *Second Nature*, 190.

42. Bitgood, Paterson, and Benefield, "Exhibit Design and Visitor Behavior," 480–81.

43. On the social construction of nuisance animals in the urban milieu, see Colin Jerolmack, "How Pigeons Became Rats: The Cultural-Spatial Logic of Problem Animals," *Social Problems* 55 (2008): 72–94.

44. Angelo and Jerolmack, "Nature's Looking Glass"; Emily S. Rueb, "Hawk Cam Returns for Third Season," *City Room/New York Times*, 20 March 2013.

45. Marc Lacey, "Lions, Check. Giraffes, Check. Squirrels, Check. Squirrels?," *New York Times*, 26 July 2011.

46. Yi-Fu Tuan, *Dominance and Affection: The Making of Pets* (New Haven: Yale University Press, 1984), 86–87.

47. It is in this sense that zoos, like cities themselves, often accumulate new cultural signifiers of place without necessarily discarding their older identifying symbols;

see Gerald D. Suttles, "The Cumulative Texture of Local Urban Culture," *American Journal of Sociology* 90 (1984): 283–304.

48. Goffman, *Frame Analysis*, 322. Although their parents might believe otherwise, it is unlikely that the basis for these frame disputes stem from children actually *misapprehending* the distinction between live animals and their representations at the zoo. While Swiss developmental psychologist Jean Piaget once posited that young children are *animistic* (or mistakenly associate qualities of animates to inanimate objects) until adolescence, more recent research disputes this theory; see Merry Bullock, "Animism in Childhood Thinking: A New Look at an Old Question," *Developmental Psychology* 21 (1985): 217–25; Kim G. Dolgin and Douglas A. Behrend, "Children's Knowledge about Animates and Inanimates," *Child Development* 55 (1984): 1646–50. However, even children who can conceptually differentiate between real animals and their immobile representations have trouble *verbalizing* this distinction. For this reason, throughout my volunteer experience as a zoo docent, kindergarten and grade-school children would relentlessly ask whether the skink or lizard or snake I was handling was "alive" or "real," while Scott asked me the same thing of one of City Zoo's gorilla statues. Psychologists have long understood that when elementary school kids question whether an animal is alive or real, what they really mean is whether the animal is *lively*, or evinces activity; see S. W. Klingensmith, "Child Animism: What the Child Means by 'Alive,'" *Child Development* 24 (1953): 51–61. Of course, given that zoos virtually swim in rivers of simulated reality (from naturalistic exhibits to plant simulators to animated films), it is no wonder that the sight of *actual* animals might sometimes confuse children at the zoo.

49. Marjorie DeVault, "Producing Family Time: Practices of Leisure Activity Beyond the Home," *Qualitative Sociology* 23 (2000): 493. According to the Rutgers sociologist Eviatar Zerubavel, *Social Mindscapes: An Invitation to Cognitive Sociology* (Cambridge, MA: Harvard University Press, 1997), 14, "Young children who have not learned yet how to focus their attention in a socially appropriate manner and therefore attend to that which is supposed to be disregarded likewise remind us that ignoring the irrelevant is something we learn to do (like my friend's son, who, on his first visit to the zoo, instead of looking at the animals, kept focusing on the patterns in the chain-link fence surrounding the areas where they were kept)."

50. On the importance of play among children (and adults), see Johan Huizinga, *Homo Ludens: A Study of the Play Element in Culture* (New York: Beacon, 1955), Jeffrey L. Kidder, *Urban Flow: Bike Messengers and the City* (Ithaca, NY: ILR Press, 2011) and "Parkour, the Affective Appropriation of Urban Space, and the Real/Virtual Dialectic," *City & Community* 11 (2012): 229–53, discusses how youth subcultures appropriate walls, ledges, stairways, streets, sidewalks, and other elements of the city's built environment for their own games of athleticism and risky play.

51. Goffman, *Frame Analysis*, 43; Karen A. Cerulo, "Nonhumans in Social Interaction," *Annual Review of Sociology* 35 (2009): 531–52; Colin Jerolmack, "Humans, Animals, and Play: Theorizing Interaction When Intersubjectivity Is Problematic," *Sociological Theory* 27 (2009): 371–89.

52. Tuan, *Dominance and Affection*, 115.

53. Erving Goffman, *The Presentation of Self in Everyday Life* (New York: Anchor, 1959), 152, and *Behavior in Public Places: Notes on the Social Organization of Gatherings* (New York: Free Press, 1963), 41–42, 47.

54. On middle-class parenting styles that emphasize the "concerted cultivation" of children through supervised play and enriching activities, see Lareau, *Unequal Childhoods.*

55. Markella B. Rutherford, "Children's Autonomy and Responsibility: An Analysis of Childrearing Advice," *Qualitative Sociology* 32 (2009): 337–53.

56. Viviana A. Zeliver, *Pricing the Priceless Child: The Changing Social Value of Children* (Princeton, NJ: Princeton University Press, 1985).

57. Rutherford, "Children's Autonomy and Responsibility." Adults also confine children in private domestic spaces as well, whether in cribs, in playpens, or under a type of house arrest; see Tuan, *Dominance and Affection,* 116.

58. DeVault, "Producing Family Time," 496–99, and "The Family Work of Parenting in Public," in *At the Heart of Work and Family: Engaging the Ideas of Arlie Hochschild,* Anita Ilta Garey and Karen V. Hansen, eds. (New Brunswick, NJ: Rutgers University Press, 2011), 167, also observes this type of behavior among families at zoos.

59. Glassner, *Culture of Fear,* 29–40.

60. David M. Hummon, *Commonplaces: Community Ideology and Identity in American Culture* (Albany: State University of New York Press, 1990), discusses how suburban dwellers interpret city living in terms of danger, real or imagined. On the urban poverty experienced by residents in the West Philadelphia neighborhood of Mantua, which borders the Philadelphia Zoo, see Elijah Anderson, *Streetwise: Race, Class, and Change in an Urban Community* (Chicago: University of Chicago Press, 1990). I give thanks to Anderson for granting me permission to reveal the identity of the Mantua neighborhood, which he refers to pseudonymously as "Northton" in *Streetwise.*

61. Glassner, *Culture of Fear,* 31.

62. Rosin, "Hey! Parents, Leave Those Kids Alone," 81.

63. Other free zoos in the United States that do not charge admission to guests include the Como Park Zoo in St. Paul, Minnesota, the Henry Vilas Zoo in Madison, Wisconsin, the Salisbury Zoo in Maryland, and the David Traylor Zoo in Emporia, Kansas. On the decline of free public spaces in the contemporary American city, see Sharon Zukin, *Landscapes of Power: From Detroit to Disney World* (Berkeley: University of California Press, 1991) and *The Cultures of Cities* (Malden, MA: Blackwell, 1995); Michael Sorkin, ed., *Variations on a Theme Park: The New American City and the End of Public Space* (New York: Hill and Wang, 1992); Mike Davis, *City of Quartz: Excavating the Future in Los Angeles* (New York: Vintage, 1992); Neil Smith, *The New Urban Frontier: Gentrification and the Revanchist City* (London: Routledge, 1996); and Bryant Simon, *Everything but the Coffee: Learning about America from Starbucks* (Berkeley: University of California Press, 2009).

64. Stanley Fish, *Is There a Text in This Class? The Authority of Interpretive Communities* (Cambridge, MA: Harvard University Press, 1980). On the differentiation among interpretive communities in other contexts of urban tourism and cultural entertainment, see David Grazian, "The Symbolic Economy of Authenticity in the Chicago Blues Scene," in *Music Scenes: Local, Translocal, and Virtual,* Andy Bennett and Richard A. Peterson, eds. (Nashville, TN: Vanderbilt University Press, 2004), 31–47.

65. Robert L. Wolf and Barbara L. Tymitz, "Studying Visitor Perceptions of Zoo Environments: A Naturalistic View," *International Zoo Yearbook* 21 (1981): 49–53; J. Mark Morgan and Marlana Hodgkinson, "The Motivation and Social Orientation

of Visitors Attending a Contemporary Zoological Park," *Environment and Behavior* 31 (1999): 227–39.

# Chapter 3: Birds of a Feather: Zookeepers and the Call of the Wild

1. Everett C. Hughes, "Work and Self" (1951), in *On Work, Race, and the Sociological Imagination*, Lewis A. Coser, ed. (Chicago: University of Chicago Press, 1994), 62.

2. Goffman, *Presentation of Self in Everyday Life*, 160.

3. I certainly tried to follow the well-intentioned advice of Erving Goffman, "On Fieldwork," 125, who once explained how when one conducts participant observation among groups of people working in a different occupation than that one normally performs, "the standard technique is to try to subject yourself to their life circumstances, which means that although, in fact, you can leave at any time, you act as if you can't and you try to accept all the desirable and undesirable things that are a feature of their life." However, this still never really changed the true reality of my life's circumstances, which was that as an adult volunteer with a full-time job who lacked professional zookeeping aspirations, I *could have* showed up late to work or refused to complete certain tasks without much consequence, unlike zoo staff members who truly experienced the precariousness of the low-wage labor market. In the end, that precariousness and the anxiety it produces makes all the difference.

4. Full disclosure: I took it upon myself to taste select brands of herbivore animal food as they came across my kitchen scales at the zoo. Designed for exotic hoofstock such as zebras and rhinos, the molasses and ground aspen in Mazuri Wild Herbivore Plus give its dark pellets and beet pulp shreds a bitter aftertaste, while Primate Browse Biscuits have a fruity taste and smell *exactly* like Kellogg's Froot Loops cereal.

5. Bunderson and Thompson, "Call of the Wild," 35.

6. As I discuss later in the chapter, zookeepers earn relatively low wages, especially considering their educational attainment as college graduates. Metro Zoo's full-time keepers make between ten and twelve dollars an hour.

7. It is also noteworthy that zookeepers do not experience the same extraordinary level of stigma as janitors, chambermaids, orderlies, or nursing-home aides, particularly because while modern Americans may consider handling animal feces and vomit repugnant, neither seems as polluting as *human* wastes or bodily emissions; see Douglas, *Purity and Danger*, 35; Zerubavel, *Fine Line*, 37.

8. A specialized job among zoo workers, veterinary technicians like Heather perform a variety of specific duties that are particularly stigmatized by the general public, including euthanasia as well as handling dead animal remains; see Clinton R. Sanders, "Working Out Back: The Veterinary Technician and 'Dirty Work,'" *Journal of Contemporary Ethnography* 39 (2010): 243–72.

9. Of course, it is a sliding scale: while zookeepers may be burdened by occasional social disapproval, they hardly face the same moral disapprobation as prison guards, hit men, soldiers defending authoritative regimes, and certain kinds of corporate lawyers and defense attorneys; see Everett C. Hughes, "Social Role and the Divi-

sion of Labor" (1956), in Coser, *On Work, Race, and the Sociological Imagination*, 51–53, and "Good People and Dirty Work," *Social Problems* 10 (1962): 3–11.

10. Meanwhile, the *median* wage for nonfarm animal caretakers in 2012 was even less: the median hourly wage was $9.46, with a median annual wage of $19,690; see Bureau of Labor Statistics, U.S. Department of Labor, "Occupational Employment and Wages, May 2012," http://www.bls.gov/oes/current/oes392021.htm, accessed 4 November 2013.

11. Sutherland, *Kicked, Bitten, and Scratched*, 60.

12. On the recent feminization of veterinary medicine, see Irvine and Vermilya, "Gender Work in a Feminized Profession"; Lincoln, "Shifting Supply of Men and Women to Occupations."

13. On sex segregation in the workplace and its relationship to wages, see Barbara F. Reskin and Patricia A. Roos, *Job Queues, Gender Queues: Explaining Women's Inroads into Male Occupations* (Philadelphia: Temple University Press, 1990); Paula England, Paul Allison, and Yuxiao Wu, "Does Bad Pay Cause Occupations to Feminize, Does Feminization Reduce Pay, and How Can We Tell with Longitudinal Data?," *Social Science Research* 36 (2007): 1237–56.

14. England, Budig, and Folbre, "Wages of Virtue."

15. Not only do zoos enjoy the free labor of unpaid interns, but the growth of unpaid internships in American zoos helps keep wages low for all other employees, just as they do in more glamorous industries such as advertising, magazine publishing, television production, and the music business. See Jim Frederick, "The Intern Economy and the Culture Trust," in *Boob Jubilee: The Cultural Politics of the New Economy*, Thomas Frank and David Mulcahey, eds. (New York: Norton, 2003), 301–13; also see "Generation I," *Economist*, 6 September 2014.

16. Alexandra Stevenson, "S.E.C. Proposes Greater Disclosure on Pay for C.E.O.'s," *DealBook/New York Times*, 18 September 2013.

17. Please note that these figures do not include deferred compensation or nontaxable benefits, and so the total compensation packages for these chief executives were in fact much higher than those reported here. All salary figures were individually pulled from U.S. Internal Revenue Service tax returns (Form 990, Schedule J) made available online through GuideStar.org, which gathers and provides copies of documentation submitted to the IRS by registered nonprofit organizations.

18. On the correlates of intraprofessional status, see Andrew Abbott, "Status and Status Strain in the Professions," *American Journal of Sociology* 86 (1981): 819–35.

19. Hughes, "Work and Self," 57–66.

20. Stefan Timmermans, *Postmortem: How Medical Examiners Explain Suspicious Deaths* (Chicago: University of Chicago Press, 2006), 275–81, discusses how medical examiners approach dead bodies as routine autopsies instead of as decomposing corpses.

21. The contextual basis of pollution and disgust not only is microsituational but cuts across entire societies and cultural traditions as well; see Douglas, *Purity and Danger*.

22. Everett C. Hughes, "Mistakes at Work" (1951), in Coser, *On Work, Race, and the Sociological Imagination*, 84.

23. Blake E. Ashforth and Glen E. Kreiner, "'How Can You Do It?': Dirty Work and the Challenge of Constructing a Positive Identity," *Academy of Management Re-*

*view* 24 (1999): 413–34. I should also add that two additional sources of social support for zookeepers exist *outside* of the workplace itself. The American Association of Zoo Keepers (AAZK) holds annual conferences where zookeepers congregate, swap stories, and help members sustain a self-affirming occupational identity. In addition, zoo workers subscribe to a number of online social media sites that allow them to maintain connections with keepers from around the world.

24. While experienced keepers may approach zoo work as routine, even zookeepers have their own thresholds for disgust. According to Lauren, "Bison poop probably smells the worst, but I don't find it that gross to clean. What I think *is* gross to clean is carnivore poop. It is usually slimy and smelly and disgusting, and you usually have to pick it up—you can't use a shovel, you have to use a trash bag, so you are kind of touching it. Mustelids are the worst: otters and ferrets have the grossest, slimiest poop. Luckily, with the otters we can hose it, and then it just all goes away. But the jaguars, they eat this ground meat product, and so they have really gross poop."

25. The phrase "call of the wild" in this context comes from Bunderson and Thompson, "Call of the Wild," who make clever use of the similarly titled 1903 novel by Jack London to emphasize how zookeeping ought to be understood as what Max Weber, *The Protestant Ethic and the Spirit of Capitalism*, Talcott Parsons, trans. (London: George Allen & Unwin, 1930), describes as a "calling." In *Protestant Ethic*, 54, Weber writes, "One's duty in a calling, is what is most characteristic of the social ethic of capitalistic culture, and is in a sense the fundamental basis of it. It is an obligation which the individual is supposed to feel and does feel towards the content of his professional activity, no matter in what it consists, in particular no matter whether it appears on the surface as a utilization of his personal powers, or only of his material possessions (as capital)."

26. It is admittedly unclear whether the life trajectories of zookeepers actually follow this pattern, or whether keepers employ these kinds of cultural narratives to manufacture consistency between the past and present; see Ann Swidler, *Talk of Love: How Culture Matters* (Chicago: University of Chicago Press, 2001). Nevertheless, regarding the facts of their biographies I see no reason not to take them at their word.

27. People invest their relationships with their pets with special significance, even as such relations are laden with complexity and contradiction; see Tuan, *Dominance and Affection*; Arluke and Sanders, *Regarding Animals*; Michael Schaffer, *One Nation under Dog: America's Love Affair with Our Dogs* (New York: St. Martin's, 2010). On how individuals integrate their home and work lives and environments, see Christena E. Nippert-Eng, *Home and Work: Negotiating Boundaries through Everyday Life* (Chicago: University of Chicago Press, 1996).

28. Bunderson and Thompson, "Call of the Wild."

29. Caregiving narratives serve as powerful cultural resources that allow zookeepers, veterinary technicians, and other deliverers of care (licensed vocational nurses, home health aides, childcare workers) to explain their occupational choices to themselves and others in an ennobling manner. On the use of cultural accounts, narratives, and scripts to justify lines of action, see Swidler, *Talk of Love*.

30. On our lack of ability to ever truly understand the subjectivity of nonhuman animals, see Thomas Nagel, "What Is It Like to Be a Bat?," *Philosophical Review* 83 (1974): 435–50.

31. Human-animal relationships can obviously be rendered quite meaningful for

people despite the impossibility of mutual understanding between humans and non-human animals; see Jerolmack, "Humans, Animals, and Play."

32. Zookeepers' rejection of desk work reflects both the routinized and demoralizing nature of much white-collar work in the digital age, and the psychological and intellectual satisfactions of manual labor performed well; see Richard Sennett, *The Corrosion of Character: The Personal Consequences of Work in the New Capitalism* (New York: Norton, 1999); Simon Head, *The New Ruthless Economy: Work and Power in the Digital Age* (Oxford: Oxford University Press, 2003); David Grazian, "A Digital Revolution? A Reassessment of New Media and Cultural Production in the Digital Age," *Annals of the American Academy of Political and Social Science* 597 (2005): 209–22; Matthew B. Crawford, *Shop Class as Soulcraft: An Inquiry into the Value of Work* (New York: Penguin, 2009).

33. In his marvelous ethnography of an English exurban village, Michael Mayerfeld Bell, *Childerley*, explores how people deploy nature as a means of structuring collective identity and moral distinction in everyday life. On the use of moral boundaries in other social contexts, see Michele Lamont, *Money, Morals, and Manners: The Culture of the French and American Upper-Middle Class* (Chicago: University of Chicago Press, 1992) and *The Dignity of Working Men: Morality and the Boundaries of Race, Class, and Immigration* (Cambridge, MA: Harvard University Press, and Russell Sage, 2000); Zerubavel, *Fine Line*; Elijah Anderson, *Code of the Street: Decency, Violence, and the Moral Life of the Inner City* (New York: Norton, 1999); Mark A. Pachucki, Sabrina Pendergrass, and Michele Lamont, "Boundary Processes: Recent Theoretical Developments and New Contributions," *Poetics* 35 (2007): 331–51; and Keith R. Brown, "The Social Dynamics and Durability of Moral Boundaries," *Sociological Forum* 24 (2009): 854–76.

34. This strategy of gender work is common among women participating in animal care occupations. Like zookeepers, female veterinarians similarly distance themselves from the performance of femininity, as they associate professionalism with attributes commonly associated with hegemonic masculinity; see Irvine and Vermilya, "Gender Work in a Feminized Profession." This orientation is common among women working in previously male-dominated occupations that have recently become feminized and sex-segregated; see Reskin and Roos, *Job Queues, Gender Queues*. On the performance of femininity and gender as a social accomplishment in everyday interaction, see Candace West and Don H. Zimmerman, "Doing Gender," *Gender and Society* 1 (1987): 125–51; David Grazian, *On the Make: The Hustle of Urban Nightlife* (Chicago: University of Chicago Press, 2008), 95–103.

35. Disagreements over priorities within professional fields are common; according to Hughes, "Social Role and the Division of Labor," 55, workers in professions often complain about "differing conceptions of what the work really is or should be."

36. Lauren's complaint not only reflects the low status accorded keepers among zoo administrators, wealthy donors, and corporate representatives; it also conforms to basic norms across occupational cultures. According to Hughes, "Social Role and the Division of Labor," 55, "Perhaps the commonest complaint of people in the professions which perform a service for others is that they are somehow prevented from doing their work as it should be done. Someone interferes with this basic relation. The teachers could teach better were it not for parents who fail in their duty or school

boards who interfere. Psychiatrists would do better if it were not for families, stupid public officials, and ill-trained attendants."

37. Hughes, "Work and Self," 65; also see Howard S. Becker, *Outsiders: Studies in the Sociology of Deviance* (Glencoe, IL: Free Press, 1963); and Grazian, *Blue Chicago.*

38. These unpleasant encounters with zoo audiences drive keepers to violate occupational norms of what Arlie Russell Hochschild, *The Managed Heart: Commercialization of Human Feeling* (Berkeley: University of California Press, 1983), calls *emotion work* (or the directed control of one's emotions in the context of paid labor) in the interests of what they consider a much greater good: the call of the wild.

39. Of course, it should be noted that Metro Zoo's keepers share other things in common that likely ease interaction and the forging of social solidarity among them as well. Nearly the entire zookeeping staff is made up of white women in their twenties or early thirties, and no fewer than *four* of them (Amber, Ashley, Christina, and Krista) were sorority sisters at the same regional college. Metro Zoo also has a recruitment process in which the keepers themselves vet all job finalists and collectively decide whom to hire; consequently, new keepers may be recruited, at least in part, on the basis of shared cultural traits and their ability to negotiate the interpersonal dynamics of the larger group; see Lauren A. Rivera, "Hiring as Cultural Matching: The Case of Elite Professional Service Firms," *American Sociological Review* 77 (2012): 999–1022.

40. Although they labor in what is obviously a far more just and liberated context, singing and dancing to hit songs during the workday may serve a similar function for zookeepers as work songs, field hollers, and early blues music would have had for black sharecroppers and farmhands in the fields of southern plantations at the turn of the twentieth century; see Robert Palmer, *Deep Blues* (New York: Penguin, 1982), 23–25; Giles Oakley, *The Devil's Music: A History of the Blues* (New York: Da Capo, 1997), 35–40.

41. Charles R. Simpson, *SoHo: The Artist in the City* (Chicago: University of Chicago Press, 1981); Richard Florida, *The Rise of the Creative Class* (New York: Basic, 2002); Frederick, "Intern Economy and the Culture Trust," 301–13; Gina Neff, Elizabeth Wissinger, and Sharon Zukin, "Entrepreneurial Labor among Cultural Producers: 'Cool' Jobs in 'Hot' Industries," *Social Semiotics* 15 (2005): 307–34.

# Chapter 4: Life Lessons:
# The Zoo as a Classroom

1. The exam also asked us to name the four major groups of arthropods, differentiate the shell of a box turtle from that of other turtles, and define "ovoviviparous."

2. On the integration of home and work life, see Nippert-Eng, *Home and Work.*

3. John Wilson and Marc Musick, "Who Cares? Toward an Integrated Theory of Volunteer Work," *American Sociological Review* 62 (1997): 706, demonstrate that women are generally more likely to report that they regularly "visit and talk with friends, attend church and pray, and believe the good life demands assisting others— all factors conducive to volunteering." Women also tend to donate more to causes re-

lated to animals and/or environmental protection (such as zoos) than men; see Francie Ostrower, *Why the Wealthy Give: The Culture of Elite Philanthropy* (Princeton, NJ: Princeton University Press, 1995), 72–76. Some metropolitan zoos attract affluent women to their docent councils, as in keeping with norms of voluntary involvement and investment among other kinds of nonprofit charitable organizations; see Randall Collins, "Women and the Production of Status Cultures," in *Cultivating Differences: Symbolic Boundaries and the Making of Inequality*, Michele Lamont and Marcel Fournier, eds. (Chicago: University of Chicago Press, 1992), 213–31. However, Metro Zoo was notable for attracting a dedicated group of middle-class volunteers, while less involved but wealthier elites served on its board of directors.

4. On the pleasures of manual work, see Crawford, *Shop Class as Soulcraft*.

5. Unlike art museums, opera and symphony houses, and other elite cultural institutions bound by rigid status-based norms of formality and restraint, zoos are far more democratic, egalitarian, and child-centered spaces of public life marked by a relaxing of constraints on public behavior. On the cultivated mannerisms embedded in social realms of class distinction and elevated status, see Pierre Bourdieu, *Distinction: A Social Critique of the Judgment of Taste*, Richard Nice, trans. (Cambridge, MA: Harvard University Press, 1984); Lawrence Levine, *Highbrow/Lowbrow: The Emergence of Class Hierarchy in America* (Cambridge, MA: Harvard University Press, 1990); Paul DiMaggio, "Cultural Boundaries and Structural Change: The Extension of the High Culture Model to Theater, Opera, and the Dance, 1900–1940," in Lamont and Fournier, *Cultivating Differences*, 21–57.

6. On the use of evergreens and other cultural conventions in television production and other narrative forms of popular culture, see Laura Grindstaff, *The Money Shot* (Chicago: University of Chicago Press, 2002) 84; Grazian, *Mix It Up*, 32–37.

7. Zoos rarely discuss their *own* involvement as animal suppliers in the wild and exotic pet trade; see Linda Goldston, "Behind Those Majestic Lions and Cute Babies is a Dirty Secret," *Hartford Courant*, 23 February 1999, who reports on the practice of selling, trading, loaning, and donating unwanted (or "surplus") animals by AZA-accredited zoos. Also see Gibson, *Reenchanted World*, 178–80.

8. According to Daphne, locals would also bring animals illegally captured in the wild to the zoo. "We had somebody show up with an albino fawn at the front gate in a blanket. Wrapped up in a blanket. And we were like, 'Oh, my god, put it back. Put it back. The mother stashed it somewhere for the day. They go out and feed and then they come back. . . . Just stick it back. And just check back later and if it doesn't come back then you have to call [a local animal rescue organization]. . . .' We are not supposed to directly accept wildlife. That's against the law."

9. This reliable repertoire of familiar themes operate in much same way as the time-tested canonical standards performed by blues and jazz musicians; see Grazian, *Blue Chicago*; Robert R. Faulkner and Howard S. Becker, *"Do You Know . . . ?" The Jazz Repertoire in Action* (Chicago: University of Chicago Press, 2009).

10. This is similar to how servers at upscale urban restaurants present themselves as wine experts to unwitting customers based on a very limited knowledge of recommended house favorites; see Grazian, *On the Make*, 38–41.

11. Goffman, *Presentation of Self in Everyday Life*, 82.

12. In the last several years, psychologists and clinicians have experimented with a rich variety of animal-assisted activities (or AAA) among young people, particularly

hospitalized children suffering from severe disabilities such as Down syndrome or autism spectrum disorder, but also children with more common emotional difficulties such as anxiety or depression. Clinical trials and psychology experiments investigating the effectiveness of AAA have included patient interactions with therapy dogs, cats, horses, canaries, fish, reptiles, dolphins, and even elephants. Examples of these published experiments include Francie R. Murry and M. Todd Allen, "Positive Behavioral Impact of Reptile-Assisted Support on the Internalizing and Externalizing Behaviors of Female Children with Emotional Disturbance," *Anthrozoös* 25 (2012): 415–25; Eva Stumpf and Erwin Breitenbach, "Dolphin-Assisted Therapy with Parental Involvement for Children with Severe Disabilities: Further Evidence for a Family-Centered Theory for Effectiveness," *Anthrozoös* 27 (2014): 95–109; also see Janelle Nimer and Brad Lundahl, "Animal-Assisted Therapy: A Meta-Analysis," *Anthrozoös* 20 (2007): 225–38.

Admittedly, these studies tend to lack the methodological robustness demanded by norms of scientific inquiry. Criticisms of this research argue that its findings collectively suffer from a variety of methodological weaknesses: placebo and novelty effects, construct validity, experimenter expectancy effects, the frequent absence of control groups, and informant bias; see Lori Marino and Scott O. Lilienfeld, "Dolphin-Assisted Therapy: Flawed Data, Flawed Conclusions," *Anthrozoös* 11 (1998): 194–200; Lori Marino and Scott O. Lilienfeld, "Dolphin-Assisted Therapy: More Flawed Data and More Flawed Conclusions," *Anthrozoös* 20 (2007): 239–49; Lori Marino, "Construct Validity of Animal-Assisted Therapy and Activities: How Important Is the Animal in AAT?," *Anthrozoös* 25 (2012): S139–51; Anna Chur-Hansen, Michelle McArthur, Helen Wineield, Emma Hanieh, and Susan Hazel, "Animal-Assisted Interventions in Children's Hospitals: A Critical Review of the Literature," *Anthrozoös* 27 (2014): 5–18. On the other hand, research in this area does suggest the *potential* efficacy of AAA on special-needs children in certain cases, if only broadly and moderately so. The examples from my interviews similarly provide at least anecdotal evidence that human-animal encounters have the ability to generate positive interactive experiences for children with disabilities. (In any case, the positive impact that facilitating these kinds of encounters has on zoo *educators* seems irrefutable.)

13. On the conditions of disadvantaged urban areas in American cities, see William Julius Wilson, *The Truly Disadvantaged: The Inner City, the Underclass and Public Policy* (Chicago: University of Chicago Press, 1987) and *When Work Disappears* (Chicago: University of Chicago Press, 1996); Jonathan Kozol, *Savage Inequalities: Children in America's Schools* (New York: Harper Perennial, 1992) and *The Shame of the Nation: The Restoration of Apartheid Schooling in America* (New York: Broadway, 2005); Douglas Massey and Nancy Denton, *American Apartheid: Segregation and the Making of the Underclass* (Cambridge, MA: Harvard University Press, 1993); Anderson, *Code of the Street.*

14. Succoth is not the only holiday than attracts observant Jews to the zoo. In the days before Passover, Jewish families clear their homes of bread, cake, and other leavened foods (called *chametz*), and it turns out that in New York City some take their discarded odds and ends to the Central Park Zoo to feed the animals, to the great consternation of their keepers; see Carol Vinzant, "Monkeys Say 'Dayenu!,'" *New York Magazine*, 9 April 2006.

15. Elijah Anderson, *The Cosmopolitan Canopy: Race and Civility in Everyday*

*Life* (New York: Norton, 2011). Sociologists have long characterized cities in terms of their residential heterogeneity, widespread anonymity, and subcultural diversity; see Simmel, "Metropolis and Mental Life"; Park, "The City"; Harvey W. Zorbaugh, *The Gold Coast and the Slum: A Sociological Study of Chicago's Near North Side* (Chicago: University of Chicago Press, 1929); Wirth, "Urbanism as a Way of Life"; and Claude Fischer, "Toward a Subcultural Theory of Urbanism," *American Journal of Sociology* 80 (1975): 1319–41.

16. Among some Hasidim, even child-centered amusement parks provoke anxiety; see Samuel Heilman, *Defenders of the Faith: Inside Ultra-Orthodox Jewry* (New York: Schocken, 1992), 306.

17. Amy J. Binder, *Contentious Curricula: Afrocentrism and Creationism in American Public Schools* (Princeton, NJ: Princeton University Press, 2002).

18. Frank Newport, "In U.S., 46% Hold Creationist View of Human Origins," Gallup, 12 June 2012.

19. Hannah observes that at the natural history museum where she currently works (along with Ryan, Brooke, and Holly), staff members *are* expected to discuss evolution, suggesting how such museums hold themselves to higher scientific standards than zoos. Still, museum audiences can be as skeptical as zoo visitors. According to Hannah, coworkers reported that during presentations on the formation of the continents and the environmental changes surrounding the extinction of the dinosaurs, adults would occasionally tell their children, "Well, technically, God made the Earth."

20. Joseph Erbentraut, "Conservative Blogger Rants Against 'Propaganda' at Popular Zoo," *Huffington Post*, 6 December 2014.

21. Jane Goodall, *In the Shadow of Man* (1971; repr., Boston: Mariner, 2000), 234, 251.

22. Diamond, *Third Chimpanzee*, 23.

23. Goodall, *In the Shadow of Man*, 251; Diamond, *Third Chimpanzee*, 23. Diamond, 24, also argues that "humans have had only a short history as a species distinct from other apes, much shorter than paleontologists used to assume," given that chimpanzees diverged from our common ancestor only perhaps three million years ago; also see Zerubavel, *Time Maps*, 73–77.

24. As Zerubavel, "Lumping and Splitting," 424, observes, "It is such inflation of intercluster mental gaps that lead us to perceive chimpanzees 'closer' to chipmunks than to humans, with whom chimpanzees share 99 percent of their genes." Also see Zerubavel, *Fine Line*, 24–32; and *Time Maps*, 73–81.

## Chapter 5: Bring on the Dancing Horses: American Zoos in the Entertainment Age

1. Wolf and Tymitz, "Studying Visitor Perceptions of Zoo Environments," 50.

2. On zoo visitors' emphasis on recreation over education, see ibid.; Morgan and Hodgkinson, "Motivation and Social Orientation."

3. Tuan, *Dominance and Affection*, 74; Croke, *Modern Ark*, 131–32.

4. Croke, *Modern Ark*, 132–33; Hancocks, *Different Nature*, 9–12.

5. Hancocks, *Different Nature*, 13.

6. Phineas T. Barnum, *Struggles and Triumphs, or, Sixty Years' Recollections of P.T. Barnum, Including His Golden Rules for Money-Making* (Buffalo: Courier Company, 1889), quotes appear on 57, 290; Tuan, *Dominance and Affection*, 78. On the exhibition of "mutant animals" as well as humans from bearded women to dwarves in American mid-nineteenth-century museums, see Paul DiMaggio, "Cultural Entrepreneurship in Nineteenth-Century Boston, Part 1: The Creation of an Organizational Base for High Culture in America," *Media, Culture and Society* 4 (1982): 33–50. On the human zoos of the nineteenth and early twentieth centuries, see Pascal Blanchard et al., eds., *Human Zoos: Science and Spectacle in the Age of Colonial Empires* (Liverpool: Liverpool University Press, 2008).

7. Kisling, "Origin and Development of American Zoological Parks," 115–16. Although chartered in 1859, the Philadelphia Zoo would not officially open its doors until 1874.

8. Ibid., 120.

9. Douglas Brinkley, *The Wilderness Warrior: Theodore Roosevelt and the Crusade for America* (Harper Perennial, 2010), 276.

10. Kisling, "Origin and Development of American Zoological Parks," 123.

11. Phillips Verner Bradford and Harvey Blume, *Ota: The Pygmy in the Zoo* (New York: St. Martin's, 1992).

12. John Sedgwick, *The Peaceable Kingdom: A Year in the Life of America's Oldest Zoo* (New York: William Morrow, 1988), 88.

13. Bradford and Blume, *Ota*, 185.

14. Like cities themselves, zoos often accumulate new entertainment attractions without necessarily discarding their older relics from the past, and therefore serve as monuments to their own cultural histories; see Suttles, "Cumulative Texture of Local Urban Culture."

15. The content of hip-hop and rap music is often unfairly maligned as dangerous or harmful, especially to young people; see Amy Binder, "Constructing Racial Rhetoric: Media Depictions of Harm in Heavy Metal and Rap Music," *American Sociological Review* 58 (1993): 753–67. I can assure readers that a local alternative rock station was just as likely to broadcast objectionable content as the hip-hop station, if not more so, especially during the morning rush hour.

16. Schaller, *Serengeti Lion*, 107–8.

17. While visitors often mistake animal inactivity for boredom, zookeepers explain that creatures are often able to rest and sleep only when they are comfortable and free from stress, whereas a truly bored animal paces in its enclosure or displays other kinds of stereotypic behavior.

18. For example, see Arluke and Sanders, *Regarding Animals*; Jerolmack, "Humans, Animals, and Play"; and Cerulo, "Nonhumans in Social Interaction."

19. Another example of misreading animal cues: according to Herzog, *Some We Love*, 63, when an alpha male baboon yawns in the presence of humans at the zoo, he isn't bored, but only displaying his teeth in order to best communicate to passersby, "I can rip your face off."

20. Unlike nearly every other exhibit at City Zoo, the petting yard was a singular case in which ordinary visitors were permitted to enter an enclosure with unleashed animals. The fence of the enclosure obviously provided a physical threshold separat-

ing these ruminating mammals (and petting yard patrons) from nonparticipating zoo visitors. But perhaps it also suggested a *figurative* boundary between nature and civility as well, the transgression of which might then explain why squealing children would chase each around the yard with abandon as their parents lost all hope at controlling their children (and occasionally themselves) while inside its wooden gates.

21. Paul G. Cressey, *The Taxi-Dance Hall* (Chicago: University of Chicago Press, 1932); Viviana A. Zelizer, *The Purchase of Intimacy* (Princeton, NJ: Princeton University Press, 2005), 119–29; Elizabeth Bernstein, *Temporarily Yours: Intimacy, Authenticity, and the Commerce of Sex* (Chicago: University of Chicago Press, 2007).

22. Mercifully, the penguin didn't spit (or urinate) on me.

23. In point of fact, City Zoo did allow donors a twenty-minute private session with an "adopted" zoo animal and its primary keeper for a $150 contribution. Of course, these arranged visits were not permitted in all cases: while the Galapagos tortoise, Humboldt penguin, and reticulated giraffe were all considered safe for paid up-close encounters with members of the public, others were decidedly not, including African lions, Amur leopards and tigers, western lowland gorillas, polar bears, and king cobras.

24. I note that the couple is childless because it is impossible to imagine being able to engage in such an involved project with young kids in tow at the zoo: the time it would take to set up the tripod, attach lenses, and essentially stay in one spot for more than a few minutes would be utterly incompatible with the demands of accompanying small children to the zoo, given their frenetic energy and short attention span.

25. Moynihan, "Man Mauled by Tiger at Zoo Is Charged."

26. Susan G. Davis, *Spectacular Nature: Corporate Culture and the Sea World Experience* (Berkeley: University of California Press, 1997), 56. In 2013 SeaWorld parks experienced a drop in attendance and stock price, largely due to bad publicity stemming from accusations of animal cruelty, the unnecessary endangerment of trainers, the forced separation of mothers from their calves, and questions of whether orcas can even be humanely held in captivity—all alleged in the Gabriela Cowperthwaite documentary *Blackfish*, released in July of that year; see William Alden, "Oscar Snub Is Applauded by SeaWorld Investors," *DealBook/New York Times*, 16 January 2014.

27. Killer whales also attract humans for reasons that go beyond their awesome size and strength. Although ferocious and supersized, orcas also possess neotenic characteristics associated with human babies, just as pandas do, including a rounded body and giant spots that make their eyes seem even larger than they already are.

28. This is in keeping with more general efforts among animal theme parks to exploit the growing culture of enchantment surrounding the human experience of animals and the natural living world for profit; see Gibson, *Reenchanted World*, 178–79.

29. As Davis, *Spectacular Nature*, 217–18, observes of SeaWorld's Shamu spectaculars, "Although the show text talks about 'communication' in terms of relationships and feelings, the communicative actions performed are instrumental: a signal leads to the performance of a task or behavior, which leads to a reward from the trainer. Training results in familiar routines, although the behaviors may be startling to the audience. Communication is the flow of command given and received, movements carefully coordinated, and actions directed. But the spoken text that raises the

idea of language conflates command with communication's ability to create connection and community; the mentions of language and communication overlay commanded action with the appearance of sharing and volition."

30. Specifically, these techniques are crucial for training zoo and aquarium animals to safely receive veterinary care; see Tanya Paul, "Animal Welfare and Sustainability: Training as a Tool" (2011 AZA Annual Conference, Atlanta, GA).

31. Not coincidentally, Bon Iver's "Holocene" also appears in Cameron Crowe's 2011 film adaptation of the Benjamin Mee memoir We Bought a Zoo (New York: Weinstein, 2008).

32. On the staged theatricality of high-concept urban entertainment spaces, see Grazian, On the Make.

33. Similarly, as Herzog, Some We Love, 1–2, 180, 195, observes throughout his book, many self-described vegetarians eat fish and seafood without moral trepidation. Of course, this may simply stem from a lack of knowledge about the social lives of these interesting creatures. According to Keven N. Laland, Culum Brown, and Jens Krause, "Learning in Fishes: From Three-Second Memory to Culture," Fish and Fisheries 4 (2003): 199–202, fish are capable of using tools, building complex nests, and exhibiting long-term memories. Moreover, scientific research demonstrates that "the learning abilities of fishes are comparable to land vertebrates, and whether one considers the neural circuitry, psychological processes or behavioral strategies, fish learning appears to rely on processes strikingly similar to that of other vertebrates." Fish can also discriminate among other fish within their own species, recognize kin, and generate "long-standing 'cultural traditions' for particular pathways to feeding, schooling, resting or mating sites"; quotes appear on 202.

34. According to Herzog, Some We Love, 75, as of 2009 Americans owned 180 million fish as house pets.

35. In fact, these stunning tanks are sometimes specifically designed to create merely the illusion of spaciousness. Nature makers sometimes design large aquarium tanks with a giant block in their center; this directs fish to swim as close to the glass walls of the tank as possible, since they literally have nowhere else to go. (As one zoo professional explained to me, "It's open on the ends, and blocked in the middle. It's like a racetrack.") Jellyfish tanks may be similarly designed, with only a few inches of space between a well-camouflaged interior wall and the tank's more visible glass exterior. On the management of aquatic zoo habitats, see Daryl J. Boness, "Water Quality Management in Aquatic Mammal Exhibits," in Kleiman et al., Wild Mammals in Captivity, 231–42.

36. Grandin and Johnson, Animals Make Us Human, 264–65.

37. Like SeaWorld's animal encounters, so-called backstage tours at zoos and aquariums are never held in actual backstage spaces, but areas designed (or at least cleaned up) for public viewing; see Goffman, Presentation of Self in Everyday Life; Dean MacCannell, "Staged Authenticity: Arrangements of Social Space in Tourist Settings," American Journal of Sociology 79 (1973): 589–603; David Grazian, "The Production of Popular Music as a Confidence Game: The Case of the Chicago Blues," Qualitative Sociology 27 (2004): 137–58.

38. Mircea Eliade, The Sacred and the Profane: The Nature of Religion, Willard Trask, trans. (New York: Harcourt, 1959), 130–31.

39. Again, as repositories of culture, zoos and aquariums often exhibit creative

works typically associated with art museums; on the artificial designation of both art objects and artistic exhibition spaces, see Becker, *Art Worlds*.

40. Durkheim, *Elementary Forms of Religious Life*, 129.

41. The penchant among American and European sociologists and anthropologists for making other cultures and folkways seem more exotic than the mundane facets of our own is hilariously satirized by Horace Miner, "Body Ritual among the Nacirema," *American Anthropologist* 58 (1956): 503–7.

42. On the use of live animals in the entertainment landscape of Las Vegas, see William L. Fox, *In the Desert of Desire: Las Vegas and the Culture of Spectacle* (Reno: University of Nevada Press, 2005).

43. Edward W. Said, *Orientalism* (New York: Pantheon, 1978).

44. While Anandapur and its Royal Forest are quite obviously imaginary places (although in point of fact it turns out there is at least one real Indian town and a village in Bangladesh named Anandapur), sociologists of culture observe that *all* places are imagined: after all, *place* is nothing more than abstract geophysical *space* rendered culturally meaningful through collective memory and shared identification; see Yi-Fu Tuan, *Topophilia: A Study of Environmental Perception, Attitudes, and Values* (New York: Columbia University Press, 1974) and *Space and Place: The Perspective of Experience* (Minneapolis: University of Minnesota Press, 1977); Zerubavel, *Fine Line*; Edward W. Soja, *Thirdspace: Journeys to Los Angeles and Other Real-and-Imagined Places* (Cambridge, MA: Blackwell, 1996); Thomas F. Gieryn, "A Space for Place in Sociology," *Annual Review of Sociology* 26 (2000): 463–96; Wendy Griswold, *Regionalism and the Reading Class* (Chicago: University of Chicago Press, 2008).

45. Of course, the human sciences could easily be accused of similar sins by portraying aboriginal or else racially or ethnically marginalized people in homogenizing and morally charged terms. For elaborations of this critique of the "savage slot" in anthropology, see Michel-Rolph Trouillot, *Global Transformations: Anthropology and the Modern World* (New York: Palgrave Macmillan, 2003); Alcida Rita Ramos, "The Politics of Perspectivism," *Annual Review of Anthropology* 41 (2012): 481–94. A similar critique of the ethnographic portrayal of African Americans in sociology can be found in Mitchell Duneier, *Slim's Table: Race, Respectability, and Masculinity* (Chicago: University of Chicago Press, 1992).

46. Like Harambe, Disney's Animal Kingdom Lodge creates a pan-African imaginary that tends to blow aside differences among African countries, regions, and cultures. For example, at its main restaurant, Boma, the dinner buffet includes dishes like fufu (West and Central Africa); peanut soup (West Africa); pap with chakalaka, sambal chili sauces, and bobotie (Southern Africa); couscous (West and North Africa); and falafel and yogurt (North Africa). At Jiko, the fine-dining restaurant at the hotel, dishes include African fusion concoctions such as the crispy beef bobotie roll with cucumber raita, green mango atjar, and honey roasted groundnuts, and the grilled wild boar tenderloin with mealie pap, chakalaka, white truffle oil, and micro cilantro. On the proliferation of global fusion cuisines that mix ethnic traditions, see David Grazian, "I'd Rather Be in Philadelphia," *Contexts* 4 (2005): 71–73; also see *On the Make*.

47. Alan Beardsworth and Alan Bryman, "The Disneyization of Zoos," in *Between the Species: Readings in Human-Animal Relations*, Arnold Arluke and Clinton Sanders, eds. (Boston: Allyn & Bacon, 2009), 225–34. Specific critiques of Disney's theme parks include Zukin, *Landscapes of Power* and *Cultures of Cities*; Sorkin, *Varia-*

*tions on a Theme Park*; Mark Gottdiener, *Postmodern Semiotics: Material Culture and the Forms of Postmodern Life* (New York: Wiley-Blackwell, 1995); Henry A. Giroux and Grace Pollock, *The Mouse That Roared: Disney and the End of Innocence* (Lanham, MD: Rowman & Littlefield, 2010).

48. On the commodification of music from Africa, the Caribbean, Latin America, and other Global South outposts as "world music," see Timothy Brennan, "World Music Does Not Exist," *Discourse* 23 (2001): 44–62.

49. While intended as flattery, these mystical characterizations deemphasize both the disturbing history of the European conquest, forced migration, and genocide of Native Americans and the urgent socioeconomic and political inequities faced today by Native American peoples in contemporary U.S. society; see Gibson, *Reenchanted World*, 35–36.

50. The passage as it appears on the zoo placard is excerpted from Helen Curry, *The Way of the Labyrinth: A Powerful Meditation for Everyday Life* (New York: Penguin, 2000).

51. Jared Diamond, *Guns, Germs, and Steel: The Fates of Human Societies* (New York: Norton, 1997), 309–10; Bill Gammage, *The Biggest Estate on Earth: How Aborigines Made Australia* (Sydney: Allen and Unwin, 2011).

52. Diamond, *Guns, Germs, and Steel*, 394.

53. Jared Diamond, *Collapse: How Societies Choose to Fail or Succeed* (New York: Penguin, 2005), 140–41. On the other side of the continent, environmental historian William Cronon, *Changes in the Land: Indians, Colonists, and the Ecology of New England* (New York: Hill and Wang, 1983), 13, points out that Native Americans in New England successfully "burned forests to clear land for agriculture and to improve hunting," although their well-developed practices were far less destructive than those of European colonists.

54. Neil Postman, *Amusing Ourselves to Death: Public Discourse in the Age of Show Business* (New York: Penguin, 1985), 143–44; also see Neal Gabler, *Life the Movie: How Entertainment Conquered Reality* (New York: Vintage, 2000).

55. On the devolution of substance, attentiveness, and deep thinking in the digital age, see Todd Gitlin, *Media Unlimited: How the Torrent of Images and Sounds Overwhelms Our Lives* (New York: Holt, 2002); Nicholas Carr, *The Shallows: What the Internet Is Doing to Our Brains* (New York: Norton, 2011).

56. For example, according to *Wired* contributing editor Steven Johnson, *Everything Bad Is Good for You: How Today's Popular Culture Is Actually Making Us Smarter* (New York: Riverhead, 2005), contemporary media diversions from independent film to concept-driven television programs to sophisticated videogames offer audiences a kind of interactive engagement conducive to collateral learning by facilitating the development of cognitive skills and competencies among active participants. Also see Edward Castronova, *Synthetic Worlds: The Business and Culture of Online Games* (Chicago: University of Chicago Press, 2005); and Grazian, *Mix It Up*, 197–98.

57. Dan Gross, "Ghost Hunters Investigating/Shooting at Philadelphia Zoo," *Philadelphia Daily News*, 6 April 2010.

58. All quotes are my transcriptions taken from "America's First Zoo" (*Ghost Hunters*, season 6, episode 612, SyFy, 1 September 2010). It is obviously possible that Baker protested the findings much more vigorously on camera, and his remarks were simply edited out of the eventually aired episode. Still, the quote speaks for itself.

59. Frank Newport, "What if Government Really Listened to the People?," Gallup, 15 October 1997; David W. Moore, "Three in Four Americans Believe in Paranormal," Gallup, 16 June 2005; Linda Lyons, "One-Third of Americans Believe Dearly May Not Have Departed," Gallup, 12 July 2005.

60. Mooallem, *Wild Ones*, 79.

61. Richard Conniff, *Swimming with Piranhas at Feeding Time: My Life Doing Dumb Stuff with Animals* (New York: Norton, 2009), 161–64.

62. James West, "Drugs, Death, Neglect: Behind the Scenes at Animal Planet," *Mother Jones*, 21 January 2014.

63. Only six animals were rescued and taken to the Columbus Zoo and Aquarium: three leopards, two Celebes macaques, and a young grizzly bear; see Greg Bishop and Timothy Williams, "Police Kill Dozens of Animals Freed on Ohio Reserve," *New York Times*, 19 October 2011.

# Chapter 6: Simply Nature: Zoos and the Branding of Conservation

1. Jonathan Franklin, " 'Frog Hotel' to Shelter Panama Species from Lethal Fungus," *National Geographic News*, 2 November 2006; also see Jeannie Kever, "Threatened Frogs Get VIP Treatment at Hotel," *Houston Chronicle*, 28 October 2006.

2. http://www.aza.org/annual-report-on-conservation-and-science/.

3. Other American zoos founded during the late nineteenth century include the Cleveland Metroparks Zoo (1882), Atlanta Zoo (1889), San Francisco Zoo (1889), Denver Zoo (1896), and Pittsburgh Zoo (1898).

4. Theodore Roosevelt, "Publicizing Conservation at the White House" (1908), in *American Environmentalism: Readings in Conservation History*, 3rd ed., Roderick Frazier Nash, ed. (New York: McGraw-Hill, 1990), 87. On Roosevelt's conservationist legacy, see Brinkley, *Wilderness Warrior*; for a feminist critique of the lionization of Roosevelt as our "naturalist president," see Donna Haraway, "Teddy Bear Patriarchy: Taxidermy in the Garden of Eden, New York City, 1908–1936," in *Primate Visions: Gender, Race, and Nature in the World of Modern Science* (New York: Routledge, 1989), 26–58.

5. Clarence S. Stein, *Toward New Towns for America* (Cambridge, MA: MIT Press, 1957); Kenneth T. Jackson, *Crabgrass Frontier: The Suburbanization of the United States* (New York: Oxford University Press, 1985), 236; Gibson, *Reenchanted World*, 93–96.

6. Nicholas Lemann, "When the Earth Moved," *New Yorker*, 15 April 2013, 73.

7. Hanson, *Animal Attractions*, 162–71.

8. Todd Bayma, "Rational Myth Making and Environment Shaping: The Transformation of the Zoo," *Sociological Quarterly* 53 (2012): 116–41, observes that although organizational ambitions like conservation and species preservation may appear on zoos' mission statements, such institutional goals often operate as "rational myths" decoupled from their everyday activities and priorities; see John W. Meyer and Brian Rowan, "Institutionalized Organizations: Formal Structure as Myth and Ceremony," *American Journal of Sociology* 83 (1977): 340–63.

9. http://www.aza.org/species-survival-plan-program/.

10. Robert Loftin, "Captive Breeding of Endangered Species," in *Ethics on the Ark: Zoos, Animal Welfare, and Wildlife Conservation,* Bryan G. Norton, Michael Hutchins, Elizabeth E. Stevens, and Terry Maple, eds. (Washington, DC: Smithsonian Institution Press, 1995), 164–80; Jonathan D. Ballou, Devra G. Kleiman, Jeremy J. C. Mallinson, Anthony B. Rylands, Claudio B. Valladares-Padua, and Kristen Leus, "History, Management, and Conservation Role of the Captive Lion Tamarin Populations," in *Lion Tamarins: Biology and Conservation,* Devra G. Kleiman and Anthony B. Rylands, eds. (Washington, DC: Smithsonian Institution Press, 2002), 95–114; Leslie Kaufman, "To Save Some Species, Zoos Must Let Others Die," *New York Times,* 27 May 2012.

11. Friese, "Classification Conundrums," and *Cloning Wild Life: Zoos, Captivity, and the Future of Endangered Animals* (New York: New York University Press, 2013); Paul A. Rees, *An Introduction to Zoo Biology and Management* (Oxford: Wiley-Blackwell, 2011), 289–91.

12. http://www.iucnredlist.org.

13. William G. Conway, "The Practical Difficulties and Financial Implications of Endangered Species Breeding Programmes," *International Zoo Yearbook* 24/25 (1986): 210–19; Kaufman, "To Save Some Species." AZA-accredited zoos are also limited in their ability to provide the extensive physical space required to successfully implement captive propagation programs; this is especially the case for city zoos confined to dense urban areas.

14. Croke, *Modern Ark,* 174–80; Conniff, *Swimming with Piranhas at Feeding Time,* 117; Leslie Kaufman, "Date Night at the Zoo, If Rare Species Play Along," *New York Times,* 4 July 2012.

15. Rees, *Introduction to Zoo Biology and Management,* 285.

16. Ibid., 285; David Owen, "Bears Do It," *New Yorker,* 2 September 2013, 26–31.

17. Jill D. Mellen and Sue Ellis, "Animal Learning and Husbandry Training," in Kleiman et al., *Wild Mammals in Captivity,* 91; David E. Wildt, "Male Reproduction: Assessment, Management, and Control of Fertility," in Kleiman et al., *Wild Mammals in Captivity,* 429–50; Rees, *Introduction to Zoo Biology and Management,* 285–86; Owen, "Bears Do It."

18. Michael Hutchins and William G. Conway, "Beyond Noah's Ark: The Evolving Role of Modern Zoological Parks and Aquariums in Field Conservation," *International Zoo Yearbook* 34 (1995): 117–30, quote appears on 127; also see William K. Stevens, "Zoos Find a New Role in Conserving Species," *New York Times,* 21 September 1993.

19. Julie Larsen-Maher, "Back from the Brink: Blue Iguanas," *Connect,* January 2012, 14–15; Carlos C. Martinez Rivera, "Amphibian Conservation Strategy for Haiti and the Dominican Republic," *Connect,* March 2013, 18–19; Della Garelle, "Quarters for Conservation: One Million Dollars, One Year, One Big Idea," *Connect,* March 2013, 35.

20. Association of Zoos and Aquariums, "2012 Annual Report on Conservation Science," 3.

21. Richard Bergl, Tara Harris, Stuart Wells, and Joanne Harcke, "Science Saves Species," *Connect,* May 2012, 21.

22. Jenny Barnett, "Conservation in Your Backyard: A Small Zoo's Perspective," *Connect*, January 2014, 16–17.

23. Sindya N. Bhanoo, "Zoos and Aquariums on Oil Spill Alert," *Green/New York Times*, 25 May 2010; Campbell Robertson and Clifford Krauss, "Gulf Spill Is the Largest of Its Kind, Scientists Say," *New York Times*, 2 August 2010.

24. These data, including examples of specific species conservation efforts, come from the Association of Zoos and Aquariums, "2012 Annual Report on Conservation Science," 4.

25. These large institutions also dedicate a greater percentage of their budgets to conservation efforts, as do a surprising number of smaller, lesser-known zoos such as Zoo Boise in Idaho, the Gladys Porter Zoo in Brownsville, Texas, and the Bramble Park Zoo in Watertown, South Dakota.

26. Paul Boyle, "New Definition of Field Conservation" (2011 AZA Annual Conference, Atlanta, GA).

27. Wilson, *Diversity of Life*, 133, insect species counts appear on 139, 198.

28. Hancocks, *Different Nature*, 154–55.

29. However, while ecologically responsive gardening has typically involved the cultivation of native species, today some argue that changes in climate due to global warming may call for a more diverse mix of native and nonnative plants to ensure a garden's resilience, given the instability of local weather patterns in the Anthropocene; see James Barilla, "Gardening for Climate Change," *New York Times*, 2 May 2014.

30. On the human activities surrounding city pigeons that habituate in New York neighborhoods and other urban outposts, see Jerolmack, *Global Pigeon*.

31. Visitors are encouraged to seek out small local birds in the wild rather than at the zoo's Avian Center, given the ethics of capturing and caging healthy wild birds native to the region. Notably, AZA zoos like the Philadelphia Zoo limit their aviary collections to large regional birds that have sustained injury in the wild, or else exotic nonnative species. Still, it is telling that visitors to the Philadelphia Zoo devote time to viewing *animated* fowl on film when the zoo displays such luminous live creatures as the Micronesian kingfisher and Mariana fruit dove just down the hall from the Avian Center's 4D theater.

32. Chris Waldron and Andrew J. Baker, "Achieving Sustainability Goals Through Partnerships" (2009 AZA Annual Conference, Portland, OR).

33. Cell phones contain toxic chemicals and metals that contaminate the environment, such as antimony, arsenic, beryllium, cadmium, copper, and lead. In addition, the mining of certain metallic ores used in cell phone manufacturing, such as Coltan, leads to habitat destruction in vulnerable places like the Congo, where endangered elephants and gorillas live.

34. This is not necessarily an evaluation of the program's ultimate efficacy or success, but simply a sign of how distracted guests can be during zoo visits.

35. In fact, according to the International Union for Conservation of Nature (IUCN) between 20,000 and 25,000 polar bears remain in the wild, and their population is decreasing, in part due to the rapid disappearance of their Arctic habitat due to global warming. The IUCN Red List assesses the endangered status of polar bears as vulnerable.

36. On the history of organized campaigns denying the scientific consensus surrounding climate change, see James Hoggan with Richard Littlemore, *Climate Cover-Up: The Crusade to Deny Global Warming* (Vancouver: Greystone, 2009); Naomi Oreskes and Eric M. Conway, *Merchants of Doubt: How a Handful of Scientists Obscured the Truth on Issues from Tobacco Smoke to Global Warming* (New York: Bloomsbury, 2010). Gallup figures are from Rebecca Riffkin, "Climate Change Not a Top Worry in U.S.," Gallup, 12 March 2014; Jeffrey M. Jones, "In U.S., Most Do Not See Global Warming as Serious Threat," Gallup, 13 March 2014; Andrew Duggan, "Americans Most Likely to Say Global Warming Is Exaggerated," Gallup, 17 March 2014; Lydia Saad, "A Steady 57% in U.S. Blame Humans for Global Warming," Gallup, 18 March 2014.

37. "IPCC, 2013: Summary for Policymakers," 4.

38. Ibid., 17.

39. Naomi Oreskes, "The Scientific Consensus on Climate Change," *Science* 306 (2004): 1686.

40. In fact, zoos are highly responsive to their audience's self-reported experiences, and many invest in collecting quantitative data to measure visitor evaluations. One major zoo asks its guests to assess its exhibits, staff interactions, and amenities (parking, retail shops, refreshment stands) on a seven-point Likert-type scale through an online survey instrument.

41. Leslie Kaufman, "Intriguing Habitats, and Careful Discussions of Climate Change," *New York Times*, 26 August 2102. As Kaufman reports, "Surveys show that American zoos and aquariums enjoy a high level of public trust and are ideally positioned to teach. Yet many managers are fearful of alienating visitors—and denting ticket sales—with tours or wall labels that dwell bleakly on damaged coral reefs, melting ice caps or dying trees."

42. Again, according to the IUCN, between 20,000 and 25,000 polar bears remain in the wild, and their population is decreasing.

43. Quoted in Nette Pletcher, "Changing Communication in a Changing Climate," *Connect*, February 2014, 9. The AZA Position Statement on Climate Change was approved by the AZA Board of Directors on 20 July 2013.

44. "IPCC, 2013: Summary for Policymakers," 11–12.

45. Ibid., 11–12.

46. Robertson and Krauss, "Gulf Spill Is the Largest of Its Kind." BP changed its name from British Petroleum in 1998 as a result of its merger with Amoco, but one also might suspect the name change was motivated to avoid the stigma associated with fossil fuels.

47. In 1995 ARCO agreed to settle with the Federal Trade Commission over charges that it falsely promoted the so-called environmental benefits and safety of one of its products, Sierra antifreeze; in its complaint, the FTC's Bureau of Consumer Protection found ARCO's claims unsubstantiated.

48. See http://www.peri.umass.edu/toxic100_20081/.

49. DiMaggio, "Cultural Entrepreneurship in Nineteenth-Century Boston, Part 1"; Paul J. DiMaggio and Helmut K. Anheier, "The Sociology of Nonprofit Organizations and Sectors," *Annual Review of Sociology* 16 (1990): 141; Ostrower, *Why the Wealthy Give*; James E. Austin, "Business Leaders and Nonprofits," *Nonprofit Management & Leadership* 9 (1998): 39–51. On the cultural diffusion of rituals of perfor-

mance among corporate firms, see Meyer and Rowan, "Institutionalized Organizations"; and Paul J. DiMaggio and Walter W. Powell, "The Iron Cage Revisited: Institutional Isomorphism and Collective Rationality in Organizational Fields," *American Sociological Review* 48 (1983): 147–60.

50. Sarah Banet-Weiser, *Authentic™: The Politics of Ambivalence in a Brand Culture* (New York: New York University Press, 2012), 149–54, discusses corporate greenwashing and the public relations strategies deployed by petroleum giants ExxonMobil, Chevron, and BP. Formerly British Petroleum, BP embarked on a eco-friendly marketing campaign in 2001, spending $200 million to change its logo and rebrand its initials to signify "Beyond Petroleum."

51. Karen Allen, "Putting the Spin on Animal Ethics: Ethical Parameters for Marketing and Public Relations," in Norton et al., *Ethics on the Ark*, 294–95.

52. Allen, "Putting the Spin on Animal Ethics," 295, emphasis added.

53. Hoggan and Littlemore, *Climate Cover-Up*, 12–13.

54. Ibid., , 82–84.

55. Aron Pilhofer, "Big Oil Protects Its Interests" (Center for Public Integrity, 15 July 2004).

56. National Response Team, U.S. Coast Guard, DHS, "On Scene Coordinator Report: Deepwater Horizon Oil Spill" (September 2011), http://www.uscg.mil/foia /docs/dwh/fosc_dwh_report.pdf, accessed 4 March 2014; Bhanoo, "Zoos and Aquariums on Oil Spill Alert."

57. The Rainforest Action Network also launched a campaign in 2009 during the Easter season, "Hershey's Week of Action," in which consumers were asked to go to markets and stamp Hershey Easter candy with stickers reading "Warning: Product May Contain Rainforest Destruction." Hershey claims innocence on this issue because they purchase their palm oil from suppliers that are members of the Roundtable on Sustainable Palm Oil (RSPO), but that organization is actually funded and run by the very same private industries that benefit from palm oil's cultivation and production. According to William F. Laurance et al., "Improving the Performance of the Roundtable on Sustainable Palm Oil for Nature Conservation," *Conservation Biology* 24 (2010): 378, as of October 2009 the RSPO's 312 members included 206 oil palm growing, processing, and trading corporations, and representatives from banking, investment, and other corporate sectors, while conservation or social development groups made up only 6.7 percent of the member organization body. Meanwhile, according to the biologists who authored the article, "Becoming an RSPO member is too easy. The RSPO allows palm-oil producers and processors to become ordinary members without actually having their operations certified, so long as they are putatively working toward certification and abiding by a rather loophole-filled code of conduct. In reality, this diminishes the significance of being an RSPO member and provides a false imprimatur of legitimacy for members that are performing poorly."

58. Quoted in Hoggan and Littlemore, *Climate Cover-Up*, 67.

59. Hancocks, *Different Nature*, 151.

60. In addition, at zoo and aquarium entrances photographers often bully visitors to have their obligatory picture taken, which the zoo then tries to sell back to the customer for an inflated price.

61. On how plastic toys and other polymer-based products wind up in our oceans, see Weisman, *World without Us*, 140–61.

62. Many such gift shops sell the exact same items (only branded with different logos) because zoos rely on vendors that operate retail stores at multiple AZA institutions around the country. One such company, Event Network, operates the gift shops at the Brevard Zoo in Florida, Cleveland Metroparks Zoo, Indianapolis Zoo, Oregon Zoo, Philadelphia Zoo, Phoenix Zoo, and Seattle's Woodland Park Zoo. Wildlife Trading Company partners with the Roger Williams Park Zoo in Providence, Rhode Island, the Reid Park Zoo in Tucson, Arizona, and Florida's Naples Zoo and Palm Beach Zoo.

63. On the inhumane trapping practices of indigenous animal poachers in the African forests of Rwanda, see Fossey, *Gorillas in the Mist*.

64. Jeffrey Gettleman, "Elephants Dying in Epic Frenzy as Ivory Fuels Wars and Profits," *New York Times*, 3 September 2012, and "Coveting Horns, Ruthless Smugglers' Rings Put Rhinos in the Cross Hairs," *New York Times*, 31 December 2012.

65. A strange irony here is that some species conservation efforts in Africa do, in fact, rely on the establishment of *trophy* hunting programs, as they make predatory wildlife more valuable to farmers alive than dead; see Conniff, *Swimming with Piranhas at Feeding Time*, 108–14, who discusses controlled cheetah hunting in Namibia as a conservation strategy.

66. Paul Rogers, "California Drought: Why Is There No Mandatory Water Rationing?," *San Jose Mercury News*, 15 February 2014.

67. Although in 2014 California governor Jerry Brown signed legislation into law aimed toward this end, University of California, Davis environmental law professor Richard Frank quickly referred to the new law's schedule for implementation as "rather languid," while Rebecca Nelson, an Australian researcher who leads the Comparative Groundwater Law and Policy Program at Stanford University, remarked to *National Geographic* that "California and the rest of the West are really ignoring groundwater's environmental role"; see Michelle Nijhuis, "Amid Drought, New California Law Will Limit Groundwater Pumping for the First Time," *National Geographic*, 17 September 2014.

68. Aldo Leopold, *A Sand County Almanac* (New York: Oxford University Press, 1949), 210.

69. There is some debate over the efficacy of this particular policy fix. According to Richard York, "Fossil Fuel Use and the Displacement Paradox" (School of Social Science, Institute for Advanced Study, 13 March 2014), consumers may in fact respond to the increased fuel efficiency of cars by driving *more*, thereby canceling out the benefits of higher mpg standards.

70. Stephen Pacala and Robert Socolow, "Stabilization Wedges: Solving the Climate Problem for the Next 50 Years with Current Technologies," *Science* 305 (2004): 968–72; also see Thomas L. Friedman, *Hot, Flat, and Crowded: Why We Need a Green Revolution—And How It Can Renew America* (New York: Picador, 2008), 257–61.

71. Friedman, *Hot, Flat, and Crowded*, 260; Justin Gillis, "Climate Panel Cites Near Certainty on Warming," *New York Times*, 19 August 2013.

72. Michael Maniates, "Going Green? Easy Doesn't Do It," *Washington Post*, 22 November 2007; also see Friedman, *Hot, Flat, and Crowded*, 253–54.

73. McKibben, *Eaarth*, 186. On Stanford University's divestment of coal stocks,

see Michael Wines, "Stanford to Purge $18 Billion Endowment of Coal Stock," *New York Times*, 6 May 2014.

74. "Turning the Tide: The State of Seafood" (Monterey Bay Aquarium, 2009), 50–51. In addition, the Monterey Bay Aquarium not only has distributed over forty million Seafood Watch pocket guides to consumers, but also partners with hundreds of restaurants and other businesses and institutions to promote sustainable fish and seafood production. Their corporate partners include the nation's two largest food service management companies, Compass Group and the Aramark Corporation.

75. On the other hand, while the Philadelphia Zoo's eagle exhibit asks audiences to "help protect open space, especially around lakes and rivers," it does not explain how one might collectively or politically organize to do so. Meanwhile, other recommendations imparted to visitors emphasize consumerist and individual behaviors— "Limit your use of pesticides and fertilizers"—without citing the six companies the Pesticide Action Network claims control 75 percent of the global pesticide market: Monsanto, Syngenta, Dow, DuPont, Bayer, and BASF. In addition, while visitors could theoretically learn more about the history of the issue by purchasing a copy of *Silent Spring*, they would have to leave the zoo grounds to do so, since (as of this writing) the Philadelphia Zoo does not actually carry the book in any of its three gift shops.

76. U.S. Department of Justice, Bureau of Justice Statistics, "Correctional Populations in the United States, 2012" (December 2013), 3.

# Chapter 7: Wrestling with Armadillos: Animal Welfare and the Captivity Question

1. On the contradictory sets of relationships we form with nonhuman animals, see Tuan, *Dominance and Affection*; Herzog, *Some We Love*.

2. Brown, Dunlap, and Maple, "Notes on Water-Contact"; Hancocks, *Different Nature*, 105.

3. In fact, according to Fossey, *Gorillas in the Mist*, xvii, "contrary to popular opinion, gorillas greatly enjoy basking in the sun."

4. During summer months City Zoo visitors also routinely complained that the lions looked particularly hot, although in the wild lions live in warm tropical savannas and grasslands in Eastern and Southern Africa.

5. William Conway, "Zoo Conservation and Ethical Paradoxes," in Norton et al., *Ethics on the Ark*, 3. Data on factory farming and zoos come from Tom Regan, *Empty Cages: Facing the Challenge of Animal Rights* (Lanham, MD: Rowman & Littlefield, 2004), 94–99; Jonathan Safran Foer, *Eating Animals* (New York: Back Bay, 2009), 15, 27, 48, 170–71, 187; Herzog, *Some We Love*, 169–70; https://www.aza.org /zoo-aquarium-statistics/. On the particularly careless treatment of animals in industrial factory-farm food production, also see Singer, *Animal Liberation*; Eric Schlosser, *Fast Food Nation: The Dark Side of the All-American Meal* (New York: Perennial, 2002); Michael Pollan, *The Omnivore's Dilemma: A Natural History of Four Meals* (New York: Penguin, 2006).

6. The visibility of zoos may also be related to their public presence in densely populated cities like New York, Chicago, Philadelphia, and San Francisco. On the relationship between stigma and visibility, see Erving Goffman, *Stigma: Notes of the Management of Spoiled Identity* (New York: Simon & Schuster, 1963), 48. Data on American vegetarianism rates come from Frank Newport, "In U.S., 5% Consider Themselves Vegetarians," Gallup, 26 July 2012. Fur data come from Regan, *Empty Cages*, 108; the Humane Society figures come from their website.

7. On the human ability to feign ignorance of known transgressions (or selectively ignore them) in the face of evidence to the contrary, see Eviatar Zerubavel, *The Elephant in the Room: Silence and Denial in Everyday Life* (New York: Oxford University Press, 2007); Ari Adut, *On Scandal: Moral Disturbances in Society, Politics, and Art* (New York: Cambridge University Press, 2008).

8. Robert Strauss, "After 132 Years, Philadelphia Will Send Off Its Elephants," *New York Times*, 5 November 2005; Robert Moran, "Former Philadelphia Zoo Elephant Dulary Dies at Tennessee Sanctuary," *Philadelphia Inquirer*, 25 December 2013. Of the three remaining elephants, Petal died at the Philadelphia Zoo in 2008, while Bette and Kallie were sent to the International Conservation Center, a breeding facility run by the Pittsburgh Zoo. Kallie was eventually transferred to Cleveland Metroparks Zoo in 2011, while Bette remains at the Pittsburgh-run facility. See Sandy Bauers, "Petal, Oldest Elephant at the Philadelphia Zoo, Dies at 52," *Philadelphia Inquirer*, 10 June 2008.

9. "Fire at the Philadelphia Zoo Kills 23 Primates," *New York Times*, 25 December 1995.

10. Bauers, "Petal, Oldest Elephant at the Philadelphia Zoo, Dies at 52."

11. In addition, Sophia organizes a local Meetup group dedicated to animal advocacy and activism. When I attended a gathering of the group at a vegetarian restaurant in downtown Philadelphia, I observed an interesting symmetry between the assembled members and the communities of animal lovers who work at City Zoo and Metro Zoo whom I had come to admire. In keeping with national and historical trends among animal rights movement adherents and participants, the group was largely female (by more than a two to one margin), just like Metro Zoo's teams of zookeepers, educators, and docents; see Charles W. Peek, Nancy J. Bell, and Charlotte C. Dunham, "Gender, Gender Ideology, and Animal Rights Advocacy," *Gender & Society* 10 (1996): 464–78; Rachel L. Einwohner, "Gender, Class, and Social Movement Outcomes: Identity and Effectiveness in Two Animal Rights Campaigns," *Gender & Society* 13 (1999): 56–76; Julian McAllister Groves, "Animal Rights and the Politics of Emotion: Folk Constructs of Emotions in the Animal Rights Movement," in *Passionate Politics: Emotions and Social Movements*, Jeff Goodwin, James M. Jasper, and Francesca Polletta, eds. (Chicago: University of Chicago Press, 2001), 212–29. Also, like the volunteers at Metro Zoo and City Zoo, these women activists were generationally bifurcated by age, with local university students and recent Penn, Temple, and Drexel graduates at one end of the age spectrum, and middle-aged animal advocates toward the older end.

12. Note that Sophia—an attorney by profession—draws not on her emotions to make her case against zoos, but on rational, scientific, and philosophical arguments that employ the language of "interests" and rely on "peer-reviewed studies." According

to Groves, "Animal Rights and the Politics of Emotion," career-oriented women involved in the contemporary animal rights movement generally draw on these kinds of rational and philosophical (rather than emotional) arguments as a way of maintaining their professional identities and social status, especially given the stigma unfairly attached to the performance of emotion among women in positions of authority and expertise.

13. Singer, *Animal Liberation*, 6.

14. Ibid., 18–19.

15. In fact, Sophia herself points out at least one major zoo she believes is heading in the right direction: "Places like the Oakland Zoo, they really try very hard to give their elephants a good life. They still don't have enough space, but they do things like they put new browse out every two hours; they're constantly doing things to make their lives as good as possible; and that's what sanctuaries do. They try to make life as good as possible for the animals that are already here [in U.S. zoos]. . . . They really try hard to give the elephants enrichment, social groupings, and things like that." Sophia also carries a great deal of respect for Ron Kagan, CEO and executive director of the Detroit Zoo, whom she admits represents a force for positive change within the zoo industry:

> Another zoo is the Detroit Zoo. I don't know much about their physical [facilities], but the fact that they made a decision in the best interest of the elephants against the recommendation of the AZA, stood up for them and sent them to a sanctuary, that speaks volumes. I met Ron Kagan—he's the Detroit Zoo's director—at an elephant event in California a few years ago, and I saw him speak, and *he* said there are problems with zoos. He admitted it, he said one of the problems is the traveling exhibits that some zoos still do—they take animals like a bird or a lizard or a tortoise or something, and bring them to a birthday party, or a Target—I saw them do it here at a Target store opening—or they do it at schools, and call it educational. Ron Kagan at the Detroit Zoo says that's wrong. He said, first of all, that's very stressful on the animal, and secondly, it's counter-educational. You're bringing this totally stressed animal into the classroom and letting the kids look at it—it's sending the wrong message that these animals are here for us to do with them what we want. And he gets that. He said we don't allow that. Another one is respect for families of animals, like don't split up the mother and the babies to send the babies to another zoo, so he's on the right track. He also admits that we can do a lot better with inspiring people to do the right thing. Because there were a lot of activists in the audience, and he's like, "You have ideas about how to educate people and how to get people to work and do things, and we have the audience, we have the people coming, and we need to do a better job of getting the two together, getting the people who are coming to the zoos to see the animals, getting them to actually do something." He admits that zoos don't do a good job of that.

16. Dale Jamieson, "Against Zoos," in *Morality's Progress: Essays on Humans, Other Animals, and the Rest of Nature* (New York: Oxford University Press, 2002), 173. In addition to critiquing the educational and research agendas of zoos, Jamieson ar-

gues that the conservation of biodiversity and endangered species in zoos leads to an irrational outcome: the preservation of genetic material at the expense of flesh-and-blood animals themselves. As he cleverly puts it, "This smacks of sacrificing the lower-case gorilla for the upper-case Gorilla"; see 173.

17. Tom Regan, *The Case for Animal Rights* (Berkeley: University of California Press, 1983) and *Empty Cages*.

18. Tom Regan, "Are Zoos Morally Defensible?," in Norton et al., *Ethics on the Ark*, 45–46. In 2013, legal scholar Steven Wise took animal rights advocacy to an entirely new level by filing lawsuits in a New York State courtroom on behalf of four captive chimpanzees, demanding they be released from their captors on account of their right to bodily liberty as legal persons; see Charles Siebert, "The Rights of Man . . . and Beast," *New York Times Magazine*, 27 April 2014, 28.

19. According to Allen, "Putting the Spin on Animal Ethics," 291, "Public support for the work of zoos and aquariums hinges on marketing efforts."

20. Scott Higley, "Caring Together We Can Make a Difference," *Connect*, June 2014, 9.

21. Julia McHugh, "Zoo Opposition: Understanding the Mindset" (2010 AZA Annual Conference, Houston, TX).

22. Jill Mellen, "Animal Welfare," *Connect*, January 2014, 28. It should be noted that not all zoo industry representatives depict their relationship with zoo critics and animal activists as necessarily adversarial. In Terry Maple, Rita McManamon, and Elizabeth Stevens, "Defining the Good Zoo: Animal Care, Maintenance, and Welfare," in Norton et al., *Ethics on the Ark*, 229, the authors—all Zoo Atlanta executives—recommend, "Rather than seeing the community, including animal welfare activists, as outsiders, the good zoo should solicit its input in establishing what the institution stands for. Zoo managers should constantly play devil's advocate with themselves, questioning their chosen option for any animal (space allotment, euthanasia, social grouping, etc.) by testing it against the various perspectives represented in the community. The zoo should be viewed as the center of a network of participating partners in the community."

23. In this sense, zoo industry personnel collectively participate in what Goffman, *Presentation of Self in Everyday Life*, 160, refers to as a shared *community of fate*, as colleagues who maintain similar professional identities, and concerns over social status and boundaries of morality.

24. While Louise jokes that animal advocates "think that carnivores should be fed vegetables," I am reminded of an incident described to me by Krista, the Metro Zoo keeper, concerning a visitor that became upset upon seeing Scout, the alligator, fed a frozen breeder rat. According to Krista,

> I came out like, "Oh, we're going to feed Scout, the alligator!" Everyone wants to watch. So I throw [the rat] through the door, and I hear this woman scream. Like, full-on scream. And I was like, "I should go check what that was." So I go out, and this woman is up against the hallway wall, panting. And I asked her, "Are you okay? What happened?" And she's says, "I'm fine, I just wasn't expecting that!" And so I asked, "Well, what were you expecting?" And she responded, "I don't know. Like carrots and celery!" And I was like, "No, it's an *alligator*! It's got big teeth!"

25. In fact, demographic research suggests that captive animals experience mortality rates equal to those living in the wild; see Kohler, Preston, and Lackey, "Comparative Mortality Levels."

26. In fairness to animal advocates, they were far more likely to direct their wrath toward zoo directors and other executives and administrators, rather than keepers themselves.

27. Like political dictators, powerful drug dealers and other organized crime bosses sometimes collect dangerous carnivores like big cats as a means of expressing machismo, as perhaps most famously exemplified by Colombian drug kingpin Pablo Escobar; see Mark Bowden, *Killing Pablo: The Hunt for the World's Greatest Outlaw* (New York: Penguin, 2002), 27.

28. While volunteering at City Zoo, one of my routine tasks was trimming the toenails of four ferrets. Given how much they disliked this particular adventure in animal hygiene, I was instructed to do so while feeding them from a tube of CAT LAX (commonly used to remove and prevent hairballs in cats), which they apparently found tasty enough to endure the procedure. (Its ingredients include caramel, malt syrup, and cod liver oil.)

29. Yann Martel, *Life of Pi* (2001; repr., Orlando, FL: Harcourt, 2011), 15–18.

30. On the growing popularity of this domineering method of animal training among companion dogs, see Malcolm Gladwell, "What the Dog Saw," *New Yorker*, 22 May 2006, 48–57; Schaffer, *One Nation under Dog*, 180–200.

31. Goffman, "On Fieldwork," 128, advises that when conducting participant observation "you have to be willing to be a horse's ass"; or perhaps in my case, one has to be willing to let a horse make you look like one. In this instance Tony was challenging not only my working knowledge of animal husbandry as a white-collar city slicker, but also my lack of what sociologist Matthew Desmond, *On the Fireline: Living and Dying with Wildland Firefighters* (Chicago: University of Chicago Press, 2007), 30, refers to as a *country-masculine habitus*, a set of dispositions associated with male-dominated (and often but not exclusively working-class) rural environments.

32. I participated in this brand of animal wrangling when I assisted a veterinarian in rounding up a goat named Blossom whose body I physically held down despite her protests while he took a blood sample and injected her with antibiotics. While the zoo industry tolerates these sorts of handling practices to deal with uncontrollable hooved mammals such as goats and elk, volunteers were always warned to keep these techniques out of public view. According to one zookeeper, "Never do it when visitors are around," lest they interpret such measures as unnecessarily abusive.

33. The contemporary curriculum in undergraduate zoo management programs emphasizes the importance of animal training, as illustrated by its inclusion in two recently published textbooks: Geoff Hosey, Vicky Melfi, and Sheila Pankhurst, *Zoo Animals: Behavior, Management, and Welfare* (New York: Oxford University Press, 2009); Rees, *Introduction to Zoo Biology and Management*.

34. Sutherland, *Kicked, Bitten, and Scratched*, discusses the recent diffusion of exotic animal training techniques such as operant conditioning and positive reinforcement techniques among zoos as well as animal shelters and wildlife sanctuaries; also see Alex Halberstadt, "Zoo Animals and Their Discontents," *New York Times Magazine*, 6 July 2014.

35. Temple Grandin, Matthew B. Rooney, Megan Phillips, Richard C. Cambre,

Nancy A. Irlbeck, and Wendy Graffam, "Conditioning of Nyala (*Tragelaphus angasi*) to Blood Sampling in a Crate with Positive Reinforcement," *Zoo Biology* 14 (1995): 261–73; Megan Phillips, Temple Grandin, Wendy Graffam, Nancy A. Irlbeck, and Richard C. Cambre, "Crate Conditioning of Bongo (*Tragelaphus eurycerus*) for Veterinary and Husbandry Procedures at the Denver Zoological Gardens," *Zoo Biology* 17 (1998): 25–32; Grandin and Johnson, *Animals Make Us Human*, 269–75; Hosey, Melfi, and Pankhurst, *Zoo Animals*, 501–2.

36. Scientific research confirms what Lauren claims: that positive reinforcement training can potentially reduce zoo animals' fear of humans and therefore contribute to agreeable keeper-animal relations, lower levels of stress in animals, and improved animal welfare; see Samantha J. Ward and Vicky Melfi, "The Implications of Husbandry Training on Zoo Animal Response Rates," *Applied Animal Behaviour Science* 147 (2013): 179–85.

37. In fact, it is to Lenny's credit that the female keepers who work for him at Metro Zoo defend his capacity for change, as perhaps an outlier among those in his generational cohort of zookeepers. According to Lauren,

> A lot of those old-school keepers have transitioned really well with it. Like Lenny—when he started, the way you interacted with the animals, how they would work directly with their big cats, or put their baby lions in harnesses to walk them around, or just crazy things that you would never do today, or being rough with the animals to get them to do what you wanted—whereas now we train them to do what we want. Lenny has changed with the times. He has done a really great job. He sees where the research has shown that things are better this way, and he has seen that it's better to do things this way. And he always wants the zoo to progress, and as a curator he is progressing with those new developments.
>
> So I think that gap is only there for the really stubborn ones because they got into this job when it was a labor job, and gosh darn it, they are close to retirement, and they've always known how to do it, and it's always worked for them, so we don't need to change it. But you can't just assume that they're all like that, because you have a lot of keepers like Lenny, who got into the job because it was a labor job and sees where the changes are happening and wants to keep up with the changes and progress and make sure that we keep the animals to the best of our ability, and to the best of our knowledge.

38. Kohler, Preston, and Lackey, "Comparative Mortality Levels."

39. Sedgwick, *Peaceable Kingdom*, 190.

40. Ibid., 194.

41. On the euthanasia practices performed in American zoos, see Leslie Kaufman, "When Babies Don't Fit Plan, Question for Zoos Is, Now What?," *New York Times*, 2 August 2012.

42. Nelson D. Schwartz, "Anger Erupts after Danish Zoo Kills a 'Surplus' Giraffe," *New York Times*, 9 February 2014. In many European zoos, so-called surplus zoo animals—typically nonendangered species whose genetic composition or reproductive fitness makes them a low priority for captive breeding or management pro-

grams—are sometimes culled for purposes of population control. According to Kaufman, "When Babies Don't Fit Plan," AZA spokespeople reported that while euthanasia is permitted in its accredited U.S. zoos under its regulatory guidelines, it is generally reserved for ill or elderly animals, while the use of contraception as a population control measure is more prevalent in North American zoos than abroad. On the ethics of culling surplus zoo animals, see Robert Lacy, "Culling Surplus Animals for Population Management," in Norton et al., *Ethics on the Ark*, 187–94; Donald Lindburg and Linda Lindburg, "Success Breeds a Quandary: To Cull or Not to Cull," in Norton et al., *Ethics on the Ark*, 195–208.

43. Ann's boss Barbara, the zoo's vice president for public affairs, confirmed that press releases announcing a recent death are usually reserved for "animals that are iconic or well known." According to Barbara, "If a snake dies, it's not going to be news, but it's a judgment call based on the popularity of an animal."

44. "Saying Good-bye to Petal the African Elephant" (Philadelphia Zoo, 9 June 2008).

45. Stefan Timmermans, "Death Brokering: Constructing Culturally Appropriate Deaths," *Sociology of Health and Illness* 27 (2005): 993–1013.

46. On the concept of the "natural" and "good" death, see ibid.

47. "Saying Good-bye to Twigga the Giraffe" (Philadelphia Zoo, 20 June 2008).

48. Timmermans, "Death Brokering," 999, discusses the ideals surrounding the "dignified" death.

49. "Pattycake, First Gorilla Born in New York City, Dies at 40 Years Old at WCS's Bronx Zoo" (Wildlife Conservation Society, 31 March 2013).

50. "WCS's Central Park Zoo's Polar Bear, Gus, Dies at 27" (Wildlife Conservation Society, 28 August 2013).

51. On the brokerage of "death without dying" as an idealized death, see Timmermans, "Death Brokering," 996.

52. Actually, *these* are the facts: according to *The Princeton Encyclopedia of Mammals*, David W. Macdonald, ed. (Princeton, NJ: Princeton University Press, 2013), 629, *wild* lions have a life expectancy of eighteen years, while captive lions like Sleepy typically enjoy a longer average life span of twenty-five years.

53. Douglas W. Maynard, *Bad News, Good News: Conversational Order in Everyday Talk and Clinical Settings* (Chicago: University of Chicago Press, 2003), 160–84.

54. In this manner, media relations offices benefit from what Goffman, *Presentation of Self in Everyday Life*, 49, refers to as "audience segregation," in which cookie-cutter performances succeed because they are delivered to distinct (in this case, temporally distant) audiences, and thus "the routine character of the performance is obscured."

55. "Denver Zoo Mourning the Loss of Polar Bear Olaf" (Denver Zoo, 5 May 2008).

56. "Denver Zoo Mourning the Loss of Polar Bear 'Voda'" (Denver Zoo, 28 April 2010), emphasis added.

57. "Denver Zoo Mourning the Loss of Polar Bear 'Frosty'" (Denver Zoo, 24 June 2010), emphasis added.

58. Such attachments are hardly uncommon among people with an affinity for animals and the outdoors; for example, Paul Colomy, "Losing Samson: Nature,

Crime, and Boundaries," *Sociological Quarterly* 51 (2010): 355–83, investigates how a community experiences the poaching of a beloved animal—in this case, a large wild elk living in the Colorado Rockies—through shared feelings of sentiment and loss.

59. Even in the case of nonhuman animal deaths, dedications and other ritual ceremonies of mourning and remembrance help communities feel a sense of emotional closure after experiencing the loss; see Colomy, "Losing Samson," 367–68.

60. Of course, these instances of handling the departed can also be fraught with great anguish, as when Heather, Metro Zoo's only veterinary technician, had to perform a necropsy on Seamus to determine cause of death, as noted in chapter 3. In order to collect brain samples, she had to cut the jaguar's head off with an electric saw. I asked her how she was able to perform the procedure, given her obvious grief. She explained, "Well, at that point, I had already cried for four hours, so I just wanted to find out what the hell was wrong with him."

# Chapter 8: The Urban Jungle: The Future of the American Zoo

1. Chakrabarty, "Climate of History."

2. For example, regarding the intelligence of chimpanzees, Goodall, *In the Shadow of Man*, 234, argues, "One of the most striking ways in which the chimpanzee biologically resembles man lies in the structure of his brain. The chimpanzee, with his capacity for primitive reasoning, exhibits a type of intelligence more like that of man than does any other mammal living today. The brain of the modern chimpanzee is probably not too dissimilar to the brain that so many millions of years ago directed the behavior of the first ape men."

3. William G. Conway, "How to Exhibit a Bullfrog: A Bed-Time Story for Zoo Men," *International Zoo Yearbook* 13 (1973): 221–26.

4. Diane Ackerman, *The Zookeeper's Wife: A War Story* (New York: Norton, 2007).

# Index

Abbott, Andrew: on the correlates of intraprofessional status, 289n18; on fractals, 273n4

Adams, Ansel, 44

Adut, Ari, on the human ability to feign ignorance of known transgressions, 308n7

Adventure Aquarium (Camden, New Jersey), 31, 155; New Age cinematic soundscapes at, 155–56; owner of (Herschend Family Entertainment), 155; petting tanks at, 151; shark tunnel at, 156; statue of Gill the shark mascot at, 45; tank dedicated to displaying the clown anemonefish (*Amphiprion ocellaris*) and the Pacific regal blue tang (*Paracanthurus hepatus*) at, 47

African tribes, 166; depiction of at zoos, 166

Akeley, Carl, dioramas of at New York's American Museum of Natural History, 277n2

Alden, William, 297n26

Allen, Karen, 310n19; on corporate sponsorship of zoos, 201, 202

Allison, Paul, on sex segregation in the workplace and its relationship to wages, 289n13

Amazon rainforest, Mesoamericans' planting and cultivation of, 6

American Association of Zoo Keepers (AAZK), 290n23; Bowling for Rhinos campaign, 92

American conservation movement, 145, 181–82

American Museum of Natural History, 145

amphibians, 265; and the chytrid fungus (*Batrachochytrium dendrobatidis*), 117, 179

Anderson, Elijah: on the conditions of disadvantaged urban areas in American cities, 294n13; on "cosmopolitan canopies," 134; on the urban poverty in the West Philadelphia neighborhood bordering the Philadelphia Zoo, 287n60

*Animal Liberation* (Singer), 221–22

animal mascots, and corporate images, 45

Animal Planet, ZOOTUBE kiosks of, 173

animal poaching, in Africa, 206

animal rights movement: women's involvement in, 308n11, 308–9n12; zoo directors, executives, and administrators as the targets of wrath of, 311n26

animal sanctuaries, 219, 261, 263, 311n34

animal training: and positive reinforcement, 241–44, 261, 312n36; training zoo and aquarium animals to safely receive veterinary care, 92, 227, 243–44, 261, 298n30

Animal Welfare Act (1966), 160, 182

animal-assisted activities (AAA), research on, 293–94n12; published criticisms of, 294n12; published experiments on, 294n12

Anthony, Lawrence, 283n8

Anthropocene, the, 7–8

aquariums. *See* aquatic zoos

aquatic zoos, 154–60; and the design of aquarium tanks, 156, 298n35; exhibition of creative works typically associ-

Environmental Protection Agency, 182
Erickson, Clark, 6
Escobar, Pablo, 311n27
ethnographic fieldwork, 275n21, 288n3, 311n31
Event Network, zoo gift shops operated by, 306n62
evergreens (recyclable topics and plot lines), 116, 293n6
exotic and wild animal collections: of ancient rulers, 46; of modern-day dictators and autocrats, 283n8; of the Tudors, 143–44
exotic and wild pet trade, 118; involvement of zoos in, 293n7
extinction rates, 7, 274n16
Exurban County Zoo, 235–36
ExxonMobil: contributions of to conservative groups responsible for publicizing climate change skepticism, 202; contributions of to Dallas-area urban institutional anchors, 200; and the *Exxon Valdez* oil spill, 199; as the second biggest air polluter in the United States in 2010, 200

factory farms, 217–18; cows, 218; pigs, 218; poultry, 218
*Fast Food Nation* (Schlosser), 218
Faulkner, Robert R., 293n9
Federal Environmental Pesticide Act (1972), 182
*Finding Nemo* (2003), 47
Fine, Gary Alan, 278n8; on the cultural and social construction of nature, 274n10; on "naturework," 278n7
Finely, Bill, 284n24
Fischer, Claude, 295n15
fish: alligator gar, 50; capabilities of, 298n33; characteristics of that makes it easy to naturalize their captivity, 156; as house pets, 298n34; learning abilities of, 298n33; onespot frigatehead, 40
Fish, Stanley, on "interpretive communities," 77
Fisher, Mathew, 179

flamingos, and beta-keratin, 35, 259, 281n47
Foer, Jonathan Safran, 219
Fossey, Dian, 25; on gorillas' enjoyment in basking in the sun, 307n3; on the inhumane trapping practices of indigenous animal poachers in Rwanda, 306n63
Foucault, Michel, 276n29
Fox, Megan, accusations of against the Brookfield Zoo, 137
Fox, William L., on the use of live animals in the entertainment landscape of Las Vegas, 299n42
fractals, 273n4
Frank, Richard, 306n67
Frederick, Jim, 289n15
Freud, Sigmund, 55
Friedman, Thomas L., 264
Friese, Carrie, 183, 285–86n29

Ganz, Sayaka, 41–42
gardening, 192; and climate change due to global warming, 303n29
gastropods, 285n36
Geertz, Clifford, on the cultural and social construction of nature, 274n10
Georgia Aquarium (Atlanta), 157; Ocean Voyager exhibit, 157; shark tunnel at, 157–58
Getty, J. Paul, 200
*Ghost Hunters* (TV show), 169–72
giant pandas: and captive breeding programs, 185; eating of bamboo by, 51; giant pandas on loan from the People's Republic of China, 50–51; habitat of in the wild (the mountain forests of southwestern China), 50; neotenic traits of, 50, 297n27
Gibran, Kahil, 159
Gibson, James William, 300n49; on a "culture of enchantment," 44, 277n2, 282n2, 297n28
Gieryn, Thomas F., 299n44
Giroux, Henry A., critique of Disney's theme parks, 300n47
Gitlin, Todd, on the devolution of sub-